AN ILLUSTRATED COLOUR TEXT

Clinical Bacteriology, Mycology and Parasitology

*To my late great Chief, Mr. Glen Buckle, who inspired me
to study Microbiology; to my patients, especially in
Bangladesh, who inspired me to study Infectious Diseases;
and to my students and colleagues, who inspire me still.*

Commissioning Editor: Timothy Horne
Project Development Manager: Jim Killgore
Project Manager: Nancy Arnott
Designer: Sarah Russell
Page make-up: Kate Walshaw

AN ILLUSTRATED COLOUR TEXT

Clinical Bacteriology, Mycology and Parasitology

W. John Spicer

MB.BS(Melbourne) FRACP FRCPA FACSHP FASM DTM&H (Sydney) Dip.Bact (London)

Senior Consultant in Infectious Diseases,
Alfred Hospital, and Austin and Repatriation Medical Centre,
Senior Consultant in Microbiology, Alfred Hospital,
Associate Professor of Microbiology, Monash University,
Melbourne, Australia

Illustrated by Peter Lamb

CHURCHILL
LIVINGSTONE

EDINBURGH   LONDON   NEW YORK   PHILADELPHIA   ST LOUIS   SYDNEY   TORONTO   2000

CHURCHILL LIVINGSTONE
An imprint of Harcourt Publishers Limited

© Harcourt Publishers Limited 2000

 is a registered trademark of Harcourt Publishers Limited

The right of W. John Spicer to be identified as author of this work has been
asserted by him in accordance with the Copyright, Designs and Patents Act 1988

First published 2000

ISBN 0443 04365 5

British Library Cataloguing in Publication Data
A catalogue record for this book is available from the British Library

Library of Congress Cataloging in Publication Data
A catalog record for this book is available from the Library of Congress

> **Note**
> Medical knowledge is constantly changing. As new information becomes
> available, changes in treatment, procedures, equipment and the use of
> drugs become necessary. The author and the publishers have, as far as it
> is possible, taken care to ensure that the information given in this text is
> accurate and up to date. However, readers are strongly advised to confirm
> that the information, especially with regard to drug usage, complies with
> the latest legislation and standards of practice.

The
publisher's
policy is to use
**paper manufactured
from sustainable forests**

Printed in China

PREFACE

All medical students, and all nurses, general practitioners and medical specialists seeing patients, need to understand the basics of microbiology and infections. Infection cannot be understood without knowing some microbiology, and infection is one of the major mechanisms causing disease, and can affect every body tissue and organ.

This book aims to present, clearly, concisely and memorably, the clinically relevant basic facts and processes in the twin disciplines of microbiology and infectious diseases. The range of the book is wide, including fungi and parasites as well as bacteria and their consequent infections.

The organisation of any book on microbiology and infectious diseases depends particularly on one question — to present the twins integrated in each chapter, or separately? If all infections were like tetanus or anthrax, where one organism causes only one disease which is in turn caused only by that organism, the integrated approach would be obvious and easy. But life, apparently, was not meant to be easy. Consider the amazing Group A streptococcus (*S. pyogenes*, pp. 30-31), causing numerous diseases by numerous mechanisms in numerous organs; or the versatile 'golden staph' (*S. aureus*, pp. 28-29), causing infections ranging from a trivial pimple through cellulitis to serious deep abscesses and bone infections, to endocarditis and overwhelming septicaemia with septic shock.

Conversely, many common disease presentations such as pneumonia, urinary infections and gut infections have more than one possible causative organism. So I have chosen, after an introductory section on general characteristics of the organisms and a second section on their attack on us and our intrinsic defences, to present the third section on specific organisms in more detail (microbiology), and then the fourth section on clinical infections in each body organ system and some special categories, with extensive cross-references between the third and fourth sections. The fifth and final section is on the extrinsic defences we have invented, such as vaccines, asepsis and antisepsis, and antimicrobials.

The format of the book is exciting, being specifically designed to help both initial learning and revision. Each subject is presented in an 'easy learning' module complete on a single or double page spread, with numerous coloured photographs, coloured graphics, integrated tables, and a 'key point' summary in the bottom right hand corner, useful both for initial orientation and for revision.

In a book of this size it is obviously impossible to give extended detail, or cover all organisms and diseases, yet I have intentionally included rarer and 'tropical' infections. Every GP and Emergency Department clinician must be able to recognise the rare meningococcal bacteraemia or meningitis, which they may never have seen before, yet the patient's life depends on early recognition and emergency treatment by the first doctor to see them. Similarly, any GP or other 'front line' doctor in a temperate industrialised country may be confronted by a sick patient with malaria or typhoid fever in a returned traveller or immigrant, needing early recognition and urgent relevant investigation and treatment. I aimed to make this book useful in both industrialised and developing countries.

I have enjoyed writing this book, and I hope you enjoy and profit from reading it. I will be interested to hear your opinions and suggestions.

2000   W. J.S.

ACKNOWLEDGEMENTS

While many of the slides illustrating this book came from my own collection of the last 35 years, I am indebted to many colleagues for filling gaps with particular slides I needed. My own staff in the Departments of Infectious Diseases and Microbiology, Alfred Hospital, were always wonderfully helpful, especially Associate Professor Denis Spelman and Dr. Andrew Fuller, also Drs. Ashley Watson, Sally Roberts, Adam Jenney, and numerous other registrars and residents in the ID Unit; Mr. Grant Perry, Ms Clare Franklin, Mrs. Jenni Williams, and the late Mr. Glen Buckle in Microbiology; and Ms Glenys Harrington in Infection Control. I am very grateful to all the patients who gave permission to be photographed, and to Alfred Hospital and Monash University Departments and individual colleagues who willingly provided particular slides from their patients in Melbourne or overseas: in the Departments of Radiology and Nuclear Medicine, Associate Professors Nina Sacharias and Victor Kalff, and Drs. Chris O'Donnell and Stephen Booth; in the Department of Anatomical Pathology, Professor John Dowling, the late Associate Professor Brian Essex, and Drs. Shant Khan, S K Tang, the late Ross Anderson, and Mr. John Hall; in Monash Microbiology, Drs. Geoff Cross, Ian Denham, Mrs. Lyn Howden and Mrs. Jan Savage. Individually, Professor Hugh Taylor, Associate Professors Suzanne Garland, Geoff Hogg, John Kelly, Hector McLean, John Murtagh, Alison Street, and Dan Sexton (Duke University, USA), and Drs. David Abell, Barry Elliott, Reuben Glass, Tony Hall, Robin Hooper, Don Jacobs, David Looke (Brisbane), Fiona McCurragh, Hugh Newton-John, Jo Sabto, Jack Swann, Hugo Standish, the late Brian Smith, Richard Stawell, Peter Thompson, Jonathan Tversky, John Waterston, Robert West and Bob Zacharin and also Ms Karen Flett and Mr Guy Brown. If I have unknowingly been provided with previously published slides, I apologize in advance. Several slides from the former Fairfield Hospital and the Royal Victorian Eye and Ear Hospital Slide Libraries were used with permission, and my thanks to Ms Caroline Hedt, Mr. Gavin Hawkins and Mr. Cam Harvey in the Alfred Visual Communications Department who were always helpful in duplicating slides, with which the Monash University Department at Alfred also assisted at times.

In Churchill Livingstone, now part of Harcourt Publishers, I received particular help, expert advice, and unfailing courtesy from Mr. Jim Killgore and Mr. Timothy Horne, with much help behind the scenes from Dr. Jane Ward (copy editor) and Dr. Laurence Errington (Indexer), and Peter Lamb who transformed my line drawings, examples and directions to the excellent figures you see.

Finally I thank my wife Heather and my family, who are almost as pleased as I am to see such a beautiful book emerge at last!

2000   W. J.S.

CONTENTS

MICROBES

2

GENERAL PRINCIPLES 2
Characteristics of bacteria 2
Characteristics of fungi 6
Characteristics of protozoa 10
Multicellular parasites 12

MICROBIAL ATTACK AND CONTROL BY INTRINSIC DEFENCES

14

Host-Microbial relationships 14
Normal flora and opportunistic infections 16
Establishment of disease: attack 18
Non-specific defences 20

Specific (immune) defences 22
Immune disorders/evading host defences 24
General body response to infection 26

SPECIFIC PATHOGENS

28

BACTERIA 28
Staphylococci 28
Streptococci and Enterococci 30
Gram-positive rods: Cornybacteria, Listeria, Bacillus 32
Clostridia 34
Neisseria, Branhamella, Kingella and Acinetobacter 36
Haemophilus, Bordetella and Legionella 38
Enterobacteriaceae 40
Vibrio, Campylobacter, Helicobacter, Aeromonas, Plesiomonas 44
Pseudomonas and rare Gram-negative rods 46
Bacteroides, Fusobacteria and other Anaerobes 48
Zoonotic bacteria 50
Spirochaetes: Treponemes, Borrelia, Leptospires 52
Mycobacteria 54
Actinomyces, Nocardia and rare Gram-positive bacilli 56
Mycoplasma and Ureaplasma 57
Chlamydia 58
Rickettsia, Coxiella, Rochalimaea and Ehrlichia 60

FUNGI 62
Aspergillus and Candida 62
Cryptococcus and Histoplasma 64
Blastomyces, Coccidioides, Paracoccidioides 66
Mycoses of skin and adjacent tissues 68
Fungi causing Zygomycosis 70

ARTHROPODS 71
Arthropods 71

PARASITES 72
Sporozoa: Plasmodia, Toxoplasma, Cryptosporidium 72
Amoebae: Entamoeba, Naegleria, Acanthamoeba 74
Intestinal and vaginal flagellates and ciliates 75
Blood and tissue flagellates 76
Intestinal nematodes 78
Tissue nematodes 80
Cestodes 82
Trematodes 84

86 | **MICROBIAL ATTACK SUCCEEDS**

CLINICAL DISEASE 86

Acute meningitis 86

Chronic diffuse CNS infections 88

CNS abscess and other focal lesions 90

Nervous system: tropical and rare infections 92

Otitis, mastoiditis and sinusitis 94

Superficial ocular infections 96

Tropical ocular infections 98

Deep eye infections 100

Stomatitis 101

Dental and periodontal infections 102

Throat infections 104

Epiglottitis and diphtheria 106

Tropical and rare oro-facial infections 108

Laryngitis, tracheitis and pertussis 109

Bronchial infections 110

Pneumonia in the normal host 112

Pneumonia in the abnormal host 114

Lung abscess and empyaema 116

Tuberculosis and atypical mycobacterial infections 118

Tropical or rare respiratory infections 120

Suppurative thrombophlebitis/lymphangitis/ lymphadenitis 122

Myocarditis, pericarditis and rheumatic fever 124

Infective endocarditis 126

Bacteraemia, septicaemia and fungaemia 128

Tropical systemic infections 130

Rarer systemic infections 132

Pyrexia of unknown origin (PUO) 134

Diarrhoeal diseases I: general features 136

Diarrhoeal diseases II: pathogens 138

Peritonitis and intra-abdominal abscesses 142

Biliary and hepatic infections 144

Tropical and rare abdominal infections 147

Urinary tract infections: cystitis and pyelonephritis 148

Renal and perinephric abscesses; prostatitis 150

Tropical and rare urinary infections 152

Urethritis 154

Cervicitis 156

Salpingitis and pelvic inflammatory disease 158

Epididymitis, orchitis and balanitis 160

Vaginitis and vulvo-vaginitis 161

Tropical and rare sexually transmitted diseases 162

Streptococcal skin and soft tissue infections 164

Staphylococcal skin and soft tissue infections 166

Gas-forming and gangrenous infections 168

Wound, bite and burn infections 170

Fungal infections of the skin, hair or nails 172

Tropical and rare bacterial skin and soft tissue infections 174

Tropical and rare fungal and parasitic skin and soft tissue infections 176

Osteomylitis 178

Special, tropical and rare bone infections 180

Joint infections 182

Zoonoses 184

Infections of mother, fetus and newborn child 186

Travellers and recent immigrants 188

Hospital-acquired infections 190

Infections in compromised patients 192

General practice patients 194

Infections in elderly patients 195

196 | **MICROBE CONTROL BY EXTRINSIC DEFENCES**

Sterilisation and disinfection 196

Anti-microbials: general properties 198

Anti-microbials: specific anti-bacterials 200

Anti-microbials: special anti-microbials 204

Vaccines and immunisation 206

Clinical and laboratory I: microbial detection and identification 208

Clinical and laboratory II: antibody response and guiding therapy 210

Index 212

CHARACTERISTICS OF BACTERIA (1)

NOMENCLATURE

All living things, hence all bacteria, have two names, firstly their generic (genus) name, e.g. *Staphylococcus*, then their specific (species) name, e.g. *aureus*. The generic name is often abbreviated to the initial letter(s), e.g. *S. aureus*.

Special additions to this universal scheme may include:

- a third name to distinguish several varieties within one species, e.g. *Acinetobacter calcoaceticus* var *anitratus* (!)
- a common, non-scientific, historical name, e.g. pneumococcus for *Streptococcus pneumoniae*, gonococcus for *Neisseria gonorrhoeae*, meningococcus for *Neisseria meningitidis*
- a serological group name, e.g. *Streptococcus pyogenes* is also called 'the group A streptococcus'
- a toxin product name, e.g. *Clostridium perfringens* type A.

CLASSIFICATION

Increasing knowledge of the properties of bacteria, fungi, protozoa and algae has necessitated revision of the simple division of living things into two kingdoms: plants and animals. One classification has two super-kingdoms: procaryotes (bacteria and blue-green algae) and eucaryotes.

Eucaryotes include four kingdoms, Protista (Protozoa and Algae), Fungi, Animalia and Plantae. Viruses are not included because they do not have the essential characteristics of living organisms (capable of independent replication or survival).

Bacteria are fundamentally different from all other living things in being *procaryotes*, distinguished by:

- DNA in a double-stranded loop, not within a nuclear membrane
- small ribosomes free in the cytoplasm; there is no endoplasmic reticulum
- the absence of mitochondria or other membrane-enclosed organelles
- a complex peptidoglycan-protein cell wall (absent in *Mycoplasma*).

Bacteria are classified by several criteria (see Table 1):

- shape
- stain
- ability to grow with or without oxygen
- size

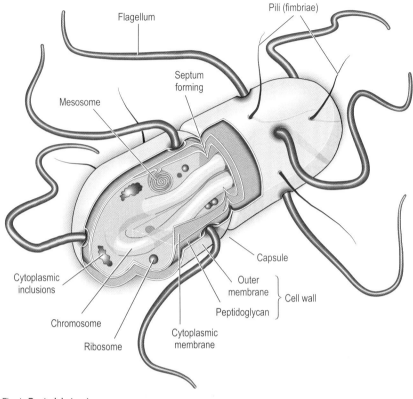

Fig. 1 **Bacterial structure.**

Table 1 **Major criteria for classifying medically important bacteria**

Criteria	Groups	Examples
Shape	Cocci (spherical) Bacilli (rods) Spirilla (curved or spiral rods)	*Staphylococcus, Neisseria* *Bacillus, Listeria, Salmonella* *Vibrio, Campylobacter*
Stain	Gram stain positive Gram stain negative Acid-fast stains Special stains for specialised structures	*Staphylococcus, Streptococcus, Bacillus* *Haemophilus, Escherichia, Salmonella* *Mycobacterium* *Clostridia* spores
Gas requirements	Strict aerobes Facultative anaerobes Strict anaerobes Capnophiles (need high CO_2)	*Pseudomonas aeruginosa* *Escherichia coli* *Clostridium* spp. *Neisseria* spp.
Specialised features	Spores Enzymes Antibiotic resistance Antigens	*Clostridium* spp. Coagulase-producing staphylococci Methicillin-resistant *Staphylococcus aureus* (MRSA) *Streptococcus* (Lancefield groups), *Chlamydia*
Nucleic acids	DNA probes DNA amplification	Enterotoxic *Escherichia coli*; rapid detection of meningococcus infection *Mycobacterium leprae*

- growth characteristics
- DNA content (G + C content) and homology.

The Gram stain is the most important staining procedure in medical microbiology. Gram-positive organisms retain the purple of crystal violet after iodine fixation and alcohol washing, whereas Gram-negative organisms lose the colour with alcohol and need counterstaining with a pink dye. Special stains are needed for organisms with unusual cell walls, e.g. acid-fast stains for mycobacteria.

Although staining and growth characteristics have formed the basis of diagnostic microbiology, the availability of techniques such as DNA probes and amplification, and polyclonal and monoclonal antibodies have greatly increased the speed, range and sensitivity of diagnostic testing.

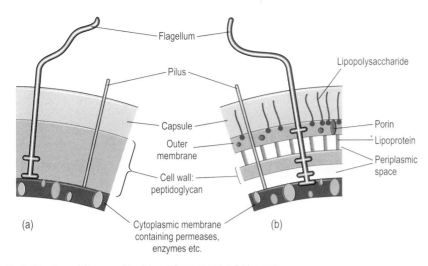

Fig. 2 **Structure of Gram-positive (a) and Gram-negative (b) bacteria.**

SIZE AND STRUCTURE (Fig. 1)

Pathogenic bacteria vary widely in size: from mycoplasma (0.2–0.8 μm diameter) to enteric Gram-negative rods (0.5–6.0 μm). Gram-positive and Gram-negative bacteria differ in their cell wall composition but have typical procaryote internal structure.

Cell membrane

The cell (cytoplasmic) membrane is made of protein and phospholipids, but (except in *Mycoplasma*) not the sterols that are found in eucaryotes. It is the osmotic barrier between cell and environment, and its essential functions include electron transport, enzyme systems (as in eucaryotic mitochondria), solute transport and cell product transport.

Cell wall (Fig. 2)

The cell wall has numerous functions reflected in its structure:

- protects the cell membrane by its rigidity from osmotic or mechanical rupture
- contains numerous characteristic antigens, important both in bacterial virulence and endotoxins, and in host antibody formation
- provides a firm base for pili (fimbriae) and flagella.

Gram-positive cells

The cell wall of Gram-positive organisms (Fig. 2a) consists mainly of many layers of *peptidoglycan* (murein), a complex polymer of long glycan (sugar) chains of alternating *N*-acetylglucosamine and *N*-acetyl muramic acid with short pentapeptide side chains cross-linked to each other by peptide bonds between the lysine of one and the D-alanine of the other, giving a rigid polar wall. Other polymers in the wall include teichoic

acids and chains of glycerol or ribitol linked by phosphodiester bonds.

Gram-negative cells

The cell wall of Gram-negative organisms (Fig. 2b) consists of a thinner layer of the same *peptidoglycan* but the cross-linking is between D-amino pimelic acid and D-alanine. This peptidoglycan layer is in a *periplasmic space* between the inner cytoplasmic membrane and a unique bi-layered phospholipid *outer membrane*, with lipoprotein on the inner surface binding to the peptidoglycan, and a special *lipopolysaccharide (LPS)* on the outer.

The three components of LPS are the lipid (lipid A, the active component of endotoxin, very important in causing septic shock), a core polysaccharide and a variable carbohydrate chain, which is the distinctive O antigen detected serologically. Its hydrophobicity gives some antibiotic and bile salt resistance.

Mycobacteria

The cell wall of mycobacteria and other acid-fast organisms contains characteristic *waxes*, complex long-chain hydrocarbons with sugars. This almost impervious coat prevents stains being removed by acid

and gives resistance to desiccation and many disinfectants and antibiotics. It also slows the entry of nutrients.

Capsule

A capsule protects the cell wall of many bacteria, particularly in adverse conditions; this mucoid polysaccharide layer may be lost in laboratory cultures. In infections, it resists phagocytosis by white blood cells and aids adherence to tissues, catheters and prostheses.

Pili

Pili (fimbriae) are hair-like in appearance and are of at least two types:

- *Sex pili* are specialised structures that enable DNA transfer by conjugation (literally, 'joined with')
- *Common pili* are shorter and aid attachment to host cells, are anti-phagocytic and by rapid changes in their antigenic protein (pilin) avoid host antibody response.

Flagella

Flagella are much longer than pili and give motility to bacteria, which may be *monotrichous* (one flagellum at one or both ends), *lophotrichous* (many flagella at one or both ends), or *peritrichous* ('covered with hair'). Coherent counter-clock-wise flagellar rotation, because of the counter-clockwise helical pitch of the flagellar protein, gives a straight swimming motion, as in chemotaxis toward an attractant, while clockwise rotation gives tumbling motility, seen particularly with repellents.

Spores

Spores, formed especially by *Clostridium* and *Bacillus* spp., are concentrated bacterial DNA surrounded by an extremely tough protective coat. The cell is metabolically inert and survives drying, heat and most chemical agents for months, years or more.

Characteristics of bacteria (1)

- The first name of a bacterium is its genus, and the second its species.
- Bacteria were initially classified by their shape, Gram stain and oxygen requirements, supplemented by biochemical and serological characters. Genetics is now being used to establish fundamental relationships.
- Bacteria are procaryotes with free circular DNA, ribosomes, no mitochondria and a peptidoglycan cell wall.
- Bacterial structure includes:
  – internal structure: nuclear material, ribosomes, cell membrane
  – cell wall: peptidoglycan (plus periplasmic space and outer membrane in Gram-negative bacteria)
  – external structures: capsule, pili, flagella.
- Spores are metabolically inert in a protective coat.

CHARACTERISTICS OF BACTERIA (2)

ENERGY, NUTRITION AND GROWTH

Energy sources and processes

Bacteria use three sources for their energy requirements: chemical reactions (*chemotrophy*), light (*phototrophy*) or the host cell (*paratrophy*). If the energy-yielding reactions use organic compounds they are termed organotrophy; those using inorganic compounds are termed lithotrophy. So energy derived from light using organic hydrogen donors would be described as photo-organotrophy (in some anaerobic bacteria).

Nutrition

Bacteria show a wide variety of nutritional requirements. *Autotrophs* can live in an entirely inorganic environment; they are free living and rarely are of medical importance. *Heterotrophs* need an exogenous supply of one or more essential metabolites. Most bacteria of medical importance come into this group and they vary from those with great synthetic capacity (e.g. *Escherichia coli*) to those pathogens that require exogenous supplies of growth factors such as vitamins to grow (e.g. some streptococci). Finally, some of the parasitic and pathogenic species can only survive intracellularly (*autotrophy*); they possess DNA and RNA, but only a certain amount of independent metabolic activity.

Growth

The doubling time or the generation time is the time between bacterial divisions (by binary fission) and ranges from about 20 minutes for *E. coli* and similar bacteria provided with rich nutrients like laboratory media, to 24 hours for tubercle bacilli. *Balanced growth* occurs when all necessary nutrients are supplied, while *unbalanced growth* is more usual in nature where changing environments and unbalanced nutritional supplies will occur. The phases in the growth of bacteria introduced into a nutrient-rich environment are shown in Fig. 1. Cells in log phase are most virulent.

METABOLISM

This is best considered in three stages:

- *Intermediary metabolism*: the utilisation of a carbon source, e.g. glucose, to provide energy and small molecules such as pyruvate.
- *Biosynthetic pathways*: to build these small molecules with nitrogen, sulphur and other minerals into amino acids, purines, pyrimidines, polysaccharides and lipids.
- *Genetic instructions*: to build these into macromolecules and bacterial organelles ready for cell division.

Intermediary metabolism

Bacteria have adapted to use a vast range of substances as energy sources and are capable of widely differing metabolic activities. Glucose is used by almost all bacteria and the inter-linking of pathways is common to all regardless of their metabolic capacity (Fig. 2). Two central metabolic pathways are:

- *Embden–Meyerhof–Parnas glycolytic pathway (EMP)*, which converts glucose to pyruvate with the formation of ATP as an energy source.
- *Tricarboxylic acid (Krebs) cycle* then uses pyruvate and ATP both to provide further energy and to form intermediate substances for amino acid and fatty acid biosynthesis.

Additional pathways are used by various bacteria, including:

- *Pentose phosphate cycle* to utilise pentoses (instead of hexoses like glucose) to provide energy for ATP and NADPH formation, and carbon compounds for nucleotide (purine and pyrimidine) synthesis.
- *Entner–Doudoroff anaerobic pathway* to convert glucose to pyruvate via 6-phosphogluconic acid instead of by the aerobic EMP pathway. It is also used by pseudomonads, important in hospital infections, to metabolise gluconate and related compounds extracellularly, using enzymes in their cell membrane and thus avoiding energy use on hexose transport.
- *Conversion of 4-carbon to 3-carbon compounds*, e.g. malate to pyruvate, and acetoacetate to pyruvate via phosphoenolpyruvate.

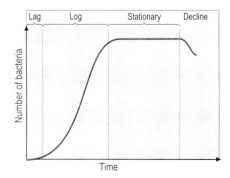

Fig. 1 **The phases of a bacterial growth curve.**

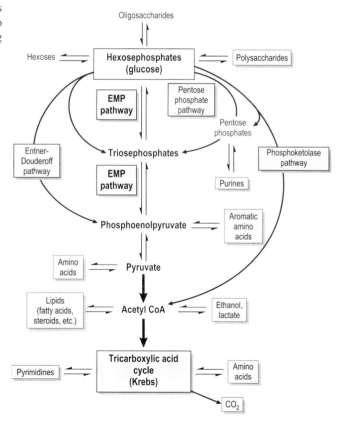

Fig. 2 **Intermediary metabolism and major biosynthetic pathways.**

- *Phosphoketolase pathway*, fermenting glucose or pentoses to pyruvate, lactate, acetate or ethanol when some enzymes of the EMP glycolytic pathway are lacking.
- *Pyruvate metabolism* to lactate, succinate, propionate, acetone, acetate, acetoin, butanol and ethanol.

Biosynthetic pathways

Purine synthesis
Purines are synthesised from ribose 5-phosphate produced in the pentose phosphate cycle or the phosphoketolase pathway.

Pyrimidine synthesis for RNA
The amino group of glutamine plus CO_2 and aspartate form carbamyl aspartic acid, which loses water to become di-hydroorotic acid, the precursor for the pyrimidines, uridine and cytidine.

Amino acid synthesis
The essential feature of each amino acid is its carbon skeleton; biosyntheses occur from four intermediates of metabolism (Fig. 3) into four families.

Utilisation of complex substrates
Bacteria can use many large molecules as substrates, decomposing them by oxidation, reduction and by specific enzymatic attack to recycle carbon and nitrogen and produce energy. Bacteria have a greater range of such processes than fungi and both perform chemical feats impossible for plants and animals.

Bacteria and fungi can use not only nitrate but also nitrite, ammonia and even nitrogen to provide their nitrogen needs. Some (chemolithoautotrophic) bacteria even reverse this process, forming nitrogen from nitrate.

Genetics
The *genome* (genetic material) of bacteria is DNA contained in one long continuous ('circular') chromosome, tightly supercoiled by the enzyme DNA gyrase. The chromosome consists of 3000 to 6000 *genes*, which are specific DNA sequences coding by *codons* of three nucleotide base pairs to messenger RNA for specific amino acids to form polypeptides. A few genes are structural, coding for ribosomal and transfer RNA. There are no introns (sequences between genes) as in eucaryotes. The full expression of the genome is the *genotype* but what is apparent – the *phenotype* – is often less, because of latent, unexpressed qualities.

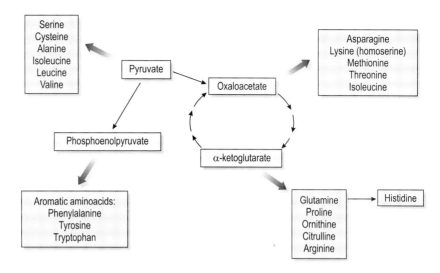

Fig. 3 **Biosynthesis of amino acids.**

Genetic changes in bacteria
As bacteria are haploid, there is no genetic exchange by meiosis and zygote formation as in eucaryotes. Genetic change comes only by random mutations or from one of three types of gene exchange. The random mutation rate is about 1 in 10^7 to 10^8 cell divisions and can lead, for example, to changes in colony colour, loss of a biochemical activity, or resistance to an antibiotic. The result of some donor bacterial genome (exogenote) entering an intact recipient is a merozygote leading to *recombination* of donor and recipient genes.

The three types of gene exchange are:

- *Transformation*: the transfer of a free fragment of DNA from one bacterium to another of the same genus, e.g. occurring in *Strep. pneumoniae*, *H. influenzae*.

- *Transduction*: a fragment of bacterial DNA is carried by a bacteriophage (a virus which infects bacteria) from one bacterium to another.
- *Conjugation*: transfer in bacteria carrying transmission plasmids, the fertility (F) factor. F$^+$ bacteria join to F$^-$ bacteria and transfer part of the donor cell's chromosome, about 1% per minute (interrupting this process at different times enables mapping of the donor chromosome).

Many bacteria contain plasmids, smaller circular bits of DNA containing 1000 to 25 000 base pairs. Plasmids carry much important genetic information, including antibiotic resistance. They can also become incorporated into the chromosome as transposons.

Characteristics of bacteria (2)
- Bacterial energy comes from one of three sources: light, chemical reactions (most pathogens) or a host cell.
- Most pathogenic bacteria are heterotrophs, needing an external source for some essential metabolites.
- The doubling time (time between cell divisions) for most pathogens is about 20 minutes but is 24 hours for *M. tuberculosis*.
- Bacteria stimulated to grow enter balanced growth with an initial lag phase, exponential growth and, finally, stationary phase when some essential nutrient is exhausted.
- Bacterial intermediary metabolism varies widely but is founded on two pathways basic to most living things: the EMP glycolytic path from glucose to pyruvate and the Krebs' TCA cycle.
- Bacteria have powerful and varied catabolic pathways that break down macromolecules and are especially important in recycling carbon and nitrogen.
- The bacterial genome is in a single tightly coiled chromosome without a nuclear membrane, coding by three nucleotide pairs via mRNA to amino acids.
- Genetic change occurs by mutation, by translocation of fragments of free DNA, by transduction through a bacteriophage, or by conjugation, a form of sexual transfer of DNA involving small free pieces of DNA called plasmids.

CHARACTERISTICS OF FUNGI (1)

NOMENCLATURE

Fungi, like bacteria, are named by the binomial Linnaean system with a generic name (capitalised and in italics) and a specific name (not capitalised, in italics). However all fungi reproduce asexually, giving the anamorphic state, and most also reproduce sexually giving the teliomorphic state. Unfortunately many pairs of these were described and named before it was realised that they were different forms of the one fungus, e.g. *Cryptococcus neoformans* (anamorph) = *Filobasidiella neoformans* (teliomorph). The name of the tissue form (anamorph) is used by clinicians.

Fungi can be normal inhabitants of the mouth and intestinal tract. In predisposing conditions, e.g. pregnancy, diabetes, immunodeficiency, therapy with broad-spectrum antibiotics, the fungi can proliferate and cause disease, e.g. thrush or even endocarditis.

CLASSIFICATION

The kingdom Fungi is divided into two phyla:

- *Zygomycota*, which quickly produce a diploid zygote sexually and sporangiophores asexually. Examples are *Rhizopus, Mucor* and other species producing zygomycosis (Fig. 1a).
- *Dikaryomycota*, in which the haploid nuclei do not fuse quickly, giving a prolonged dikaryotic sexual cycle.

Phylum *Dikaryomycota* has two sub-phyla: *Ascomycotina* and *Basidiomycotina*. In the subphylum Ascomycotina, which includes most fungi causing ringworm, this cycle occurs entirely *within* a sac (an ascus), producing ascospores (Fig. 1b). In the subphylum Basidiomycotina, which includes *Cryptococcus*, this begins in a bag (a basidium) and ends with maturation on the *outside* of the basidium, producing basidiospores (Fig. 1c).

Fungi for which a sexual form is not yet recognised and which, therefore, cannot be fully classified are called Fungi imperfecti. These include many pathogens: *Candida, Torulopsis* and *Epidermophyton* spp.

STRUCTURE

Fungi, being eucaryotes, have a structure (Fig. 2) differing considerably from

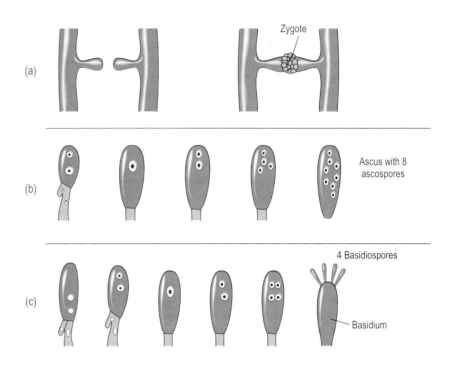

Fig. 1 **Classification by reproductive structures. (a)** Zygote formation in *Zygomycota*, e.g. *Mucor* species. (**b**) Ascus formation in *Ascomycotina*, e.g. *Trichophyton* species. (**c**) Basidium formation in *Basidiomyocotina*, e.g. *Cryptococcus*.

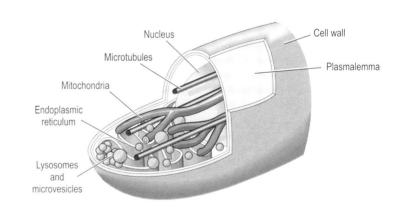

Fig. 2 **Structure of fungal cells.**

procaryotic bacteria (see pp. 2–3). Fungi have a nucleus containing their chromosomal DNA and a RNA-rich nucleolus within a nuclear membrane.

The cytoplasm contains not only ribosomes but also mitochondria, lysosomes and microvesicles, microtubules, Golgi apparatus and a double-membraned endoplasmic reticulum. Surrounding the above structures (the cytosol) is the cell membrane or *plasmalemma*, which contains not only lipids and glycoproteins but also ergosterol. Bacteria (except for *Mycoplasma*) do not contain sterols, and mammalian cells contain cholesterol rather than ergosterol, which is, therefore, a site of attack by antifungal drugs.

Outside the plasmalemma is a rigid cell wall containing a polymer of *N*-acetylglucosamine, *chitin*, on which are layers of polypeptides with complex polysaccharides including mannans and glucans. Some fungi, e.g. *Cryptococcus* ssp. have a further layer, a polysaccharide *capsule*. The cell wall and capsule have multiple functions, including protection, transport and virulence, and are involved in invoking the host response.

MORPHOLOGY

There are two major morphological forms of fungi (Fig. 3), small round yeasts and long filaments called hyphae. Both have sexual and asexual forms.

Yeasts

Yeasts are round, unicellular and multiply by budding or by fission. Some yeasts form long buds called pseudo-hyphae ('germ tubes', used to identify *Candida albicans*) but not true hyphae.

Filamentous fungi

Filamentous fungi form hyphae, long tubes which may have cross-walls called septa, or simply be multinucleate (coenocytic). A collection of hyphae is called a mycelium, which may be vegetative, growing on a nutrient surface, or extending upwards as an aerial mycelium producing conidia ('spores') which spread very easily, contaminating a laboratory if culture plates are carelessly opened! The morphology of conidia is important in classifying fungi.

Dimorphic fungi

Dimorphic fungi exist in both forms. Many pathogenic fungi are dimorphic, usually the yeast form occurring in tissues and the filamentous (mould) form in the environment or on culture at 25°C. *Candida* is an exception, forming mycelium in tissues.

DIAGNOSIS OF FUNGAL INFECTION

Yeasts and fungi grow on ordinary media but are mostly slow growing, and cultures need to be examined over 2–3 weeks. A glucose or blood agar is often used at acid pH to inhibit bacterial growth.

Identification of fungi is mainly made from morphology (Fig. 4a), and yeasts may be detected in stained films during routine examination of swabs etc. Special stains are usually needed for filamentous fungi. Antibodies (Fig. 4b), and DNA probes can also be used.

Clinical classification

Fungi are sometimes grouped by the clinical syndromes they cause:

- superficial and cutaneous mycoses, e.g. 'athletes foot' caused by *Trichophyton* spp.
- subcutaneous mycoses, e.g. sporotrichosis, ulcerative lesions caused by *Sporothrix* sp.
- systemic/deep mycoses, e.g. histoplasmosis, a pulmonary or generalized disease caused by *Histoplasma capsulatum*.

(a)

Fission   Budding (blastoconidium)   Germ tubes   Pseudo-hyphae

(b)

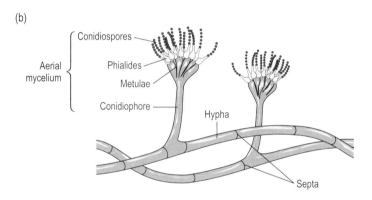

Aerial mycelium { Conidiospores — Phialides — Metulae — Conidiophore — }   Hypha   Septa

Fig. 3 **Morphology of fungi.** (**a**) Yeasts, e.g. *Candida;* (**b**) filamentous moulds, e.g. *Aspergillus.*

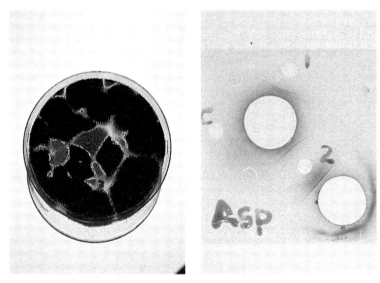

Fig. 4 **Diagnosis of *Aspergillus niger*.** (**a**) Colonies; (**b**) double diffusion plate test for aspergillus precipitins (antibodies). (Reproduced with permission from Inglis, T.J.J. 1999 Colour Guide Microbiology, 2nd edn. Churchill Livingstone, Edinburgh)

Characteristics of fungi (1)

- Nomenclature is confused because, historically, sexual and asexual forms were separately named.
- Classification depends on the sexual reproductive mode and structures.
- Structure is eucaryotic and, therefore, unlike the procaryotic bacteria, fungi have a nucleus with a nuclear membrane and the cytoplasm contains mitochondria, Golgi apparatus, lysosomes and an endoplasmic reticulum.
- Morphology is either small round yeasts or filamentous moulds with a mycelium of hyphae. Many fungi are dimorphic, with a yeast form in tissues and an environmental mycelial form.

CHARACTERISTICS OF FUNGI (2)

REPRODUCTION

Asexual reproduction is most commonly seen when haploid cells divide by mitosis to form spores, the chromosome number being unchanged. There is, by definition, no sexual mating prior to this sporulation. It is the only type of reproduction seen to date in some human fungal pathogens, which are therefore called Fungi imperfecti, sexual reproduction being considered the perfect state.

Sexual reproduction produces spores by mating when two haploid cells fuse to become diploid and then divide by meiosis. Many fungi need two colonies of the opposite mating type for sexual mating to occur (heterothallic fungi), while others need only one colony (homothallic). Spore formation occurs either within a sac called an ascus or partly on the surface of a bag called a basidium (see p. 6). The characteristic appearance of the spore-bearing structure and the spore is used in classification and hence identification of pathogens (Fig. 1).

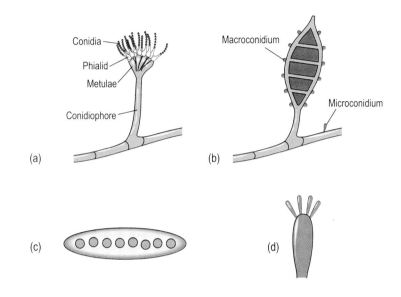

Fig. 1 **Fungal reproductive structures.** (**a**) Asexual fruiting structure. (**b**) Macroconidium and microconidium. (**c**) Ascospores in ascus. (**d**) Basidiospores on a basidium.

PATHOGENICITY

Fungal pathogenicity describes the pathogen's attack mechanisms, whereas resistance describes the host's defence mechanisms. It is essential to distinguish between:

- *primary pathogens*, i.e. fungi such as *Cryptococcus* spp. that can infect normal hosts
- *opportunistic fungi*, i.e. those only able to infect abnormal hosts with impaired defences resulting from, for example, antibiotics, cytotoxics, X-ray 'therapy', steroids and other immunosuppressant drugs, endocrine disease such as diabetes mellitus, or AIDS.

Although less is known about fungal than bacterial pathogenicity, the following mechanisms are recognised:

- mycotoxins
- hypersensitivity
- invasive infection.

Mycotoxins

Unlike bacteria, fungi are not known to produce any endotoxins. Some make exotoxins, i.e. elaborated outside the fungus. These are also only made outside the human body. There are three major groups causing mycotoxicoses:

- *Aflatoxins*, made by *Aspergillus flavus*, causing turkey X disease, and human disease via coumarin anticoagulant action. Aflatoxins are also carcinogenic.
- *Ergot alkaloids*, made by *Claviceps purpurea* infecting rye, causing St Anthony's fire or ergotism, with smooth muscle contraction, peripheral vaso-constriction and then gangrene. Ergot alkaloids are used with care in obstetrics to contract uterine smooth muscle.
- *Psychotropics*, such as psilocybin and the derivative LSD.

Hypersensitivity

Hypersensitivity results from repeated exposure to fungal spores and consequent immunoglobulin or sensitised lymphocyte production. There is no toxin production or tissue invasion. Inhalation of spores (e.g. of *Aspergillus*) causes allergic rhinitis, asthma and alveolitis, i.e. hypersensitivity pneumonitis.

Invasive infection

Colonisation is the continuing presence of the organism without disease, whereas infection means tissue invasion and damage (p. 14). Anatomically, fungal infection causes superficial, cutaneous, subcutaneous or systemic mycoses (see pp. 62–71). Tissue damage in infection can occur by at least six mechanisms (Fig. 2):

(a) direct invasion leads to distortion (e.g. fungus balls formed from *Aspergillus*, Fig. 3) hence tissue destruction and ill-understood toxic effects

(b) blood vessel wall invasion causes thrombosis, obstruction, ischaemia and infarction of tissues, e.g. with *Aspergillus* spp. and Zygomycetes (Phycomycetes) such as *Mucor*

(c) obstruction leads to secondary bacterial infection and further tissue damage

(d) embolism to distant vessels occurs, e.g. in endocarditis

(e) some fungi, e.g. *Histoplasma*, persist and even multiply in macrophages and neutrophils, being resistant to lysosomal enzymes

(f) capsule formation can occur in species such as *Cryptococcus*; this protects the fungus and may damage tissues.

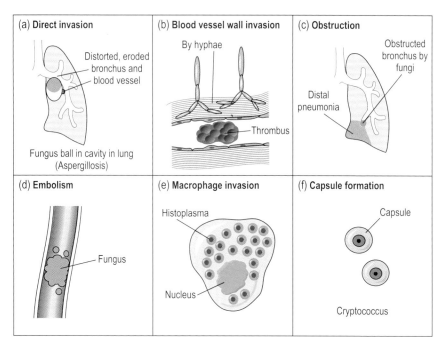

Fig. 2 **Six mechanisms of tissue damage in fungal infection.**

HOST RESISTANCE

All four lines of defence (see p. 19–23) are of some but variable importance against fungi:

1. Intact skin and mucous membranes plus the associated chemical and bacterial factors are primary barriers. Normal bacterial flora compete with fungi for nutrients, the balance being upset by antibiotics.

2. Non-specific inflammatory reactions occur, though neutrophil phagocytosis and macrophage activity is often less against fungal infection.

3. Antibodies and complement can kill *Aspergillus* and *Candida* spp., though many antibodies are not protective.

4. Cell-mediated immunity is the most important defence, and its loss, in diseases such as AIDS, causes a multitude of serious and often fatal fungal infections.

ECOLOGY

The ecology of fungi include their **reservoirs** in the environment and the **sources** of infection.

Environmental reservoirs

This is the natural ecology of many fungi, e.g. free living in soil or air.

These reservoirs are the usual source of most human pathogenic fungi, infection occurring after inhalation or implantation. Geophilic ('soil loving') dermatophytes (pp. 68–9) live in the soil.

Some have particular ecological niches, for example:

- *Cryptococcus neoformans*: soil and buildings contaminated with pigeon droppings: their high urea content is a fungal nutrient
- *Sporothrix schenckii*: rose thorns and rotting vegetation
- *Histoplasma capsulatum*: soil contaminated with bat, starling and chicken droppings.

In addition, some have particular geographic distribution:

- *Blastomyces dermatitidis* is limited to North America and a few foci in Africa.
- *Coccidioides immitis* is limited to hot dry areas in south-west USA, Central and South America.
- *Paracoccidioides brasiliensis* is limited to Central and South America.

Animals

Zoophilic ('animal-loving') dermatophytes obviously live on (and infect) animals, such as cats, dogs and horses.

Humans

Two yeasts are commonly found as part of our normal flora: *Candida albicans* on skin and mucous membranes, and *Pityrosporum ovale* on skin rich in nutrient lipids from sebaceous glands. Thirdly, dermatophytes are sometimes found in the absence of symptoms, probably as colonisation (see p. 14).

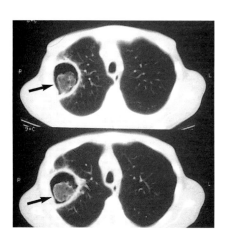

Fig. 3 **Fungus ball within cavity (CT chest).**

Characteristics of fungi (2)

- All fungi reproduce asexually with haploid cells dividing by mitosis to form spores.
- Fungi imperfecti, in which only asexual replication is known, include some human pathogens.
- Sexual reproduction by meiosis is the basis for definitive classification and naming.
- Primary pathogens are fungi able to infect normal hosts.
- Opportunist pathogens are only able to infect abnormal hosts with impaired defences.
- Fungi cause disease by mycotoxins, hypersensitivity or invasive infection with tissue damage.
- Cell-mediated immunity is the most important defence.
- Reservoirs and sources of infection are frequently the environment, sometimes animals, and commonly ourselves (*Candida*).

CHARACTERISTICS OF PROTOZOA

NOMENCLATURE AND CLASSIFICATION

Protozoa are unicellular eucaryotic organisms that were initially classified in the kingdom Animalia but are now usually considered in the kingdom Protista along with Algae.

There are three phyla of medical importance:

- Apicomplexa, containing the Sporozoa such as *Plasmodium* and the Coccidia such as *Toxoplasma*. Their more complicated life cycles are described on pages 72–3.
- Sarcomastigophora, containing the amoebae such as *Entamoeba histolytica* and the flagellates such as *Giardia* and *Trypanosoma*. Division is by binary fission, and locomotion by pseudopodia (amoebae) or whip-like flagellar movement (pp. 74, 75, 76, 77).
- Ciliophora, the ciliates, such as *Balantidium coli*, which divide by binary fission or by conjugation with nuclear exchange. Locomotion is by co-ordinated movement of the rows of hair-like cilia (p. 75).

The four medically important groups are the Sporozoa including the Coccidia, the Amoebae, the Flagellates and the Ciliates.

STRUCTURE

All protozoa are unicellular (Fig. 1). All have a *trophozoite* form with one or more nuclei containing nucleoli or karyosomes and bounded by a nuclear membrane, and the usual eucaryotic cytoplasmic organelles including mitochondria, ribosomes and endoplasmic reticulum. Vacuoles are often prominent and specialised structures such as sucking discs (*Giardia*) or a mouth-like cytosome (*Balantidium*) are found in the larger, more complex protozoa. Trophozoites have a cell membrane but no cell wall. Most intestinal protozoa also develop *cysts* that are more resistant than the fragile trophozoites to drying, cold or other environmental stresses.

REPRODUCTION

Amoebae replicate either by simple binary fission of the trophozoite or by formation of trophozoites within the multinucleate

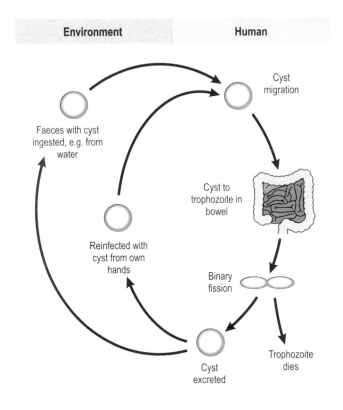

Fig. 1 **Typical life cycle of a simple protozoan.**

cyst. Flagellates and ciliates multiply by longitudinal binary fission.

Sporozoa and coccidia have complex asexual and sexual life cycles, which are described on pages 72–3. Hosts and generations classically alternate, e.g. the malarial asexual cycle (schizogony) in humans alternates with the sexual cycle (gametogeny) in mosquitoes.

PATHOGENICITY AND HOST DEFENCES

Pathogenicity

The pathogenicity of protozoa is not well understood and varies between genera. Details for particular organisms are given on pages 72–7. In general, however, protozoa:

- have fewer pathogenic mechanisms than bacteria
- have several surface attachment mechanisms (e.g. *Giardia*)
- are less invasive than bacteria
- have few known cytotoxins: *E. histolytica* means 'tissue lysis', and this species is an exception

- avoid host defences by several ingenious mechanisms, e.g. trypanosomes frequently alter their surface antigens, making antibody formation or vaccine development extremely difficult, while leishmaniae produce a superoxide dysmutase that protects them from macrophage superoxide so well they actually live within the macrophage!

Host defences

Host defences are similarly less well defined and variable, but in general:

- Chronic rather than acute disease and inflammation are associated with protozoan infection: again *E. histolytica* is an exception, causing acute amoebic dysentery.
- Eosinophils, so characteristic of metazoan infections, are few or absent.
- Both antibody formation and cell-mediated immunity (see pp. 22–3) are stimulated but are relatively ineffective.

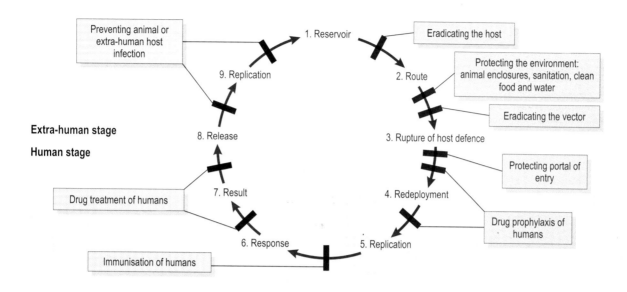

Fig. 2 **Achieving control by interrupting the life cycle (see also Fig. 2, p. 14).**

ECOLOGY: RESERVOIRS AND SOURCES OF INFECTION (Table 1)

Geographic distribution

Protozoal diseases are most common in poor and tropical communities, but this is principally because of poor control measures.

Reservoirs

Reservoirs can be:

- animal, e.g. for *Toxoplasma* and *Cryptosporidium*
- human, e.g. *Entamoeba* and *Giardia*
- soil contaminated with faeces.

Spread

Spread can occur by

- ingestion, either directly faecal-oral (or ano-oral), or by contaminated water, or unwashed food (particularly fruit and vegetables) or uncooked meat
- direct contact, either sexual or through the nose
- injection, usually by an insect vector, or, rarely, by needle or blood transfusion.

CONTROL

Theoretically, control can be achieved by interrupting the life cycle (Fig. 2) at any point. The practical difficulties are noted in the discussion of the individual species.

Table 1 **Protozoal infections**

Protozoa	Source	Spread by	Syndromes
Intestinal			
Entamoeba histolytica	Human faeces	Water, fruit, vegetables	Diarrhoeal, dysentery, dissemination
Giardia lamblia	Human faeces	As above	Diarrhoeal, malabsorption
Dientamoeba fragilis	Human pinworm eggs	As above	Diarrhoeal, discomfort (abdominal)
Balantidium coli	Pig faeces	Water and food	Diarrhoea, dysentery
Cryptosporidium spp.	Human, calf, sheep	Food and water, human sexual contact	Diarrhoea, dehydration, debility
Blood and tissue protozoa			
Naegleria fowleri	Free-living	Water	Meningitis, keratitis
Toxoplasma gondii	Cat faeces, beef, lamb, pork	Fingers, undercooked meat	Congenital, disseminated (silent)
Sarcocystis hominis	Dog faeces, beef	As above	Diarrhoea, discomfort
Leishmania spp.	Canines, rodents	Sandfly vector	Cutaneous, visceral
Trypanosoma gambiense	Human	Tse-tse fly	Encephalitis
Trypanosoma rhodesiense	Cattle	Tse-tse fly	Encephalitis
Trypanosoma cruzi	Human	Reduviid bugs	Cardiomyopathy, megacolon
Plasmodium spp.	Human	Mosquito	Disseminated

Characteristics of protozoa

- Protozoa include intestinal parasites, e.g. *Entamoeba histolytica* and *Giardia lamblia*, and blood and tissue parasites, e.g. *Toxoplasma gondii* and malarial parasites.
- Protozoa are unicellular with eucaryote cellular structures. They all have a fragile trophozoite stage and most have a resistant cyst form.
- All are small, invisible without a microscope.
- All have life cycles outside the human host, and most can multiply in humans. Infection is by ingestion, by inhalation or by insect bite.
- Their life cycles vary from direct passage of trophozoite during coitus (*Trichomonas*) or passage of cysts in faeces and subsequent ingestion (*Entamoeba* and other intestinal protozoa) to complex alternation of generations in different hosts (malaria).
- Eosinophilia is not found in protozoal infections.
- Protective immunity is poorly developed in most protozoan infections, which are very common and may be multiple.
- Drug prophylaxis and therapy are often unsatisfactory.

MULTICELLULAR PARASITES

NOMENCLATURE AND CLASSIFICATION

It is sensible but scarcely scientific to include some of these medically important parasites in Microbiology – only the eggs of 25 cm roundworms and 6 metre tapeworms are microscopic! A general overview of these varied and important parasites is given here, with details on pages 79–85.

The sub-kingdom Metazoa of the kingdom Animalia contains the **helminths** (worms) in two phyla:

- **Nematoda** are roundworms with round cylindrical bodies. Some are intestinal parasites, e.g. hookworms and thread (pin) worms, while others are blood and tissue parasites, such as filarial and guinea worms.
- **Platyhelminthes** are flatworms ('platelike') and are divided into Trematoda, comprising the leaf-like flukes such as schistosomes, and Cestoda, which are the ribbon-like tapeworms such as the beef and pork tapeworms.

A third phylum of medical importance in the Metazoa is the **Arthropoda** ('jointed limbs') comprising invertebrates with segmented bodies, paired jointed limbs, a hard exoskeleton of chitin and well-developed respiratory and digestive systems (p. 71). It includes:

- **Crustacea**, including crabs, shrimps and copepods, some of which are intermediate hosts in the life cycles of some helminths.
- **Arachnida**, including mites and ticks (vectors for some microbial diseases) plus venomous animals such as spiders and scorpions. All adults have eight legs, but no wings or antennae.
- **Insecta** comprising mosquitoes, bugs, fleas and lice (vectors for further important microbial diseases) plus venomous stinging animals such as bees and wasps. All have six legs, antennae and wings.

STRUCTURE

All the above are multicellular eucaryotes: their cells have nuclei, nucleoli, nuclear membranes, mitochondria, ribosomes, endoplasmic reticulum and other specialised structures within a cell membrane and cell wall. Their multicellular structure is variable and complex.

Nematodes (roundworms) are cylindrical, have complete digestive systems and separate sexes.

Trematodes (flukes) are flat, leafshaped worms with an oral muscular sucker and an incomplete digestive system. Schistosomes have separate sexes but other trematodes are hermaphrodites (see below).

Cestodes (tapeworms) are ribbon-like and may reach enormous size. The head (scolex) usually has characteristic hooklets and suckers, and the body is divided into segments (proglottids). There is no digestive system, nutrients being absorbed through the soft body wall from the host gut. Cestodes are also hermaphrodites, each proglottid having male and female sexual organs.

REPRODUCTION

Some parasites have very complex life cycles; a knowledge of the life cycle is essential for understanding the method of infection and the prevention of disease. In the simplest life cycle,

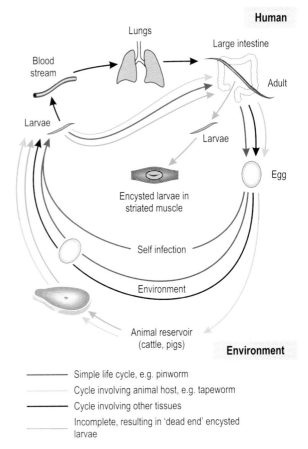

Fig. 1 **Levels of complexity in the life cycles of multicellular parasites.**

seen in pinworm and whipworm infections, eggs are ingested, hatch to larvae in the gut and develop to adult worms that produce eggs that embryonate outside the body (Fig. 1). Similarly ingested encysted larvae in meat or on vegetables develop in the gut into tapeworms or *Fasciolopsis*. In more complex cycles, larvae may pass through the lungs before migrating to the adult's niche in gut (e.g. roundworm, hookworm) or blood vessels (schistosomes). Lastly, helminths ingested, or injected by insects, migrate through tissues to live in the lungs, liver, skin, muscles, lymphatics or skin.

The part of the life cycle outside the human body also increases in complexity, from simple eggs through skin-penetrating larvae to the development of infectious larvae in insects or other animals and finally to alternation of generations when schistosomes have asexual reproduction in snails and sexual reproduction in humans. Individual life cycles are discussed below (pp. 78–85).

PATHOGENICITY AND RESISTANCE

Pathogenicity

As with protozoa, the pathogenicity of many multicellular parasites is variable and poorly understood (see pp. 78–85 for particular organisms). In general, however, these parasites:

- have fewer known pathogenic mechanisms than bacteria
- have several surface attachment mechanisms (e.g. tapeworms)

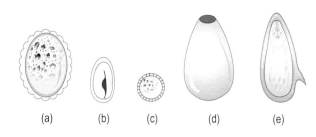

Fig. 2 **Eggs of helminths.** (**a**) Roundworm; (**b**) pinworm; (**c**) tapeworm; (**d**) intestinal fluke; (**e**) blood fluke.

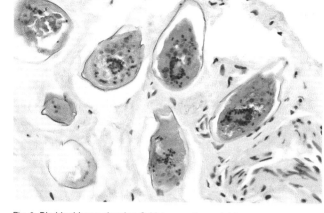

Fig. 3 **Bladder biopsy showing *Schistosoma haematobium* eggs.**

- are often less invasive than bacteria, living only in gut
- have few known cytotoxins
- avoid host defences by several ingenious mechanisms, e.g. schistosomes cover themselves with host plasma proteins, hence are antigenically invisible to host defences
- show marked but largely unexplained tissue tropism.

Resistance

Host defences are similarly less well defined and variable but, in general, metazoan infection results in:

- chronic disease and chronic inflammation more often than acute
- raised levels of eosinophils (which is characteristic of metazoan infections) occurring in response to parasite surface polysaccharides and glycoproteins. Concomitant increased IgE production with the eosinophilia assists killing of parasites
- stimulated antibody formation and cell-mediated immunity (see pp. 22–3) although these are relatively ineffective.

ECOLOGY: RESERVOIRS AND SOURCES OF INFECTION (Table 1)

Geographic distribution

Metozoan diseases, like protozoan, are commonest in poor and tropical communities, but again this is principally because of poor control measures.

Reservoirs

Reservoirs include:

- animal, e.g. *Taenia* or *Echinococcus* spp.
- human, e.g. pinworm, whipworm
- soil contaminated with faeces, e.g. roundworm.

All trematodes have at least one *intermediate* host as well as the reservoir host (pp. 84–5).

Spread

Spread can be by ingestion, either directly faecal–oral or by contaminated water or unwashed food (particularly fruit and vegetables) or uncooked meat; by direct contact, through the skin; or by injection by an insect vector.

DIAGNOSIS

Most of the helminths encountered in developed countries are intestinal parasites, and diagnosis usually depends on detection of the parasites or their eggs in faeces (Fig. 2). Eosinophilia is often used as an indicator of infection. Biopsy may be necessary for tissue parasites (Fig. 3).

CONTROL

Control, as with protozoa, can be by attack on:

- the host(s)
- the vector or route of spread
- the human infection, by immunisation (in theory) or by drug prophylaxis or treatment.

Table 1 **Features of infection involving multicellular parasites**

Group	Examples	Comment
Nematodes		
Intestinal	Pinworm, roundworm	Spread from faeces via food, water
Tissue/blood	Filariae, guinea worm	Spread via food or insect bites or by skin penetrating larvae
Animal nematodes	Dog and cat ascaris	Human is an accidental host so life cycle is not completed; disease is caused by migrating larvae
Trematodes	Liver fluke	Spread from faeces via infected food from the intermediate host
Cestodes	Beef tapeworm	Humans ingest encysted larvae or eggs from the intermediate host
Arachnida	Mites, ticks, spiders	Venomous bites, or acting as vectors for microbial disease
Crustacea	Crabs, shrimps	Intermediate hosts for helminths
Insecta	Mosquitoes, fleas	Venomous bites, or acting as disease vectors

Multicellular parasites

- Multicellular parasites include nematodes (roundworms), trematodes (flukes), cestodes (tapeworms) and arthropods.
- All are metazoa, with eucaryote cellular structures.
- All are large, visible without a microscope.
- All have life cycles outside the human host and most cannot multiply in humans. Infection is by ingestion, by skin penetration or by insect bite.
- Life cycles vary from simple embryonation outside humans, to complex alternation of generations in different hosts.
- Eosinophilia is found in almost all helminth infections.
- Protective immunity is poorly developed in most helminth infections, which are very common and often multiple.
- Drug prophylaxis and therapy are often unsatisfactory.

HOST–MICROBIAL RELATIONSHIPS

Microbes and hosts live in an uneasy peace between microbial attack and host defence. Some microbes within the normal human microflora can become pathogenic if the state of the host changes. Other microbes are present in the environment and can infect a host if the host defences are penetrated. In this and the following six sections the relationship between microbes and humans in health and disease is discussed including, here, the terminology and principles involved.

CONTAMINATION, COLONIZATION AND INFECTION

It is important to distinguish between the above three concepts though the practical distinction is not so simple!

Contamination is the transient presence of microbes, pathogenic or non-pathogenic, on our skin or other body surfaces, without any injury or invasion of our tissues.

Colonization is the continuing presence of such microbes, usually for weeks, months or even years, again without injury or invasion of our tissues.

Infection is injury to or invasion and damage of our tissues by microbes. Tissue invasion is usual in microbial attack, but cholera is an example of severe injury and disease from a bacterial toxin without any significant tissue invasion.

The microbes which can infect us fall into one of three groups (Fig. 1):

- *Normal flora* only inhabit the surfaces of the body although these can be both external, the skin, hair and nails, and internal, the mucous membranes of our digestive tract, the respiratory tract down to the larynx, the terminal urethra and the vagina.
- *Aggressive pathogens* or *primary pathogens* are those microbes which can cause disease in normal hosts, i.e. those with normal defence mechanisms.
- *Opportunistic pathogens* are those microbes which do not cause disease in normal hosts but do so in those with impaired defences (Fig. 1).

MICROBIAL ATTACK AND HOST DEFENCE

Nine steps can be distinguished in the development of infection (Fig. 2):

1. The reservoir or source of the infecting organism (p. 18).
2. The route by which the organisms spread externally to reach us.
3. The rupture of our first line of defence, the skin or mucous membrane, i.e. entry into the body.
4. The redeployment of the invading microbes, i.e. spread within the body, opposed by our second line of defence, complement and phagocytosis (pp. 20–1).
5. The replication of the invading microbes (pp. 20–1), again opposed by our second line of defence.
6. The response of our immune system (pp. 20–3) and our general responses to infection with tissue damage.
7. The result of successful microbial attack, i.e. clinical disease (pp. 86–195), opposed by our external defences.
8. The release of infectious units into the environment may follow step 7, especially for protozoan cysts and multicellular parasites (pp. 72–85) with extra-human life cycles.
9. Replication, in Step 1.

Hopefully the body's external defences can deal with the infection and the outcome is recovery rather than recumbency (chronic disease) or rigor mortis.

PATHOGENICITY, VIRULENCE AND INVASIVENESS

Pathogenicity is the ability to cause disease. It, therefore, includes virulence and toxins, microbial factors determining

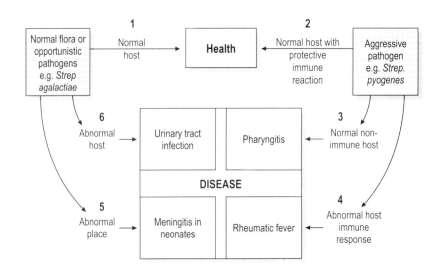

Fig. 1 **Six major microbiological principles illustrated by streptococcal interaction with host defences.**

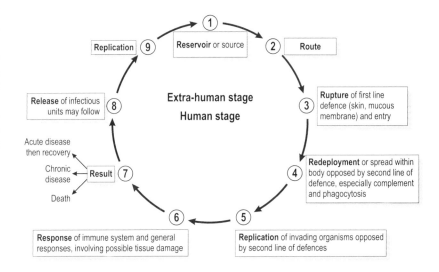

Fig. 2 **The nine steps in the development of infection.**

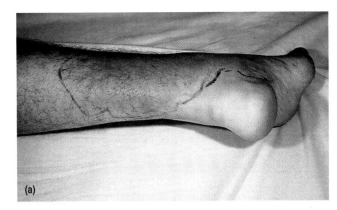

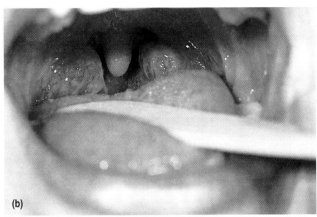

Fig. 3 *Strep. pyogenes* can cause both (a) erysipelas (local spread) or (b) tonsillitis (direct invasion).

Table 1 **Some important microbial toxins**

Toxin and organism	Result	Mechanism
Anthrax toxins		
B. anthracis		
Oedema factor	Oedema	Adenylate cyclase
Lethal factor	Pulmonary oedema	Cytotoxic
Botulinum toxin		
C. botulinum	Neurotoxins	Neuromuscular block
Cholera toxin		
Vibrio cholerae	Diarrhoea	Adenylate cyclase, CAMP activation
Clostridial toxins		
C. difficile	Diarrhoea	Membrane permeability
C. perfringens		
alpha toxin	Necrosis	Phospholipase C
beta toxin	Oedema	Capillary leak
delta toxin	Haemolysis	Haemolysin
kappa toxin	Spread, necrosis	Collagenase
mu toxin	Spread, necrosis	Hyaluronidase
Diphtheria toxin		
C. diphtheriae	Cytotoxic	ADP ribosylation of elongation factor 2
Enterotoxin		
E. coli and enteric bacilli	Diarrhoea	As cholera toxin
(Also cytotoxin)	Haemorrhage	As shigella toxin
Shigella toxin	Haemorrhagic enteritis	60S ribosome inactivation

See also Staphylococci (pp. 28–9) and Streptococci (pp. 30–31)

adherence, invasiveness (the ability to enter and spread in the body), the ease and speed of microbial replication, and their ability to impede host defences.

Two related concepts are the **infective dose**, which is the number of microbes necessary to cause infection, and the **period of infectivity**, which is the time during which a source, usually human, is disseminating organisms and hence is potentially able to cause infections in others.

Virulence factors

Virulence factors are the microbial factors essential for the development of infection and disease.

Toxins are usually protein exotoxins, though many Gram-negative organisms have a very important complex lipopolysaccharide endotoxin. Some of the more important are summarized in Table 1.

Adhesins determine adhesiveness to cells and are usually found on fibrillae, fimbriae or pili, the fine hair-like structures on the outside of many bacteria. They are known to be important in streptococci and staphylococci, in *Neisseria* spp., *E. coli*, *Pseudomonas* spp. and *Shigella* spp., and have been found in many other bacteria, in *Candida albicans* and in some protozoa and viruses.

Impedins (as the name suggests) impede host defence mechanisms (see pp. 19–23) and act against:

- *anatomical barriers:* impedins include bacteriocins and factors for direct skin penetration and mucosal invasion, and connective tissue-disrupting enzymes
- *serum factors:* impedins can act against complement ('serum resistance') and fibrinolysins
- *phagocytosis:* impedins include protective capsules, blocking of the oxidative burst, stopping phagosome–lysosome fusion and resisting lysosomal enzymes.

- *antibody production:* impedins are antibody-degrading enzymes
- *cell-mediated immunity:* impeded by mechanisms such as surface disguise (p. 25).

In addition to these microbial factors in pathogenesis, an abnormal host response may cause further damage (see p. 24).

Interplay of factors

The interplay between the virulence of the microbe (normal flora or aggressive pathogen), the invasion of the host and the state of the host defences is illustrated by the streptococci, where we see six major microbiological principles (Fig. 1).

DISEASE AND MICROBIAL SPECIES

While sometimes one species of organism only causes one disease, and one disease is only caused by one species (e.g. anthrax, tetanus), very frequently one species causes numerous diseases, e.g. *Strep. pyogenes* (Fig. 3); similarly, one disease, e.g. impetigo, can be caused by different organisms.

Host–microbial relationships

- Contamination is transient, colonization is more permanent but produces no tissue damage, while infection entails tissue damage.
- Normal flora live on our skin and mucous membranes and cause no disease in normal hosts. Aggressive or primary pathogens cause disease in normal hosts, while opportunist pathogens only cause disease in hosts with impaired defences.
- Microbial virulence factors in pathogenicity include toxins, adhesins and impedins.
- Infection depends on the balance between virulence, invasiveness, host defences and host response.
- Various microbial species cause more than one disease and many diseases can be caused by a variety of microbial species.

NORMAL FLORA AND OPPORTUNISTIC INFECTIONS

DEFINITION

Microbes that coexist with humans are adapted to life on the surface of our body including the mucous membranes lining the surface of the respiratory, digestive, urinary and genital systems, which are in continuity with the skin and actually or potentially in contact with our exterior environment. These are the normal human flora and have a typical spectrum in varying areas of the body (Table 1):

- flora of the small bowel, as expected, is intermediate between that of the mouth and colon
- vaginal flora resembles a mixture of that of the skin and colon
- nose, conjunctiva and external ear have similar flora
- *Staph. aureus, Staph. epidermidis* and *Streptococcus* spp. are widespread
- *Escherichia coli* and other aerobic Gram-negative rods and *Enterococcus* spp. are far outnumbered in the gut by anaerobic rods (*Bacteroides* and *Fusobacteria*) and anaerobic cocci (*Peptostreptococci, Peptococci* and *Veillonellae*).
- *Candida albicans* is frequently found in the mouth, colon, vagina and skin.

USEFUL ROLES

Protection from invading microbes

Normal flora form part of our first line of defence (see p. 19). There are at least two mechanisms: first their simple physical presence, often in large numbers (Fig. 1) and well established in their niche with necessary nutrients, and, secondly, the antimicrobial protein bacteriocins and even antibiotics which some produce. When part of our bacterial flora is removed by antibiotics intended as therapy, we may instead suffer from overgrowth of resistant normal flora such as *Candida albicans* or *Clostridium difficile*, causing thrush or enterocolitis, respectively. We can also be infected by a much lower dose of exogenous pathogens, e.g. *Salmonella* spp.

Immune stimulation

Our bacterial normal flora in particular stimulate production of the surface antibody IgA in our mucous membranes, presumably protecting us from invasion of deeper tissues by normal flora, or by similar exogenous species.

Human nutrition and metabolism

Normal flora in the gut, including *E. coli* and *Bacteroides* spp. (now called *Prevotella* spp.) synthesize vitamin K. In addition these bacteria make enzymes including sulphatases and glucuronidases which deconjugate bile salts and sex hormones after excretion from the liver so they can be reabsorbed in the so-called enterohepatic loop.

HARMFUL ROLES

The normal flora also have several less desirable attributes.

Table 1 **Normal flora in different regions of the body**

Organism	Skin	Conjunctiva	Nose	Mouth/oropharynx	Small bowel	Colon	Vagina
Staph. aureus	+	±	+	±	±	±	−
Staph. epidermidis	+++	++	+++	−	−	±	±
Streptococci	±	±	±	+++	−	+	±
Enterococci	−	−	−	−	+	+	±
Diphtheroids	++	−	−	−	−	−	+
Lactobacilli	−	−	−	−	++	++	+++
Haemophilus spp.	−	±	±	±	−	−	−
Moraxella spp.	−	±	±	±	−	−	−
Neisseria spp.	−	±	±	+	−	−	−
E. coli	±	±	−	±	+	+++	±
Klebsiella spp.	±	±	−	±	+	++	±
Other aerobic GNR[a]	±	−	+[c]	±	±	±	±
Bacteroides and *Fusobacterium* spp.	±	−	−	+++	+	+++	±
Veillonella spp.	−	−	−	+++	−	±	+
Clostridium spp.	−	−	−	−	+	+++	+
Peptococcus and *Peptostreptococcus* spp.	±	−	−	±	++	+++	+
Mycobacteria spp.	±	−	−	−	+	±	±
Mycoplasma spp.	−	−	−	−	−	−	±
Treponemes	−	−	−	++	−	±	−
Candida albicans	±	−	−	+	−	±	+
Other fungi	+++[b]	−	−	−	−	±	±

+++, almost always present; ++, usual; +, frequent; ±, occasional or rare.
[a] GNR, Gram-negative rods.
[b] *Pityrosporum* spp.
[c] *Pseudomonas aeruginosa* frequent in external ear.

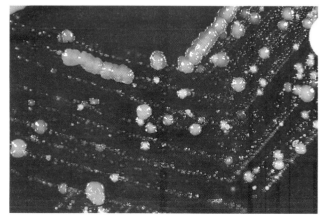

Fig. 1 **Oral flora colonies on horse blood agar.**

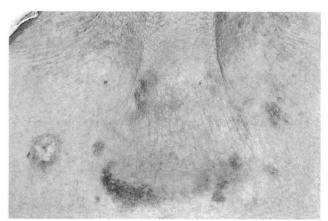

Fig. 2 **Fungal skin infection.**

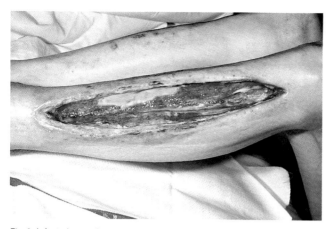

Fig. 3 **Infected wound.** *Staph. aureus* from the skin is the commonest cause.

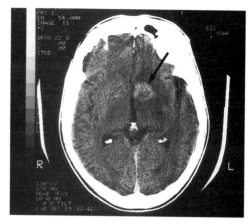

Fig. 4 **Toxoplasmosis: brain cyst (light) and oedema (dark)**.

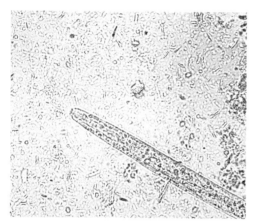

Fig. 5 *Strongyloides stercoralis* larva.

Opportunist pathogenicity

Opportunist pathogenicity occurs when the host defences are impaired and microbes which do not cause disease in normal hosts then become pathogenic. The normal flora is an obvious source of such organisms: they occur in large numbers adjacent to our tissues, well placed to invade and infect if our host defences become impaired. For example, when patients become neutropenic through disease or anti-cancer 'chemotherapy', they are frequently infected by their own gut, skin or respiratory flora unless precautions are taken (Fig. 2). Similarly, a surgical wound can give entry of the patient's skin flora into their deeper tissues, causing a surgical wound infection (Fig. 3). (Other opportunists come from hospital and natural environments, and sometimes from other people, e.g. from the skin or nasopharynx of hospital staff to immunocompromised patients.)

Possible source of carcinogens

Sources of carcinogens is a controversial topic, but it is known that enzymes from the gut flora, including sulphatases, can modify ingested chemicals to known carcinogens. Whether this is actually important in the production of colonic or bladder carcinoma is as yet unknown.

LATENT PATHOGENS

Pathogenic organisms which are *not* normal flora can become resident and lie dormant in a host with normal defences, either after subclinical infection (inapparent, asymptomatic, without symptoms) or after clinical infection with apparent recovery. Months, years or even decades later they cause clinical disease, especially when host defences become impaired ('compromised') by disease, drugs or other cause. These infections in compromised patients are discussed on pages 192–3.

Latent pathogens include:

- bacteria, particularly intracellular ones, causing brucellosis, listeriosis, melioidosis, nocardiosis, salmonellosis (especially *S. typhi*), tuberculosis and other mycobacterial infections such as that caused by the *M. avium* complex (MAC).
- fungi: *Candida*, *Pneumocystis carinii*, *Cryptococcus neoformans* and *Histoplasma capsulatum*.
- protozoa, particularly those with persistent cyst forms, as in amoebiasis (*E. histolytica*), cryptosporidiosis (*C. parvum*) or toxoplasmosis (*T. gondii*) (Fig. 4)
- helminths (worms), particularly those with a life cycle wholly in humans, e.g. strongyloidiasis from *S. stercoralis* (Fig. 5)
- viruses.

It is these infections which occur in AIDS patients who have widespread severe immune impairment. Details of these pathogens and diseases are found on the relevant spreads.

Normal flora and opportunistic infections

- Normal flora are important in:
  - protection from invading microbes
  - immune stimulation
  - human nutrition and metabolism
  - opportunist infection when host defences are impaired
  - possible production of carcinogens.
- Latent pathogens
  - lie dormant in a normal host but cause clinical disease when host defences are compromised
  - often are intracellular bacteria, tissue fungi, protozoa with cyst forms, or helminths with life cycles only involving humans
  - cause disease that is more acute, more severe and more life-threatening in the immunocompromised host than in normal hosts.

ESTABLISHMENT OF DISEASE: ATTACK

The establishment of microbial infection requires firstly a *reservoir* or source of infection, secondly a *route* of transmission and thirdly *rupture* of the non-specific surface defences of the body, providing a portal of entry.

RESERVOIRS AND SOURCES

Epidemiology is the study of the behaviour of diseases in the community rather than in individual patients. It includes the study of the reservoirs and sources of human diseases (Fig 1). Strictly speaking, **a reservoir is the organism's usual residence**, where it resides, replenishes and replicates. **The source is the site from which spread immediately occurs to the host**, directly or indirectly. For example, the soil is the reservoir of eggs or cysts of many parasites, while soil-contaminated vegetables are the source of human infection with the parasite. Sometimes the reservoir and the source are the same site, e.g. nasopharyngeal carriage of streptococci and staphylococci.

Epidemiological information can be used statistically to generate **incidence rates** (number acquiring a disease in a certain period, divided by total population) and **prevalence rates** (number of people having a disease at a specific time divided by total population). These figures can give indications of populations at-risk, e.g. using age-adjusted data the high incidence of shigellosis among children under 5 years is highlighted in what is a relatively rare disease in the population as a whole. Epidemiology can also indicate possible causes of disease and possible means of prevention, e.g. diseases transmitted by insect vectors or by food.

Reservoirs

Reservoirs can be almost anywhere on our planet, including:

- *soil:* for many parasitic infections (pp. 72–83)
- *water or other fluids:* even disinfectants! (p. 197)
- *inanimate objects:* often important in hospital-acquired infections (p. 190)
- *animals:* infections from this source are called zoonoses (pp. 184–5)
- *people:* spread their normal flora (harmless to themselves) to those with impaired defences; people called carriers may be colonized with primary pathogens against which they have protection from their own antibodies or immunization, but which cause disease in those not so protected.
  Note below four other categories of people who can be temporary sources rather than permanent reservoirs.

Sources

Sources of infections are also extremely varied, but can be divided into three groups.

- *Inanimate objects* include water and food, particularly fruit and vegetables for many parasitic and bacterial diseases. Industrial or hospital equipment may also be a temporary or immediate source as well as a permanent reservoir.
- *Animals* may be a source, e.g. some snails, crustacea or fish act as intermediate hosts between humans and another animal reservoir in many cestode infections (pp. 82–3). Uncooked or undercooked meat or fish is another source, e.g. clostridial or staphylococcal food poisoning
- *People* are the third important source: those infected with an obvious illness; those incubating an illness without developed symptoms; those with inapparent or subclinical

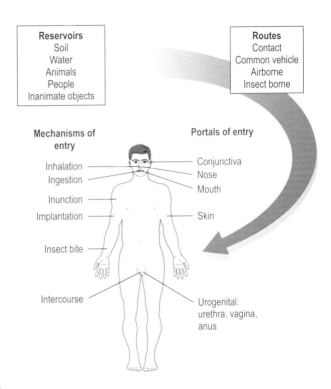

Fig. 1 **Reservoirs, routes and entry of disease.**

infection; or those improving, i.e. convalescing but still infectious to others. They may, therefore, be patients, family members, hospital and health-care staff, or daily or casual contacts, and only those obviously infected will be easily identifiable as a potential source.

ROUTES OF TRANSMISSION

There are four major ways in which microbes can move from a reservoir or source to a host.

- *Contact* can be either *direct* contact ('person-to-person') such as a boil or a skin infection transmitting directly by touch or *indirect* contact via some object. (Droplet infection is sometimes illogically classified as a contact infection.)
- *Common vehicle* transmission occurs when some object touches first the reservoir or source and then the host. This common vehicle infects many; replication of the microbe may occur in the food or liquid vehicle. Common vehicle transmission is really a special form of indirect contact, differentiated for epidemic investigations. Food and water are the commonest common vehicles, but batches of blood products, intravenous or dialysis fluids, or drugs may also be contaminated common vehicles.
- *Airborne* transmission occurs either in *droplets* larger than 5 nm, which only travel about 1 metre, or by *droplet nuclei* less than 5 nm, skin squames or dust, which can be carried for kilometres.
- *Vector-borne* transmission usually involves insects (vector means carrier). They may be passive vectors ('flying pins') like houseflies, simply carrying salmonellae externally from a reservoir to us, or be harbouring the organism internally without change in the organism (as *Yersinia pestis* within a

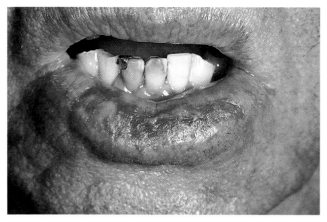

Fig. 2 **Candidiasis of the lips caused by lip licking allowing breakage of the skin barrier.**

Fig. 3 **Ixodes tick.**

plague flea), or be an active, essential part of the biological life cycle of the organism as mosquitos are in malaria.

RUPTURED DEFENCES: ENTRY

Once bacteria have been carried to the human body by one of the above four routes of transmission, how do they enter?

The four major *portals of entry* are:

- mouth and gastrointestinal tract
- respiratory tract
- skin (by three different mechanisms) (Figs 1 and 2)
- genital tract.

The eye and placenta are less common.

Exogenous infection ('arising from outside') therefore can occur by seven mechanisms:

- *ingestion:* by eating or drinking something contaminated by contact or common vehicle
- *inhalation:* usually of airborne infection, but in hospital it may be direct contact into the respiratory tract
- *inunction:* literally rubbing on, hence by direct or indirect contact to skin or conjunctiva
- *implantation:* by traumatic or surgical wounds, or by transfusion or organ transplant
- *insect bite:* in vector-borne transmission (Fig. 3)
- *intercourse:* in sexually transmitted diseases arising from some variety of sexual intercourse. While this is microbiologically only another form of direct contact, the pathogens and portals have sufficient special characteristics to be listed separately
- in utero.

Table 1 lists examples of pathogens linked to specific portals of entry.

Endogenous infections ('arising from within') occur when our normal flora become opportunist pathogens (p. 17).

THE FIRST LINE OF DEFENCES: SKIN AND MUCOUS MEMBRANES

The skin and mucous membranes enclose the body forming a physical barrier to infection; additional physical factors such as ciliary and mucus movement over the mucosal surfaces also contribute. The normal flora on these surfaces reduce the ability of pathogens to survive by competing for nutrients and by producing anti-microbial chemicals (Table 2 and page 16).

Table 1 **Pathogens using differing portals of entry**

Portal	Disease	Pathogen example
Skin		
Direct invasion	Abscesses	*Staph. aureus*
Wounds	Tetanus	*Clostridium tetani*
Bites	Plague	*Yersinia pestis*
Nose	Pneumonia	*Strep. pneumoniae*
Conjunctiva	Trachoma	*Chlamydia trachomatis*
Mouth	Cholera	*Vibrio cholerae*
	Salmonellosis	*Salmonella* spp.
	Diphtheria	*Corynebacterium diphtheriae*
Urethra	Gonorrhoea	*Neisseria gonorrhoeae*
	Urinary tract infection (UTI)	*Escherichia coli*
Vagina	Vaginitis	*Trichomonas vaginalis*
	Gonorrhoea	*Neisseria gonorrhoeae*
Placenta	Syphilis	*Treponema pallidum*

Table 2 **Chemical protection of body surfaces**

Organ/system	Origin	Protective chemicals
Skin	Sebaceous glands	Fatty acids
	Sweat	Fatty acids
Mucous membrane	Secretions	Mucus, lysozyme
Gut	Salivary glands	Thiocyanate
	Parietal cells (stomach)	Hydrochloric acid
	Hepatocytes	Bile acids
	Normal flora	Fatty acids (short chain)
Lung	A cells	Surfactant

Establishment of disease: attack

- Permanent reservoirs of infection are soil, water, inanimate objects, animals and people.
- Immediate sources of infection are usually inanimate objects, (including food and water), animals and people.
- Routes of transmission are direct and indirect contact, common vehicle, airborne and vector-borne.
- Exogenous infection enters by rupture of body defences and occurs by ingestion, inhalation, inunction, implantation, insect bites and intercourse, and in utero.
- Endogenous infection occurs when our natural flora become pathogenic.

NON-SPECIFIC DEFENCES

The non-specific defences form part of the normal constitution of the body and as such do not need prior contact with a particular microbe, i.e. they are constitutive or innate.

By contrast, the specific defences (pp. 22–3) form the *immune response*, and are *inducible*, i.e. are only present when they have been induced by the presence of a particular microbial species.

The non-specific defences include:

- normal flora (p. 16)
- physicochemical factors (p. 19)
  — skin
  — mucosal membranes with cilia and mucous secretions
- cellular factors (phagocytic cells)
  — neutrophils
  — macrophages
  — natural killer cells
- humoral factors in blood, mucosal secretions and cerebrospinal fluid
  — complement
  — opsonins
  — enzymes.

These innate defences are most useful in providing protection against pyogenic organisms, fungi and multicellular parasites.

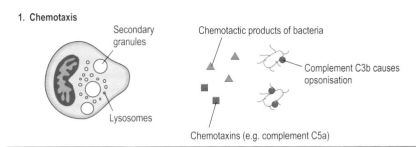

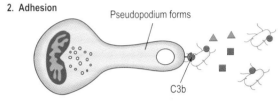

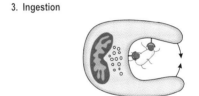

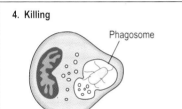

(a) Oxidative burst (oxygen dependent killing)
(b) Oxygen independent killing
(c) Death and degradation

Fig. 1 **The steps in phagocytosis.**

SECOND LINE OF DEFENCE—CELLULAR FACTORS

Phagocytosis

Phagocytosis ('the eating of cells') is the process by which neutrophils, and tissue macrophages derived from monocytes, find, engulf and kill microbes (Table 1). Neutrophils, particularly, phagocytose extracellular bacteria, while macrophages are most active against intracellular bacteria, protozoa and viruses. Eosinophils are involved in metazoan parasitic infections.

The first steps in phagocytosis are (Fig. 1):

- **Chemotaxis**. The attraction of neutrophils by chemicals, some made by bacteria, others by the host's complement cascade.
- **Adhesion**. This is facilitated by C3b from the complement system (and by specific antibodies if they exist at this time).
- **Ingestion**. This occurs when membrane activation leads to pseudopodia forming around the organism and its inclusion into a vacuole called a phagosome.

Opsonins are cofactors that coat microbes and enhance the ability of neutrophils to engulf them. Opsonins include complement C3b, C-reactive protein and antibodies. The last enables the specific immune system to activate the innate system.

Within the phagosome the microbe is killed and degraded by the following mechanisms initiated by fusion of the granules of the phagocyte with the phagosome:

- **Oxygen-dependent killing (oxidative burst)**. Active oxygen molecules – superoxide anion, hydrogen peroxide, 'singlet' activated oxygen and hydroxyl free radicals – are formed from O_2 and NADPH in the presence of cytochrome b_{245}. The active oxygen molecules kill microbes.

Table 1 **Cellular defences**

Phagocytic cell	Origin	Anti-microbial mechanism
Neutrophil	Bone marrow	Margination, diapedesis, and chemotaxis, then phagocytosis, degranulation, O_2-dependent killing (H_2O_2), O_2-independent killing, e.g. by lactoferrin, lysozyme
Eosinophil	Bone marrow	O_2-independent especially anti-parasitic
Monocyte	Bone marrow	As neutrophils except for degranulation (In addition, inducible activity, see p. 23).
Macrophage	Blood monocyte	As monocytes, major inducible activities

- **Oxygen-independent killing after phagosome–lysosome fusion with degranulation.** Lysosomes are primary granules within neutrophils that contain myeloperoxidase, lysozyme and cationic proteins which damage bacterial membranes and lead to microbial killing and digestion. In addition, the secondary 'specific' secretory granules, containing lactoferrin and lysozyme, discharge these inside the phagosome contributing to bacterial killing and digestion.
- **Degradation.** Once microbes are killed, they are degraded by hydrolytic enzymes and the products released from the neutrophils. When the granule contents are released outside the phagocyte, they can lead to tissue damage.

Natural killer cells

These are special lymphocytes which recognise virus-infected cells and release a cytolysin which kills the host cell before it can release further virus. Like interferons, which are natural anti-viral agents, they are outside the scope of this book.

SECOND LINE OF DEFENCE —HUMORAL FACTORS

Complement

The complement system is a group of about 20 serum proteins which respond to a stimulus such as invading microbes by a cascade of serial chemical reactions where the product of one reaction is the enzyme catalysing the next. There are two major pathways in the complement system, the classical, which was found first and is important in antibody action (the third line of defence, pp. 22–3), and the alternative pathway (Fig. 2), discovered later, which is important in the second line of defence together with the phagocytes.

The major actions of complement are:

- assisting phagocytosis by facilitating adherence, stimulating mast cells to release chemotaxins, stimulating chemotaxis directly, stimulating the respiratory burst
- increasing vascular permeability
- lysing organisms.

These occur by a complex series of reactions which are much simplified here.

Even the nomenclature appears complex (Table 2).

Further components

In addition to the major components of phagocytosis and complement activation, the non-specific tissue defences include:

- some simple chemicals and ions (Table 3)
- some individual proteins (Table 3)
- some complicated protein systems.

The individual proteins include the following:

- *Acute phase proteins.* These include C-reactive protein (CRP) which binds to numerous bacteria and activates the classical complement pathway independently of antibody, causing binding of C3b to the bacterial cell and hence opsonisation and phagocytosis.
- *Lysozyme.* This is an important enzyme that is widely distributed in the body. It is a muramidase which breaks the peptidoglycan in bacterial cell walls.

Coagulation, fibrinolysin and kallikrein systems. These complicated protein systems play only a small role in controlling most infections but become very important in overwhelming infections when excess activation of these systems contributes to shock and death (see pp. 26–7).

Table 2 **Nomenclature in the complement system**

C: Complement

Numbers: order of discovery rather than place in reaction series, e.g. C3

Component parts: each component splits into a smaller 'a' and a larger 'b' fragment, e.g. C3a, C3b

Complexes: formed with another component, e.g. with Factor B as C3bB

Activation: to form *active* enzyme is shown by line over the enzyme, e.g. C3bBb

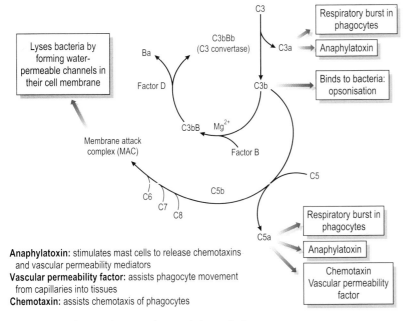

Anaphylatoxin: stimulates mast cells to release chemotaxins and vascular permeability mediators
Vascular permeability factor: assists phagocyte movement from capillaries into tissues
Chemotaxin: assists chemotaxis of phagocytes

Fig. 2 **The alternative complement pathway and phagocytosis.**

Table 3 **Chemical humoral mediators**

Substance	Origin	Anti-microbial mechanism
Simple chemicals		
Hydrogen ion	Phagocytes	Low pH is anti-microbial
Reduced oxygen	Phagocytes	Directly anti-microbial
Chloride ion	Tissue fluid	Combination with H_2O_2
Fatty acids	Metabolites	Anti-microbial at low pH
Individual proteins		
Acute-phase proteins:		
C-reactive protein	Liver cells	Activates complement
α1-Antitrypsin	Liver cells	Inhibits proteases
Fibronectin	Macrophages, fibroblasts	Opsonin for staphylococci
Interferons	Infected cells	Virus entry and virus multiplication impeded
Lactoferrin	Neutrophils	Binds iron needed for microbial growth
Lysozyme	Macrophages, neutrophils, secretions (e.g. tears, saliva)	Murein degradation
Myeloperoxidase	Neutrophils	Combines with H_2O_2 and Cl^-
Transferrin	Liver cells	Binds iron needed for microbial growth

Non-specific defences

- The first line of defences are skin and mucous membranes.
- Second line defences are provided by integrated phagocytosis and complement (alternative pathway) with some additional chemical factors.
- Phagocytosis is principally by neutrophils for most common bacterial infections and by macrophages for intracellular bacteria, protozoa and viruses.
- Complement acts by assisting phagocytosis (adherence, chemotaxis and respiratory burst), by increasing vascular permeability and hence the supply of neutrophils to infected tissues, and by lysing bacteria.
- Additional chemical factors include lysozyme and acute phase proteins such as C-reactive protein.

SPECIFIC (IMMUNE) DEFENCES

Microbes have many mechanisms for successfully overcoming or avoiding our first and second lines of defence (pp. 19–21); evidence of such success would be the ability to enter our cells or to infect again. So our third and fourth lines of defence are integrated with the initial constant defences, are rather slower to develop but are particularly effective against microbes we have experienced previously (antibody formation against specific microbes), or intracellular microbes (cell-mediated immunity, CMI). As they only develop after exposure to a microbe, they are called the **acquired immune response**.

SEROLOGIC IMMUNITY BY ANTIBODIES — THE THIRD LINE

This integrated and specific response is adroitly achieved with an antibody molecule that couples phagocytosis, complement activation *and* specific microbial recognition. Each antibody molecule, therefore, has three particular regions: two constant with the biological functions of activating phagocytes and complement, and a third, variable, recognition site to bind a specific microbe.

How can we have the myriad different antibody molecules needed for recognition of all the different microbes we meet in life? The answer is in our lymphocytes. These are of two types, **B-cells**, bone-marrow-derived and antibody producing, and **T-cells**, which are thymus processed, help B cells to make antibody and are especially important in cell-mediated immunity.

It used to be thought that B-cells used each new antigen ('any substance which causes *anti*body to be *gen*erated') as a template to make the corresponding antibody. It is now known that each B-cell is programmed to make one single antibody and no other, and has about 100 000 copies on its surface as receptors. When a microbe invading the tissues meets a B-cell with an antibody which fits a microbial cell surface antigen, the B-cell is stimulated to proliferate and then differentiate to plasma cells, which make more antibody. Thus a huge number of clones of the one **effector cell** selected are produced (as predicted by the clonal selection theory of Burnet and Fenner), resulting in an amplified amount of antibody against the invader. This takes several days and is called a **primary response**.

In addition, a smaller number of B-cells persist as **memory cells**, ready to proliferate and differentiate with large amounts of antibody if an invader returns. This explains the amplified **secondary response** seen in antibody production in a second infection (Fig. 1).

Immunisation

Immunisation (or vaccination) is based on the ability of B-cells to produce memory cells. The B-cells are primed by a harmless antigen (live–attenuated or killed) to produce memory cells ready to proliferate and differentiate when stimulated by a microbial antigen, thus producing large amounts of antibody in a short time (pp. 206–7).

How do antibodies work?

In summary, antibodies activate the classical complement pathway that then activates phagocytosis, either directly or indirectly via the mast cell. This is shown in Fig. 2 (see also Fig. 2, p. 21).

Antibody has two further benefits:

- Immunoglobulin E (see below) can bind directly to mast cells and subsequent antigen binding *directly* triggers

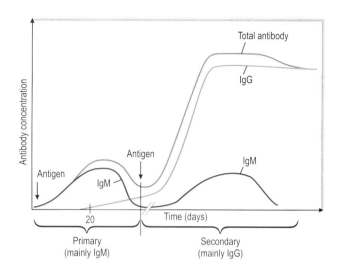

Fig. 1 **Primary and secondary antibody responses.**

release of the mediators (VPF, chemotaxins, MAC) without the need for complement activation.
- Antibody binding carries 'the bonus of multivalency', which means that while a single molecule of antibody may be insufficient to cause phagocytosis of a C3b-coated microbe, the attraction forces of multiple bonds is geometric, so three antibodies bound closely on a microbe can have 1000 times the attraction to a phagocyte. This is called high avidity multiple binding.

How does antibody structure give these functions?

There are five major classes of antibodies, called IgG, IgM, IgA, IgE and IgD. All except IgM have a similar structure (Fig. 3). Two identical heavy peptide chains are flanked by two identical light chains and all are linked by disulphide bonds. This is conventionally shown in a Y shape, but in reality it is considerably folded to expose three hypervariable regions in each adjacent light and heavy chain, which form the antigen-recognising area. This area is able to vary enormously to allow for specific recognition of innumerable antigens.

The C-terminal ends of the heavy chains by contrast are constant and are responsible for the biological functions of complement activation, and hence activation of phagocytosis, and binding to cell surface Fc receptors (the base of the Y is the Fc region).

Why do we have different classes of antibody?

The five major classes of antibodies have diverse functions:

- IgM is formed early in infection (primary response) and is composed of five Y-shaped molecules. It activates complement very efficiently.
- IgG is the major immunoglobulin in blood and tissues, is found particularly in the secondary response (Fig. 1), activates complement, binds to phagocytes and also acts with NK (natural killer) cells.
- IgA is present particularly at mucous membranes, often as a dimer with an extra 'secretory' component. It is exceptional in *not* activating complement or binding to Fc receptors, and probably acts by preventing attachment of microbes to mucous membranes.

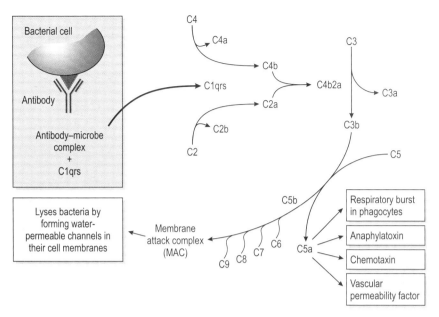

Fig. 2 **Activation of the classical complement pathway by antibody.**

- IgD is a minor Ig, probably a membrane receptor for antigen.
- IgE binds to mast cells and basophils, releasing histamine and other vasoactive mediators in the inflammatory response. It is particularly important in parasitic infections. In excess it is responsible for the symptoms of hay fever and other allergies.

CELL-MEDIATED IMMUNITY (CMI) — THE FOURTH LINE

CMI is particularly aimed at intracellular microbes and again consists of the three components, lymphocytes, specific chemical messengers, and phagocytes, but differs in that:

- the lymphocytes are T-cells, not B-cells
- the chemicals are called cytokines (meaning 'cell activators') and are made not only by the T-cells (lymphokines), but also by macrophages (monokines), neutrophils and other cells. Their aim is other cells, not antigen
- the phagocyte is the macrophage (and its predecessor, the blood monocyte).

There are different types (subsets) of T-cells, distinguished by their function and their surface markers (e.g. CD8, CD4). The important types are:

- *cytotoxic T-cells (CD8, T8)* which recognise antigen plus class I MHC (see below) in virus-infected cells and kill the cell before it can release infective virus; they also release gamma-interferon to make nearby cells resistant to infection
- *suppressor T-cells (also CD8, T8)* that decrease (downregulate) the activity of other T-cells and B-cells
- *helper/inducer T-cells (CD4, T4)* which help other T-cells to become cytotoxic, help B-cells make antibody and help macrophages kill intracellular microbes.

How does a T-helper cell help a macrophage kill a microbe within the macrophage? It must be able to do three things — recognise the macrophage, recognise that it contains microbes, and then activate the macrophage to kill the intracellular microbes. To achieve this, the T-cell recognises not one but always a pair of substances on macrophages containing intracellular microbes. First it recognises the macrophage by molecules belonging to class II of the major histocompatibility complex (MHC); these are tissue type markers originally discovered through their role in rejection of incompatible tissue or organ transplants. Secondly, it recognises adjacent antigenic fragments of the microbes which the macrophage has processed onto its surface. Then the T-helper cell releases gamma-interferon and activates the macrophage and other cells so that they can kill the intracellular parasites.

In addition, T-cells help the proliferation and differentiation of B-cells by production of B-cell stimulating factor (BSF-1), B-cell growth factor (BCGF II), and B-cell differentiation factors (BCDF-mu and BCDF-gamma), and probably other lymphokines.

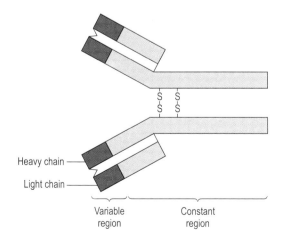

Heavy chain

Light chain

Variable region   Constant region

Fig. 3 **Stylised antibody molecule.**

Specific (immune) defences

- Antibodies link phagocytosis, complement activation and specific microbial recognition.
- Antibody activates the classical complement pathway.
- Antibody plus complement activates phagocytosis, both directly, and via mast cell activation.
- Antibody specific for one antigen is made by B-cell lymphocytes (each making only one antibody); when stimulated, this cell proliferates and differentiates to a clone of identical selected plasma cells (effector cells).
- Some lymphocytes persist (memory cells) ready for a more rapid secondary response to any subsequent invasion; this forms the basis for immunisation.
- T helper cells recognise antigen and class II MHC on the surface of macrophages containing intracellular microbes, release gamma interferon and activate the macrophage to kill the microbes.
- Soluble cytokines from T-cells stimulate B-cell proliferation and differentiation.

IMMUNE DISORDERS / EVADING HOST DEFENCES

IMMUNE DISORDERS

The immune system developed to protect the organism against infection and, probably, malignancy. Shifts to a too active state – hypersensitivity – or to a weakened state – immunodeficiency – can have devastating effects, both direct from the disarrayed immune response and indirect from the vulnerability to infection this allows.

HYPERSENSITIVITY

Hypersensitivity can occur by five basic mechanisms (Fig. 1).

'Innate' hypersensitivity

This term has been used to describe a sixth type of exaggerated response, such as septic shock in Gram-negative septicaemia (pp. 26 and 128–9).

IMMUNODEFICIENCY

When the immune system fails, a major consequence is an increased risk of infection (other alterations in immune function occur in hypersensitivity, autoimmune disease and malignancy). Immunodeficiency can be either a **primary** event caused by congenital or genetic abnormalities, or it can be **secondary** to a wide range of systemic insults including disease or medical treatment. Secondary immunodeficiency has become more common with the widespread use of drugs such as corticosteroids, cytotoxic agents, anti-cancer chemotherapy, and immunosuppressive drugs, or treatments such as X-ray therapy and organ and bone marrow transplantation. Immunocompromised patients, whether from disease such as AIDS or treatment, are vulnerable not only to serious attacks of the infections seen in normal hosts but also to unusual pathogens rare in normal hosts, and to opportunistic infections by normally harmless microbes (see p. 192).

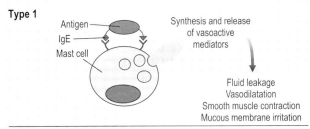

Type 1
Antigen
IgE
Mast cell
Synthesis and release of vasoactive mediators
Fluid leakage
Vasodilatation
Smooth muscle contraction
Mucous membrane irritation

Type 1 Anaphylactic hypersensitivity and atopic allergy. Here antigen reacts with IgE antibody and mast cells to cause release of vasoactive mediators such as preformed histamine, eosinophil and neutrophil chemotaxins, and platelet-activating factor from the granules, and newly synthesised leukotrienes and prostaglandins. These cause fluid leakage, vasodilatation, smooth muscle contraction and mucous membrane irritation. The result is hay fever or extrinsic asthma, depending on the site. This is also the mechanism seen in anaphylaxis to penicillins, cephalosporins and other antibiotics.

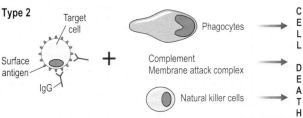

Type 2
Target cell
Surface antigen
IgG
+
Phagocytes
Complement
Membrane attack complex
Natural killer cells
CELL DEATH

Type 2 Antibody-dependent cytotoxicity. Surface antigen reacts with IgG antibody causing cell death. Cell death may occur because coating the cell with IgG (with or without C3b) increases adherence of phagocytes (this is called opsonisation), or because activation of complement to C9 causes cell lysis by the membrane-attack complex, or because of antibody-dependent cell-mediated cytotoxicity (ADCC), a direct non-specific method without phagocytosis, by phagocytes or natural killer (NK) cells. Type 2 mechanisms are not common in microbiology and infectious diseases, but do occur in *Mycoplasma pneumoniae* pneumonia with cold agglutinins, and in some drug reactions. They are very important in other areas of medicine, including Rh and ABO blood group incompatibility, organ transplants, and autoimmune diseases.

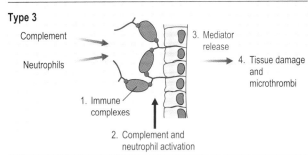

Type 3
Complement
Neutrophils
1. Immune complexes
2. Complement and neutrophil activation
3. Mediator release
4. Tissue damage and microthrombi

Type 3 Immune-complex mediated hypersensitivity. This involves the reaction of antigen with antibody (either in excess) causing tissue damage through excess activation of complement and neutrophil chemotaxis with release of vasoactive mediators, tissue-damaging enzymes and platelet-activating factor causing tissue damage and microthrombi. If excess antibody is circulating, this reaction occurs near the entry site of antigen, while in antigen excess circulating soluble complexes form and often deposit in filtering tissues like kidney or choroid plexus. Examples include farmer's lung from thermophilic actinomycetes, allergic bronchopulmonary aspergillosis, filariasis causing elephantiasis, streptococcal glomerulonephritis, malarial nephrotic syndrome, serum sickness, and reactions to chemotherapy such as erythema nodosum leprosum (ENL) in leprosy, and the Herxheimer reaction in syphilis.

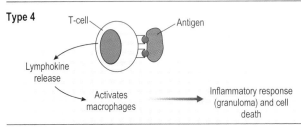

Type 4
T-cell
Antigen
Lymphokine release
Activates macrophages
Inflammatory response (granuloma) and cell death

Type 4 Cell-mediated hypersensitivity. Antigen reacts with class II MHC and primed T-cells. This causes lymphokine release that in turn causes macrophage and lymphocyte infiltration, and target cell death by the action of the killer T-cells. Note this is simply an exaggerated normal CMI response. Persisting antigen stimulation leads to chronic granuloma formation in the tissue. Type 4 is also called **delayed-type hypersensitivity** for, being cell-mediated, it takes several days to develop while the other four types are more immediate. Clinically it causes redness and swelling as in the Mantoux reaction. Other examples are the rash in smallpox and measles, and much of the pathogenesis in TB (caseation), borderline leprosy, fungal diseases including candidiasis and histoplasmosis, and parasitic diseases including leishmaniasis and schistosomiasis.

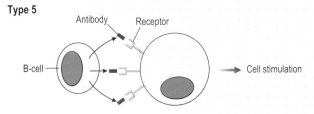

Type 5
Antibody
Receptor
B-cell
Cell stimulation

Type 5 Stimulatory hypersensitivity. This occurs when antibody reacts with surface receptors causing cell stimulation, such as that resulting in excess thyroid hormone production in thyrotoxicosis.

Fig. 1 **Types of hypersensitivity reaction.**

The defect can occur at any point in the immune system and will give rise to a distinct spectrum of disease (Table 1).

Diagnosis

Diagnostic tests in **hypersensitivity** diseases depend on the type.

Type 1 (anaphylactic) is investigated by intradermal scratch tests that provoke the release of histamine and other mediators causing an immediate wheal (oedema) and flare (erythema).

Type 2 (antibody-dependent cytotoxicity) is usually investigated with haemagglutination tests to predict and avoid Rh and ABO incompatibility. Tissue typing of the MHC antigens is used to avoid incompatible tissue transplants.

Type 3 (immune complex formation) is detected in tissues by immunofluorescence with anti-C3 and conjugated anti-immunoglobulins

Type 4 (cell-mediated) is investigated by biopsy if necessary.

Diagnostic tests in **immune deficiency diseases** depend on the defect suspected by the clinical presentation.

Complement components can be measured and in vitro function quantified.

B-cell function is assessed by measuring immunoglobulin levels, naturally occurring antibodies like A and B iso-agglutinins, and antibody response (if any) to killed vaccines.

T-cell function is assessed by skin tests to tuberculin, *Candida*, or mumps antigen, or by the reactivity of monocytes to phytohaemagglutinin. The numbers in each T-cell subset can now be measured, and this test is now widely available because of the diagnostic needs in AIDS.

EVADING HOST DEFENCES

The clinical picture of infectious illness results from the interaction of microbial factors and host factors, both the non-specific defences and the immune system.

For a pathogen to survive successfully it must reach the site where it is best adapted to survive, and there it must avoid the host defences and multiply.

Adherence. The ability of pathogens to adhere to specific tissues assists them in overcoming the non-specific defence mechanisms, including desquamation of epithelial cells and ciliary action. Specific adhesin membrane proteins bind to host cell membrane components. Some strains of bacteria show preference for particular surfaces, e.g. group A β-haemolytic streptococci from throat culture adhere better to oral epithelial cells than to skin. Many adhesins are associated with

Table 1 **Some major immunodeficiency diseases**

Defective component	Clinical disease	Defect/result
Phagocyte	Chronic granulomatous disease Myeloperoxidase deficiency 'Lazy leucocyte' syndrome	Cytochrome b$_{245}$ *Candida* survives Chemotaxis impaired
Complement	Deficiency of C1, 2, 3 or 4 Deficiency of C5, 6, 7 or 8 Hereditary angio-oedema	(SLE common) Disseminated Neisserial infection C1-inhibitor defect
B-cells	X-linked hypogammaglobulinaemia (Bruton's) Common, acquired hypogammaglobulinaemia	Immunoglobulins absent Low B-cells, poor B & T function
T-cells	DiGeorge syndrome (congenital thymic aplasia) Chronic mucocutaneous candidiasis	Absent T-cells, absent CMI MIF deficiency
Combined B- and T-cells	Severe combined immunodeficiency disease (SCID)	T- & B-cell defect, adenosine deaminase defect

CMI, cell-mediated immunity.

fimbriae or pili and this is often a determinant of virulence.

Adherence to prosthetic surfaces. The use of prosthetic devices has allowed different organisms to become pathogenic, e.g. infection of plastic intravascular devices with coagulase-negative staphylococci is a major cause of bacteraemia in hospital patients.

Capsules. The formation of a slippery mucoid capsule, e.g. in *Klebsiella pneumoniae*, prevents opsonisation, presents a relatively non-immunogenic surface and may make antibody or complement that does bind inaccessible to phagocytic cell receptors. All three properties assist in avoiding phagocytosis.

Evasion of respiratory burst. Some intracellular pathogens, e.g. *Leishmania donovani*, enter cells by binding to complement receptors and so do not activate NADPH oxidase as the Fc receptor would normally do.

Survival within the phagosome. Intracellular organisms can avoid non-oxidative killing by

- inhibiting fusion of phagosome and lysosome (*Toxoplasma gondii*)
- rupturing the phagolysosome so the microbe can multiply in the cytoplasm (*Shigella flexneri*)
- withstanding inactivation, and replicating in the phagolysosome (*Mycobacteria* spp.)

Bacterial cell wall. Surface proteins, for example the M protein in Gram-positive bacteria, can bind to molecules such as complement and prevent them activating host defences. In Gram-negative bacteria outer membrane proteins can block antibody- and complement-mediated lysis (e.g. in *Campylobacter fetus*) as can the lipopolysaccharide.

Antigenic variation. By varying the structure and antigenic composition of surface molecules, pathogens can avoid antibodies and appear to be a constantly 'new' infection, e.g. *Neisseria gonorrhoeae* varies its pilin protein. *Trypanosoma*

brucei has a variable surface glycoprotein (VSG) that limits complement activation; during infection waves of new parasites are produced every few days each with a new variant VSG.

Nutrient supply. Although microbial methods of ensuring sufficient nutrients are not directly related to avoidance of the immune system, they do assist the survival of the pathogen. Iron-scavenging mechanisms using secreted iron chelators called siderophores occur in *Escherichia coli* infections.

DAMAGING THE HOST

Pathogens damage the host in three ways:

- direct tissue injury (mechanical or chemical) or by subverting the cellular machinery so it becomes non-viable
- toxicity: exo- and endotoxins damage the host locally and at sites distant to the site of microbial growth
- immunopathogenic injuries result when the pathogen causes the host immune system to damage the host.

Immune disorders / evading host defences

- Hypersensitivity reactions occur when an overactive immune system causes injury; it can involve five mechanisms.
- Immunodeficiency can occur as a result of a primary genetic or congenital abnormality or as a result of systemic disease or medical treatment.
- Any component of the immune system can be defective in immunodeficiency.
- Pathogens have evolved mechanisms to evade each level of the immune response.
- Pathogens damage the host directly by destroying cells, through toxins and by initiating immunopathogenic mechanisms.

GENERAL BODY RESPONSES TO INFECTION

FEVER

Fever is an elevated body temperature, of 37.4°C (99°F) or higher. It is almost always present in any significant bacterial infection but is less frequent and less severe in fungal and parasitic infections. Four questions arise. How is fever produced? Is it useful? Is it harmful? How is it treated?

Production of fever

In the production of fever, five substances are important (Fig. 1):

- *endotoxin*: the lipopolysaccharide of the Gram-negative cell wall; it is composed of a long carbohydrate chain, a core polysaccharide and the active component lipid A, a unique glycophospholipid of disaccharide, short-chain fatty acids and phosphate groups (pp. 40–1)
- *peptidoglycan* (murein) in the cell walls of Gram-positive bacteria, which lack endotoxin
- *cytokines* IL-1 (interleukin-1) and tumour necrosis factor (TNF) from macrophages (p. 23)
- *acute phase reactants* including prostaglandins.

Endotoxin from Gram-negative bacteria and cell wall peptidoglycan from Gram-positive bacteria are called **exogenous pyrogens** because when infection occurs they stimulate macrophages to release the **endogenous pyrogens** IL-1 and TNF. These stimulate the acute phase response and the resultant prostaglandins stimulate the thermoregulatory centre in the hypothalamus to reset the body's thermostat higher, thus producing fever.

Is fever useful?

Perhaps the greatest use of fever is as an alerting mechanism, so that the host (or its parents!) knows it is infected and ill. In only two infections, syphilis and leishmaniasis, is there evidence that fever harms the organism directly.

Is fever harmful?

Stimulation of the thermoregulatory centre initially results in attempted heat conservation by vasoconstriction, headache and shivering, seen clinically as a *rigor*. When the fever continues, there is vasodilatation and sweating to lose heat. These are uncomfortable but not intrinsically harmful, unless fluid loss or the accompanying metabolic changes (see below) are excessive or long-continued. Temperatures above 40°C can permanently affect the brain or other organ functions. Temperatures above 43°C usually kill.

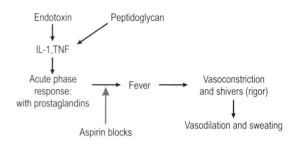

Fig. 1 **Mechanism of fever.**

Treatment of fever

Treatment of fever follows from the above: treatment of the infection, and fluids, nutrition and prostaglandin inhibitors such as aspirin for symptom control.

SHOCK

Shock, characterised by hypotension and decreased tissue perfusion, is a serious, often fatal consequence of severe sepsis, i.e. systemic infection. It is commonest in bacterial infections, infrequent in fungal infection and occurs in a few specific viral infections. It is best studied in Gram-negative infections (endotoxic shock); shock in Gram-positive infections probably shares some final pathways with endotoxic shock, though it is probably initiated by exotoxins in sepsis caused by *Staph. aureus* (e.g. pyrogenic toxin C, enterotoxin F (TSST-1, Fig. 2)), *Strep. pyogenes* (e.g. streptolysins, proteases) and *Strep. pneumoniae* (pneumolysin, purpura-producing principle, neuraminidase). Yet another mechanism, myocarditis, is the major cause of shock in meningococcal sepsis.

It is ironic that in severe Gram-negative infections endotoxic shock is a consequence of excessive, uncontrolled normal defence mechanisms (see pp. 20–3 and Fig. 3). Endotoxin in *small* amounts leads to:

- complement activation and acute inflammation
- macrophage and neutrophil activation and phagocytosis
- minor degrees of fibrinolysis and kinin activation.

All of these are beneficial defences to the host. However, endotoxin in *large* amounts leads to:

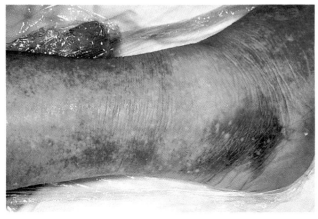

Fig. 2 **Staphylococcal toxic shock — ankle.**

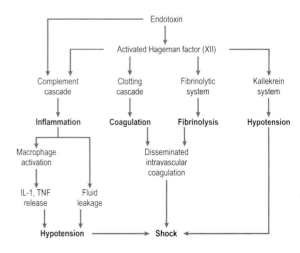

Fig. 3 **Endotoxic shock.**

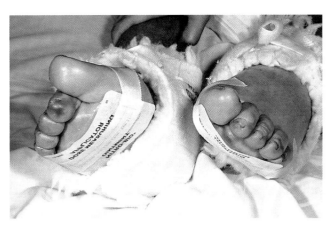

Fig. 4 **DIC with infarcts (death of tissue) on toes.**

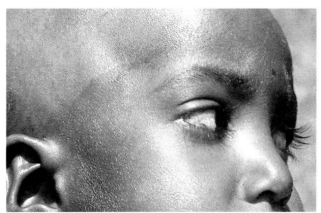

Fig. 5 **Vitamin A deficiency in malnutrition-leads to Bitot's spots (xerosis conjunctivae).**

- excessive complement activation and capillary leakage, and hypercoagulation with excess consumption (and therefore lack) of platelets and coagulation factors, hence clotting and bleeding
- excessive macrophage activation and release of IL-1 and TNF, hence hypotension
- excessive fibrinolysis, which with hypercoagulation leads to disseminated intravascular coagulation (DIC) (Fig. 4)
- excessive kinin activation and hence hypotension (Fig. 3).

The end results are severe shock, organ underperfusion and organ failure (cardiac, pulmonary, renal, cerebral) and death.

Treatment of shock attempts to prevent or reverse the pathological mechanisms. It includes early diagnosis and treatment of the causative infection, reversal of clotting abnormalities and excess fibrinolysis, and maintenance of blood pressure and organ function. Anti-endotoxin antibody is now available but its place is not established owing to debatable indications and great expense.

METABOLIC CHANGES

Changes in energy/carbohydrate, protein, fat and mineral metabolism with infection depend on the severity and duration of infection; the site of infection is also important if it diminishes food intake or increases loss of protein.

Energy/carbohydrate metabolism is increased, with glucose mobilisation from liver glycogen and other carbohydrate stores, from body fat and, in extreme situations, by gluconeogenesis (making new glucose) from body protein. Thus there is increased urinary nitrogen from amino acid destruction, and increased serum insulin, growth hormone and corticosteroids.

Protein metabolism is affected not only by the above protein breakdown (gluconeogenesis), but also by diminished albumin and transferrin synthesis. Conversely, there is use of some of the liberated amino acids in the production of some new proteins for host defence, including complement C3, C-reactive protein and fibrinogen, carrier proteins like haptoglobin and caeruloplasmin, and enzyme inhibitors like α_1-antitrypsin.

Fat metabolism is altered by both defective lipid clearance from plasma, and defective lipid uptake into storage; hence there are increased levels of serum lipids, especially triglycerides.

Mineral metabolism alters. Serum copper increases as it is bound to caeruloplasmin, which also increases. Oxidising ferrous iron for haemopoiesis increases but serum iron falls because it complexes with lactoferrin from neutrophils and is taken up by the liver; this may help to protect the host because iron is important for the pathogenicity of some bacteria. Serum zinc decreases with uptake into lymphoid cells, important in some key enzymes.

Cytokine control. The mediators of most of these changes are the now-familiar IL-1 and TNF (also called cachectin, 'substance causing wasting').

IL-1 increases carbohydrate metabolism and acute-phase protein production, decreases hepatic albumen and transferrin synthesis, depletes fat stores and moves iron and zinc into tissues from serum.

TNF induces IL-1 production, causes anorexia and weight loss, and has most of the above activities of IL-1.

The consequence of these changes is frequently malnutrition.

MALNUTRITION

There is a vicious circle that occurs in which infection leads to malnutrition, which in turn leads to impaired host defences and then to further infections. In addition, malnutrition is already a problem in poor communities where contaminated food and water are more likely to result in further infections (Fig. 5) (pp. 72–85).

We have seen above how infection leads to malnutrition through fever, anorexia, diarrhoea, increased nutritional requirements and increased catabolism. In turn, malnutrition leads to impaired host defence through skin diseases such as pellagra, or by decreased gastric and gut secretions: both impair the first-line of defence. Decreased complement activity and impaired T-cell activity, also resulting from malnutrition, impair the 2nd and 3rd lines of defence, which impair the 4th line of defence. By contrast, immunoglobulin synthesis and phagocytosis are usually normal.

General body responses to infection

- Fever is produced by the sequential action of microbial cell wall components (especially endotoxin), IL-1 and TNF and, finally, prostaglandins on the hypothalamic thermoregulatory centre.
- Fever is a useful alerting mechanism, has an effect on few pathogens and affects the host adversely if high or long continued.
- Shock is also usually endotoxin-induced, causing excessive activation of the complement, coagulation, fibrinolytic and kinin systems, resulting in hypotension, disseminated intravascular coagulation, decreased tissue perfusion and death if severe.
- Metabolic changes are predominantly adverse, and affect energy, carbohydrate, protein, fat and mineral metabolism.
- Infection leads to malnutrition, impaired host defences and further infections. In addition, malnutrition frequently co-exists with socioeconomic factors which increase infection.

STAPHYLOCOCCI

Staphylococci are ancient, common, versatile and important human pathogens: some of the osteomyelitis in Egyptian mummies is almost certainly staphylococcal. The microscopic grape-like clusters were described by Robert Koch in 1878, and grown by Louis Pasteur in 1880, who rightly said 'osteomyelitis is a boil in bone marrow'.

CLASSIFICATION AND DESCRIPTION

Staphylococcus is by far the most important genus in the family of Gram-positive cocci called Micrococcaceae. Two other genera in the family, *Stomatococcus* and *Micrococcus*, very rarely cause human infections.

Staphylococcus aureus, which produces the enzyme coagulase and usually has golden-yellow colonies, is the major human pathogen. It is pyogenic (pus-producing), causing **abscesses** in skin and most other organs, leading to **bacteraemia** and **endocarditis**, and also produces many **toxins**.

It has four special characteristics:

- virulence: causing severe disease in normal hosts
- difference: causing different disease in different sites, by different mechanisms, and involving different strains
- persistence: both in the environment, and on humans, who are frequently asymptomatic carriers
- resistance: to many antibiotics that were previously effective.

The rest of the genus are coagulase-negative staphylococci (CNS), and the most important are *S. epidermidis* (formerly *S. albus* because colonies are usually white), and *S. saprophyticus*.

CELL STRUCTURE AND FUNCTION

Staphylococci have the typical bacterial procaryotic internal structure and Gram-positive cell walls (Fig. 1). In addition to the usual peptidoglycan (murein), *S. aureus* has two special components in the cell wall:

- **Protein A** is unique to *S. aureus*. It is linked to the peptidoglycan with an outer end that surprisingly binds to the Fc receptor of IgG, protecting the microbe from opsonisation. This property is used in some serological tests for other organisms, to carry an antibody against them.
- **Teichoic acids** are polyribitol glycerophosphates found in all staphylococci and are involved in complement activation and attachment to mucosal surfaces as they

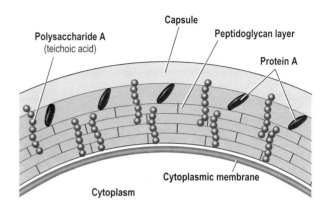

Fig. 1 **Staphylococcal structure.**

bind to fibronectin. Anti-teichoic acid antibodies are a research test for systemic staphylococcal infections.

Other components include:

- **Capsules** are found in few staphylococci in culture but may be more common in infected tissue. The capsule protects from complement, antibodies and phagocytes
- **Slime layers** are found in some coagulase-negative staphylococci, which assist adherence to synthetic catheters, grafts, and prostheses while hindering chemotaxis and phagocytosis.

CONFIRMATORY TESTS

Microscopy shows Gram-positive cocci about 1 nm in diameter, classically in clusters (Fig. 2) but often singly or in small groups in clinical specimens (Fig. 3).

Culture on blood agar shows large smooth colonies after 24 hours; these are usually golden-yellow surrounded by haemolysis for *S. aureus*, and pale yellow or white for others, but colony colour is not constant enough for speciation. Selective media may also be used, e.g. 7.5% sodium chloride in the blood agar suppresses other genera as staphylococci are unusually salt tolerant, +/– methicillin if seeking methicillin-resistant *S. aureus* (MRSA, see below). Methicillin is no longer used in patients but is standard in laboratory use.

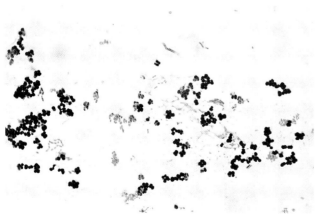

Fig. 2 **Staphylococci in grape-like clusters.**

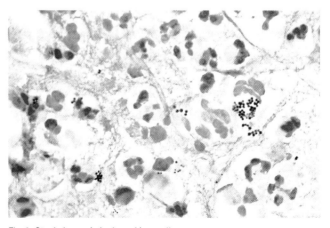

Fig. 3 **Staphylococci singly and in small groups.**

Table 1 **Clinical laboratory identification**

	Haemolysis blood agar	Coagulase	Mannitol fermentation	Novobiocin
S. aureus	+	+	+	Sensitive
S. epidermidis	+/–	–	–	Sensitive
S. saprophyticus	–	–	–	Resistant

Table 3 **Clinical syndromes**

Organism	Pathogenic mechanism	Clinical syndrome
S. aureus	Tissue destruction	Abscess, e.g. skin, joint, bone, brain, lung, etc.
	Blood spread	Bacteraemia, endocarditis
	Toxin	Scalded skin, toxic shock, food poisoning
S. epidermidis	Adhesion (slime?)	Infected prosthesis or catheter
S. saprophyticus	Adhesion	Urinary tract infection

Table 2 **Virulence factors**

Type	Virulence factors	Action
Toxins	Cytotoxins (5), alpha to epsilon (haemolysins) (epsilon = leucocidin)	Lyse neutrophils, RBC, other cells Inhibits phagocytosis
	Exfoliatin (epidermolytic toxin) A & B	Separates skin granulosum cells
	Toxic shock syndrome toxin-1 (TSST-1) (= enterotoxin F, exotoxin C)	Stimulates IL-1 release
	Enterotoxins (6), A to F.	Stimulate vomiting and peristalsis
Enzymes	Coagulase (bound form, and free)	Protects by fibrin formation
	Catalase	Protects by destroying H_2O_2 in WBC
	Fibrinolysin (staphylokinase)	Dissolves fibrin
	Hyaluronidase (spreading factor)	Hydrolyses hyaluronic acid in CT
	Lipases	Allow survival in sebaceous glands
	Nuclease	Degrades DNA
	β-Lactamases (penicillinases, cephalosporinases)	Destroy β-lactam antibiotics
Structure	Cell wall, capsules	Decrease chemotaxis, opsonisation and phagocytosis
	Slime	Facilitates adhesion to synthetics

Biochemical tests and resistance to novobiocin (an otherwise unused antibiotic) distinguish between species (Table 1). Additional specialised tests are used for confirmation, or for other coagulase-negative strains.

Further tests are used to distinguish between different strains, especially of *S. aureus* in hospital cross-infection studies. Antibiotic sensitivity patterns may be of some help, as may further biochemical tests in larger laboratories. Phage typing (susceptibility to lysis by standard panels of bacteriophages), PFGE (pulse field gel electrophoresis) or DNA typing are reference laboratory techniques.

PATHOGENESIS AND VIRULENCE

S. epidermidis is a member of the normal skin flora, and an opportunist pathogen in hosts with impaired defences. It particularly infects by attaching to foreign, synthetic materials like intravascular catheters or joint prostheses, often assisted by slime production. It lacks most of the toxins and enzymes of *S. aureus*.

S. saprophyticus is thought to have special adherence factors for urinary tract epithelium.

S. aureus, by contrast, has many virulence factors. These are toxins, enzymes and the actual structure of the organism (Table 2). Its success as a pathogen also results from its lack of antigenicity and the consequent lack of protective antibodies.

CLINICAL SYNDROMES

Staphylococcal infection presents with a wide range of syndromes affecting many tissues (Table 3) and caused by three mechanisms:

- local destruction (abscess, Fig. 4)
- blood spread
- toxin production.

CHEMOTHERAPY

Penicillinase-resistant penicillins such as oxacillin, cloxacillin and flucloxacillin are used for serious infections. First- or second-generation cephalosporins such as cephalothin, cephalexin and cefuroxime are usually safe in patients who are hypersensitive to penicillins. Vancomycin is necessary for methicillin-resistant staphylococci. Erythromycin and its newer relatives are used in milder infections.

CONTROL

Control is both important and difficult in hospitals, for staphylococci persist for months in dust, curtains and linen, and human carriage is often permanent. Reservoirs, routes of spread and ruptures of skin and mucous membranes differ, so different measures are appropriate in different circumstances; these include cleaning (+/– disinfection), air-control, decrease in direct and indirect contact, and portal protection by aseptic or no-touch technique for all invasive procedures.

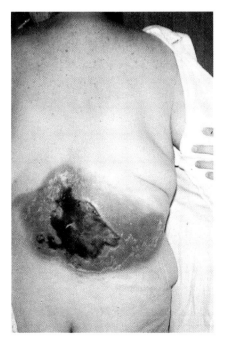

Fig. 4 **Huge abscess (carbuncle) on back with yellow pus, necrosis, ulceration and local spread.**

Staphylococci
- Staphylococci are Gram-positive cocci. They are important pathogens, particularly the coagulase-positive *S. aureus*.
- Coagulase-negative staphylococci include *S. epidermidis*, important in infecting prostheses and catheters, and *S. saprophyticus*, a cause of urinary infections.
- *S. aureus* is a virulent primary pathogen causing many different infections in many different tissues by different mechanisms: by abscess formation, blood spread (bacteraemia and endocarditis) and by toxins.
- Cell structure includes the usual Gram-positive cell wall plus protein A and teichoic acid.
- Virulence factors include many enzymes and toxins.
- Management includes antibiotics (testing for susceptibility) and minimising infection by cleansing techniques, control of air and contact, and aseptic procedures.

STREPTOCOCCI AND ENTEROCOCCI

CLASSIFICATION AND DESCRIPTION

Streptococci and enterococci have certain characteristics that contribute to their ability to cause disease:

- they have the ability to live as normal flora on our skin and mucosal surfaces, mainly in the nasopharynx, gut and vagina
- *Strep. pyogenes* and *Strep. pneumoniae* are aggressive pathogens with numerous **virulence factors**, which give the ability to adhere, invade and damage tissues.
- other strains are 'conditional or opportunistic pathogens': normal flora that can become pathogenic in abnormal sites or in abnormal hosts
- infection is followed by spread locally, to distant organs and to other people.

The organisms in these genera are:

- Gram-positive cocci (GPC), usually in chains, sometimes in pairs (Fig. 1a, b)
- non-motile, non-sporing and may be capsulated (Fig. 1c)

(a)

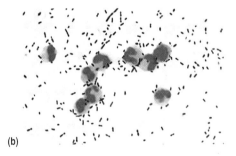

(b)

- facultatively anaerobic
- nutritionally fastidious, needing blood or other rich media, with some important strains growing only in pyridoxal-rich media
- catalase negative, unlike staphylococci

The classification of streptococci is confusing, as three separate criteria are used:

- **biochemical** into species
- **serological** into Lancefield groups based on specific polysaccharide antigens in the cell wall
- **haemolytic** by the lysis seen when streptococci are cultured on sheep blood agar: beta means a clear zone; alpha, a green zone (viridans means 'making green' in Latin) and gamma means no haemolysis (see Table 1 and Fig. 2).

Enterococci are now placed in a separate genus because of differences in their characteristics, including resistance to bile, 6.5% NaCl and antibiotics.

Streptococci that are **obligate** anaerobes are also placed in a separate genus, *Peptostreptococcus* (see p. 49).

CELL STRUCTURE AND FUNCTION

Streptococci have a complex cell wall (Fig. 3). The biological principle that structure relates to function is illustrated by the components:

- pneumococcal capsule gives resistance to phagocytosis
- lipoteichoic acid on pili helps adhesion to host cells
- type-specific M Protein in group A gives virulence
- Lancefield group-specific carbohydrate protects peptidoglycan
- linear peptidoglycan with cross-linking gives rigidity.

CONFIRMATORY TESTS

Clinical specimens (throat swabs, pus, sputum, etc.), are examined by:

- Gram stain
- culture on sheep blood agar shows small colonies, usually glistening, mucoid if encapsulated, and with haemolysis as in Table 1.
- biochemical tests (Table 2)
- serology for the development of serum antibodies, i.e. antistreptolysin O titre (ASOT) or antiDNAase B (Table 3).

PATHOGENESIS AND VIRULENCE

Virulence factors (Table 3) are found in the aggressive pathogens *Strep. pyogenes* and *Strep. pneumoniae* and are related to surface antigens and extracellular products. Some appear to help the spread of disease, but it is not yet possible to link every individual toxin with particular clinical infections.

Table 1 **Classification and normal habitat**

Species (biochemical)	Serologic Lancefield group	Haemolysis on sheep blood agar	Normal flora (nf) or asymptomatic carriage (ac)
Strep. pyogenes	A	Beta	Throat, nose (ac)
Strep. agalactiae	B	Beta (alpha, gamma)	Vagina, gut (nf)
E. faecalis	D	Gamma (alpha)	Gut, perineum (nf)
Strep. bovis, equinus	D	Gamma (alpha)	Gut, perineum (nf)
Strep. pneumoniae	Ungroupable	Alpha	Nasopharynx (ac)
Strep. viridans group (*Strep. sanguis, salivarius, mitis, 'milleri', mutans*)	Ungroupable	Alpha (gamma)	Mouth (nf)

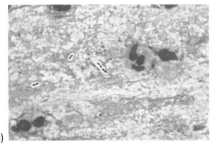

(c)

Fig. 1 **Gram stain showing *Strep. pyogenes* in chains (a) and *Strep. pneumoniae* in pairs (b) and encapsulated (c).**

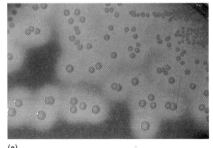

(a)

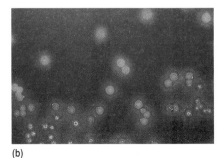

(b)

Fig. 2 **Haemolysis by (a) *Strep. pyogenes* (beta) and (b) *Strep. pneumoniae* (alpha).**

Table 2 **Identification of streptococci and enterococci**

Organism	Group	Susceptibility to: Bacitracin	Optochin	CAMP[(a,b)] test	Hydrolysis of: Hippurate[(b)]	Aesculin	Growth in: Bile	6.5% NaCl
Strep. pyogenes	A	S	R	–	–	–	–	–
Strep. agalactiae	B	R(S)	R	+	+	–	–	+(–)
Enterococci	D	R	R	–	– (+)	+	+	+
Strep. bovis, equinus	D	R	R	–	–	+	+	–
Strep. pneumoniae	–	R	S	–	–	–	–	–
Strep. viridans group	–	R(S)	R	–	–	–	–	–

S, sensitive; R, resistant; +, present; –, absent. [a]Extracellular protein giving synergistic haemolysis with β-haemolysin of *Staph. aureus*. [b]Now largely replaced by commercial latex or co-agglutination tests. ( ) less common.

Table 3 **Virulence factors of *Strep. pneumoniae* and *pyogenes***

Virulence factor	Actions
Strep. pneumoniae	
Capsular 'C' polysaccharides (the most important)	Inhibit phagocytosis and opsonisation
Pneumolysin	Beta haemolytic, dermotoxic
Neuraminidase	Splits membrane glycoproteins, may aid invasion and spread
Purpura-producing principle	Active in animals, may be in humans
Strep. pyogenes	
Structural components	
Capsule (if present)	Not a virulence factor
M-protein	Anti-phagocytic, anti-complementary (i.e. blocks action of complement)
Lipoteichoic acid	Adheres to epithelial cells
Toxins and enzymes	
Streptolysin O (oxygen labile)	Lyses red cells, white cells, tissue cells and platelets, releasing cell enzymes
Streptolysin S (oxygen stable)	Action as for streptolysin O
DNAase, type B	Depolymerises DNA in pus
Streptokinases	Lyse clots, help bacterial spread
Hyaluronidases	Solubilize collagen in tissues, may help bacterial spread
Erythrogenic toxins	Mediate rash in scarlet fever

Table 4 **Streptococcal organisms and associated clinical syndromes**

Organism	Direct invasion and inflammation	Local spread	Distant spread	Distant toxin effects	Immune mechanisms
Strep. pyogenes	Throat, wound and burn infections Puerperal sepsis	Erysipelas	Septicaemia	Scarlet fever	Rheumatic fever
Strep. agalactiae	Neonatal pneumonia Puerperal sepsis	Abscess	Neonatal meningitis		
Enterococci	Urinary tract infection	Abscess	Endocarditis Septicaemia		
Strep. pneumoniae	Bronchitis	Pneumonia	Septicaemia Meningitis		
Strep. viridans	Caries		Endocarditis Bacteraemia		

CLINICAL SYNDROMES

Disease is caused by:

- direct invasion and inflammation
- local spread
- distant spread
- distant toxin effects, e.g. scarlet fever
- immune mechanisms, e.g. rheumatic fever.

There is therefore a wide range of clinical syndromes associated with streptococcal infections (see Table 4):

- *Strep. pyogenes*: throat infections, wound and burn infections, puerperal sepsis, scarlet fever, rheumatic fever, septicaemia etc.
- *Strep. agalactiae*: neonatal pneumonia and meningitis, puerperal sepsis
- enterococci: urinary tract and wound infections, endocarditis, septicaemia
- *Strep. pneumoniae*: bronchitis, pneumonia, bacteraemia, meningitis
- *Strep. viridans* group: caries, endocarditis, bacteraemia.

CHEMOTHERAPY

In general, streptococci are very sensitive to penicillin, while enterococci are quite resistant to most antibiotics, except ampicillin or vancomycin. However, pneumococci with partial or complete resistance to penicillin (p. 113), and vancomycin-resistant enterococci, are increasingly common.

CONTROL

A multivalent pneumococcal vaccine is used to protect those particularly at risk, including splenectomised and immunocompromised patients. Locating carriers (nose, throat, skin, perineal carriage) is important in controlling outbreaks, especially in hospitals or closed communities.

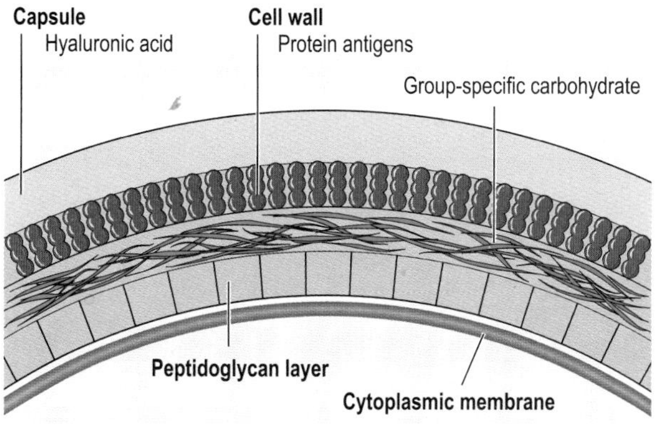

Capsule
Hyaluronic acid

Cell wall
Protein antigens

Group-specific carbohydrate

Peptidoglycan layer

Cytoplasmic membrane

Fig. 3 **Structure of *Strep. pyogenes*.**

Streptococci and enterococci

- Streptococci are Gram-positive cocci, usually growing in chains, facultative anaerobes, nutritionally fastidious and catalase negative.
- Enterococci are more resistant than streptococci to bile, salt and antibiotics.
- Identification depends on Gram stain, haemolysis, biochemical tests and Lancefield grouping.
- The complex cell wall and enzymes and toxins have important functions, including adhesion, virulence and spread.
- *Strep. pyogenes* and *Strep. pneumoniae* are aggressive pathogens, invasive and virulent even in normal hosts.
- Other streptococci are opportunistic pathogens, i.e. normal flora that cause disease in abnormal sites or abnormal hosts.
- Disease is caused by invasion and spread, toxin effects and immune mechanisms.

GRAM-POSITIVE RODS: CORYNEBACTERIA, LISTERIA, BACILLUS

CORYNEBACTERIA

Corynebacteria are aerobic (facultatively anaerobic) Gram-positive rods. They are non-spore-bearing, non-motile, catalase positive, and ferment various carbohydrates producing lactic acid. The appearance of stained films is said to resemble Chinese letters (Fig. 1) because of incomplete fission initially during multiplication; metachromatic granules occur in *C. diphtheriae* (granules staining a different colour from the rest of the cell).

Only *C. diphtheriae* causes a major disease, diphtheria, which is now rare where immunisation is effective, for only humans are hosts. Other *Corynebacteria* are our normal flora or have animal hosts. *C. minutissimum* causes a superficial skin infection (erythrasma) that appears similar to the mycoses.

Pathogenesis and virulence

Corynebacteria provide a spectrum of pathogenicity from the aggressive primary pathogen *C. diphtheriae* through the normal flora which occasionally become opportunist pathogens (*C. haemolyticum*, *C. pseudodiphtheriticum*, *C. xerosis*, and *C. jeikeium*, formerly known as group JK) to accidental human infections from animal reservoirs.

Virulence in *C. diphtheriae* is caused by the diphtheria exotoxin, which interferes with protein synthesis; this is only produced when the *tox* gene is transduced into the corynebacterium by a lysogenic bacteriophage. Toxin is only produced when the concentration of iron in the medium is low. All exotoxin is antigenically the same, so one (monovalent) antitoxin is sufficient for treatment.

The second toxin, dermonecrotic toxin, is a sphingomyelinase acting on vascular endothelial cells to increase vascular permeability.

The role of a third, haemolytic toxin is uncertain.

Confirmatory tests

Laboratory tests can take a week, so the initial diagnosis of diphtheria must be clinical.

- Gram stain: throat swab often negative
- culture on blood agar is more reliable: three variants occur, *mitis* (small, black, smooth colonies), *intermedius* and *gravis* (large grey-black, dull colonies), not correlated with virulence; tellurite is used in selective media

- biochemical tests: *C. diphtheriae* ferments maltose
- toxin production: Elek's immunodiffusion method shows a line of precipitation occurring where toxin from a streak of organisms meets antitoxin from a filter paper on an agar plate.

Clinical syndromes and management

C. diphtheriae causes pharyngeal, nasopharyngeal and largyngeal diphtheria with a *pseudo-membrane*, and rarely cutaneous diphtheria (Veldt sore, Barcoo rot). The other corynebacteria may cause opportunistic infections and rarely pharyngitis (pp. 106–192).

Diphtheria is completely preventable by immunisation, usually as 'triple antigen'. Antitoxin is essential for clinical diphtheria and must be given as early as possible without waiting for confirmatory tests. Penicillin is used in addition to kill the bacteria.

LISTERIA MONOCYTOGENES

Listeria monocytogenes is another aerobic (facultatively anaerobic) non-spore-forming Gram-positive rod. Animals are the reservoir. It survives pasteurisation of milk, and cheesemaking.

L. monocytogenes causes meningitis, bacteraemia and endocarditis; it is a particular problem in mothers, babies and the immunosuppressed.

Pathogenesis and virulence

There are no known virulence factors. The important feature in its pathogenesis is its ability to survive protected from host defences as an **intracellular pathogen** in macrophages and monocytes.

Confirmatory tests

- Microscopy: typical rods can be overlooked as commensal corynebacteria; numbers in CSF may be too low for detection in meningitis.
- Culture on blood agar: zone of beta-haemolysis often surrounds colonies (Fig. 2) grown at 25°C.
- Culture in liquid media: unusual tumbling motility is diagnostic.
- Biochemical tests: catalase (+), H_2S production (−).

Fig. 1 *C. diphtheriae* with Gram stain showing 'Chinese letter' arrangement.

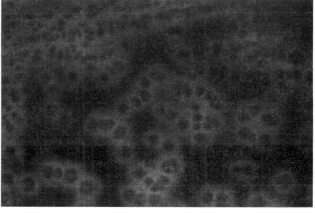

Fig. 2 *L. monocytogenes* culture on blood agar showing beta-haemolysis.

Clinical syndromes and chemotherapy

Pregnant women may develop bacteraemia, rarely meningitis. Neonates infected in utero develop early-onset **granulomatosis infantiseptica** with disseminated abscesses; if infected after birth, later-onset meningitis develops. Immunosuppressed patients develop meningitis, bacteraemia or endocarditis. Sporadic adult cases of meningitis may be related to cheese or milk products as the organism is common in cattle and goats. The drug of choice is ampicillin, with cotrimoxazole or gentamicin in severe cases.

BACILLUS SPP.

The genus *Bacillus* consists of aerobic (facultatively anaerobic), Gram-positive rods distinguished by forming spores. The important human pathogens are *B. anthracis* and *B. cereus*. Other *Bacillus* spp. are rare opportunistic pathogens.

B. anthracis

B. anthracis is a large non-motile spore-forming rod. As spores persist in the soil, animals and their products are the usual source of human infection.

Characteristic **virulence factors** are the capsule, which is anti-phagocytic, and the anthrax toxin. This has three components, the oddly-named protective factor that with the oedema factor causes severe oedema, and with the lethal factor causes death in untreated animals and humans.

Laboratory diagnosis is by finding a large Gram-positive rod, usually single or paired, with a capsule but without spores in clinical specimens but in long 'bamboo' chains with spores and no capsule in culture. *B. anthracis* differs from other *Bacillus* spp. in that the rapidly growing colonies adhere to the media, are non-haemolytic and are 'curly' like a 'medusa head'.

The **clinical syndrome** is of malignant pustule progressing to massive swelling, systemic symptoms and death (p. 168). **Chemotherapy** is by penicillin.

B. cereus

Soil, rice or other foods are the usual source of human infection with *B. cereus*. Four characteristic **virulence factors** are known: a heat-stable and a heat-labile enterotoxin (necrotic toxin), cerelysin (a haemolysin) and phospholipase C (a lecithinase). **Laboratory diagnosis** is by

culture (Fig. 3) of incriminated foods in food poisoning, and of the eye, blood or other infected tissues.

Clinical syndromes. These include:

- short-incubation short-duration emetic-type food poisoning in reheated or improperly stored rice (heat-stable toxin)
- long-incubation longer-duration diarrhoeal food poisoning (from heat-labile toxin affecting ion transport) from soil-contaminated, undercooked meat or vegetables
- post-traumatic pan-ophthalmitis, usually causing blindness
- rare opportunistic bacteraemia, pneumonia or meningitis.

Chemotherapy is unnecessary for food poisoning, but urgent and difficult because of multi-resistance in the other syndromes: vancomycin or clindamycin, and gentamicin are used empirically until sensitivity results are available.

ERYSIPELOTHRIX RHUSIOPATHIAE

This organism is an aerobic (facultatively anaerobic and micro-aerophilic) non-spore-forming small, slender rod with typical Gram-positive cell structure. It is pleomorphic and often filamentous in culture. Colonies are α-haemolytic, small and grey.

Many animals, especially pigs and fish, are sources of human infection which occurs through a skin abrasion (often in abattoir workers, fishmongers, etc.) producing a distinctive purplish-red spreading skin lesion (erysipeloid). Culture of a biopsy specimen is required as surface swabs are usually negative. Endocarditis is very rare, but fatal in 30% of patients. Chemotherapy is with penicillin or a cephalosporin.

Gram-positive rods

Corynebacteria

- *C. diphtheriae* is an aerobic, non-spore-forming, non-motile, catalase-positive Gram-positive rod.
- There are three colonial types: gravis, intermedius and mitis.
- It causes acute, potentially fatal diphtheria by a potent exotoxin, so urgent treatment is essential with anti-toxin and penicillin. Immunisation prevents the disease.
- Rarely, other corynebacteria cause opportunistic disease, e.g. bacteraemia from I.V. catheters.

Listeria

- *L. monocytogenes* is an aerobic (facultatively anaerobic) Gram-positive rod. It is non-spore-forming, catalase positive, with characteristic tumbling motility at 25°C. Many animals form the reservoir, and milk or cheese are often sources.
- It is an intracellular organism which particularly infects special groups – pregnant women, newborns, and immunosuppressed patients – though sporadic infections, probably from food, occur.
- It causes meningitis, multiple abscesses, bacteraemia or endocarditis.
- Ampicillin, with cotrimoxazole or gentamicin, is the treatment of choice.

Bacillus

- These are aerobic (facultatively anaerobic) Gram-positive rods distinguished by spore formation. Soil and animals are the reservoirs.
- *B. anthracis* has an anti-phagocytic capsule and three toxins: protective factor (!), oedema factor and lethal factor. It causes anthrax, with malignant pustule, massive oedema and septicaemia.
- *B. cereus* has four toxins and causes two types of food poisoning (emetic and diarrhoeal), serious eye infections and rare opportunistic infections, as may other *Bacillus* spp.

Erysipelothrix rhusiopathiae

- This is an aerobic micro-aerophilic Gram-positive rod found in animals and fish which causes erysipeloid: it is treated with penicillin.

Fig. 3 ***Bacillus cereus* colonies.** Culture on horse blood agar.

CLOSTRIDIA

Clostridia are Gram-positive, strictly anaerobic, spore-forming rods. They are mainly free-living soil microorganisms. There are four major human pathogens, but numerous other species occasionally cause infections. The four important pathogens are:

- *C. perfringens*, causing skin, soft tissue, and muscle infections, ranging from simple cellulitis to gas gangrene, and also causing food poisoning and enteritis necroticans ('pig-bel')
- *C. tetani*, the cause of tetanus
- *C. difficile*, implicated in antibiotic-associated colitis
- *C. botulinum*, the cause of botulism.

Features used to distinguish species include colony appearance, the shape and position of spores, motility, biochemical tests and toxin production.

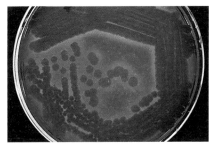

Fig. 1 *C. perfringens* culture on blood agar.

Fig. 2 **Gram stain of clostridia showing short Gram-positive rods.**

C. PERFRINGENS

C. perfringens (formerly *C. welchii*) is a large, spore-forming Gram-positive rod which is an obligate anaerobe, though aerotolerant for up to 72 hours. Unlike most other clostridia it is non-motile, although colonies spread rapidly on agar plates (Fig. 1). It is haemolytic, and metabolically active, doubling in only 8 minutes in ideal conditions! It is found in soil, the gut, the female genital tract and nearby skin.

The occurrence of gas gangrene in war wounds, where extensive tissue damage is associated with impaired blood supply and contamination with soil and other foreign matter, led to extensive studies of anaerobic bacterial growth.

Pathogenesis and virulence
C. perfringens produces a huge range of toxins and extracellular enzymes, as might be expected of such a fearsome pathogen. Some of their properties are shown in Table 1 and correlate with five major strain types, A to E: the table is intentionally incomplete, yet still forbidding!

- Alpha toxin is the most important, lysing red and white blood cells, platelets and endothelium, causing severe haemolysis, bleeding and tissue destruction.

- Beta toxin causes vascular leakage and is also important in necrotising enteritis ('pig bel') (p. 138).
- Delta toxin haemolyses red cells.
- Theta toxin causes haemolysis, pulmonary oedema and cardiac arrhythmia.
- The enzymes help the organism spread rapidly through tissues.
- The enterotoxin is quite different, acting like cholera toxin on the adenylate cyclase system of ion transport to cause fluid loss and diarrhoea.

Confirmatory tests
Microscopy shows large Gram-positive rods (Fig. 2) usually without spores in swabs, fluid and tissues but with terminal or sub-terminal spores in media. At times it may stain poorly, even appearing Gram-negative. Appearance is suggestive but not pathognomonic.

Culture should be both anaerobic on blood agar and a cooked meat or thioglycollate broth, and aerobic to detect other pathogens. Rapid, spreading, haemolytic growth is characteristic.

Biochemically, full identification rests on five sugar fermentations and nitrate reduction, but presumptive identification is on two tests: *C. perfringens* causes rapid stormy digestion with acid and gas in litmus milk medium, and the α-toxin (a phospholipase) causes visible opacity on egg yolk medium that is inhibited by specific antiserum, the so-called Nagler reaction (Fig. 3).

Clinical syndromes
In skin, soft tissues and muscle there is a spectrum from **cellulitis** to **gas gangrene** (Fig. 4) and rapid death (p. 169). In the gut

Table 1 **Virulence factors produced by Clostridia**

Virulence factor (toxins and enzyme)	Activity	Strain type				
		A	B	C	D	E
Alpha[1]	Phospholipase, haemolytic	++	+	+	+	+
Beta[1]	Capillary leakage, necrosis		++	++		
Delta[1]	Haemolysin		+	+		
Epsilon[1]	Capillary leakage, necrosis		+		++	
Iota[1]	Capillary leakage					+
Theta	Capillary leakage, cardiotoxia	++	+	++	+	
Kappa	Collagenase	++				
Lambda	Protease					
Mu	Hyaluronidase					
Nu	Deoxyribonuclease	+	+	+	+	+
Neuraminidase	Hydrolyses serum protein					
Enterotoxin	Destroys gut ion transport	++				

[1]The first five listed are lethal for experimental animals.

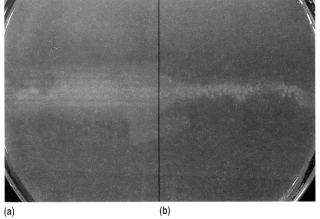

(a) (b)
Fig. 3 **Nagler plate.** (a) No antitoxin; (b) with antitoxin.

Fig. 4 **Clostridial infection.**

there are two different conditions, infective 'food poisoning', and necrotising enteritis (pig bel) (p. 138).

Management
Penicillin is given for the tissue infections, but urgent surgery and consultation concerning anti-toxin and hyperbaric oxygen are essential. Antibiotics are not needed for food poisoning and are of little help in pig-bel.

C. TETANI
C. tetani is a large anaerobic Gram-positive rod; because the spores, when present, are terminal, it resembles a drumstick. It is particularly found in soil.

Tetanus, which is caused by the toxin produced by growing *C. tetani*, has been known since antiquity, particularly related to war injuries. Tetanus only occurs when a wound is contaminated with tetanus spores if the tissue conditions are suitable for their germination, i.e. necrosis and anaerobiosis occurs.

Pathogenesis and virulence
Germinating growing organisms produce **tetanospasmin**, one of the two most potent poisons known. It is a heat-labile neurotoxin, of 150 kDa (in a heavy 100 000 kDa chain and a light chain), which is released, binds to peripheral nerve membranes, then moves by retrograde neuronal transport to anterior horn cells where it blocks the release of inhibitory neurotransmitters, thus causing spasms and spastic paralysis.

Confirmatory tests
Diagnosis is clinical, as *C. tetani* may be isolated from contaminated wounds in which it has not released toxin and, conversely, often cannot be found in the wound (which may be trivial but must be anaerobic) causing tetanus. It may be found in the umbilical cord remnant in neonatal tetanus.

Clinical syndrome and management
The almost unmistakable clinical disease tetanus is described on page 93. The

therapeutic drug of choice is penicillin, plus anti-toxin and supportive measures. Since patients with severe injuries usually receive antitoxin, most cases of tetanus arise from relatively trivial injuries involving contamination with soil or foreign bodies. Active immunisation with tetanus toxoid before injury or after recovery gives excellent immunity.

C. BOTULINUM
C. botulinum is an anaerobic, motile, spore-forming Gram-positive rod with oval, subterminal or central spores. The spores from soil or vegetables are relatively heat resistant.

Pathogenesis and virulence
Like tetanus, disease is produced in adults not by infection but by intoxication with an extremely potent heat-labile toxin, usually types A, B or E, in uncooked or improperly cooked foods (the toxin is destroyed by boiling at 100°C for 10 minutes). The heavy chain of the toxin binds to cholinergic nerves, blocking acetylcholine release and hence blocking transmission.

Infant botulism *is* an actual infection caused as the organisms (often from honey) multiply in the gut and liberate toxin there.

Wound botulism is very rare and is produced by toxin from multiplying organisms in an infected wound.

Confirmatory tests
Diagnosis is primarily clinical but may be confirmed by toxin detection in a mouse assay, or by growth of the organism (made easier by heating the specimen to 80°C for 10 minutes to kill other contaminating vegetative organisms).

Clinical syndromes and management
The three forms are food-borne adult botulism, infant botulism and wound botulism (pp. 138, 170).

Penicillin treatment is used, but early antitoxin before the toxin all binds to nervous tissue is necessary, and ventilatory support may be needed. Control depends on education in correct cooking and home bottling methods.

C. DIFFICILE
C. difficile was only recognised as a pathogen in the early 1970s. It is an obligate anaerobe with typical Gram-positive structure, and resistant spores. It is more antibiotic resistant than other clostridia. It is part of normal bowel flora in most children and some

adults. Its spores persist in hospital and other environments, and some infections are exogenous.

Pathogenesis and virulence
C. difficile produces two toxins, an **enterotoxin** (toxin A) that causes secretory, haemorrhagic diarrhoea and a **cytotoxin** (toxin B) that causes a destructive cytopathic effect in tissue culture cells. Both cause changes in experimental animals, and each appears important in pathogenesis when the balance of normal flora in the colon is upset by antibiotic therapy.

Confirmatory tests
C. difficile is isolated from stools by anaerobic culture on special selective antibiotic-containing media. Detection (in a tissue culture test) of the cytotoxins is even better evidence of clinical relevance.

Clinical syndrome and management
C. difficile produces antibiotic-associated diarrhoea, varying in severity from several loose stools daily to severe pseudomembranous colitis (p. 137). Withdrawal of the causative antibiotic is sufficient in mild cases, but oral metronidazole, vancomycin or bacitracin are effective in more severe cases, though relapse is common. Control is by judicious selection of antibiotics, and care in hospital practice to avoid cross-infection, e.g. with sigmoidoscopes.

OTHER CLOSTRIDIA
Other species including *C. septicum, C. novyi* and *C. tertium* can cause cellulitis and gangrene similarly to *C. perfringens.* Apparently spontaneous infection with *C. septicum* is often a sign of undiagnosed colonic cancer.

Clostridia
- Clostridia are anaerobic spore-forming Gram-positive rods.
- They produce severe disease by numerous very potent toxins.
- *C. perfringens* causes a range of skin, soft tissue and muscle disease, from simple cellulitis to fatal gas gangrene; it also causes food poisoning.
- *C. tetani* causes tetanus.
- *C. botulinum* causes food-borne, infant and wound botulism.
- *C. difficile* causes antibiotic-associated diarrhoea and pseudomembranous enterocolitis.

NEISSERIA, BRANHAMELLA, KINGELLA AND ACINETOBACTER

These genera contain aerobic Gram-negative cocci including:

- **primary pathogens** like *Neisseria meningitidis*, causing acute and often over-whelming septicaemia and meningitis; *N. gonorrhoeae*, the cause of gonorrhoea, and *Moraxella catarrhalis*, causing respiratory infections
- **opportunistic pathogens** among our normal flora, rarely causing disease, such as *N. subflava*, *N. sicca*, *N. mucosa*, *K. kingii* and *Actinobacter* spp.

NEISSERIA

N. GONORRHOEAE (GONOCOCCUS)

N. gonorrhoeae are Gram-negative cocci which are oval or bean shaped and occur often in pairs with their long sides parallel (Fig. 1). They are fragile and fastidious, dying on drying, and are aerobic and capnophilic (growth is enhanced by 5% CO_2). They are oxidase positive and die rapidly in the environment. The only known host is mankind.

Pathogenesis and virulence

Figure 2 shows structural features of importance. Protein I has 16 serotypes related to virulence; protein II is found in less virulent strains. Exotoxins are unknown. Gonococci can colonise mucosal surfaces in the genital tract and cause asymptomatic infections, especially in women. They can pass through the mucosal cells, invade, and cause inflammation in the underlying tissues. Many strains are killed by serum factors, including IgG, IgM and complement, so disseminated infection to joints, tendons and skin is unusual. No significant immunity develops, so repeat infections easily occur.

Confirmatory tests

Microscopy is very important. Diagnosis is almost 100% certain if the Gram stain shows intracellular Gram-negative diplococci. False negatives are more common in women and asymptomatic patients.

Culture. Urethral, cervical, rectal and pharyngeal swabs are plated immediately onto warm media. The fragile gonococcus is inhibited by some components of usual media, so isolation is improved by using a selective antibiotic-containing medium (e.g. Thayer-Martin medium) to suppress contaminants, and a rich non-selective medium such as chocolate agar to ensure growth of oxidase positive (Gram-negative) cocci. Joint aspirate may be positive, but skin and blood even in disseminated disease are usually negative.

Biochemical tests show acid production from glucose only. Serology is useless.

Clinical syndromes

The gonococcus is the cause of the venereal disease gonorrhoea (pp. 154–160). It infects organs alternately: urethra not vagina, cervix not uterus, fallopian tube not fimbriae (primarily), ovary not posterior abdominal wall. Salpingitis (Fallopian tube infection) leads to tubal blockage and pelvic inflammatory disease (PID). Urethritis and epididymo-orchitis are the common infections in males. Rectum and pharynx are infected by direct contact with infectious discharge, usually urethral. Disseminated infection is uncommon. Asymptomatic patients are a reservoir of infection, which occurs by direct invasion and local spread.

Chemotherapy

Penicillin G or amoxycillin is still highly effective for sensitive strains, while ceftriaxone or ciprofloxacin is necessary for β-lactamase positive strains (penicillinase-producing *N. gonorrhoeae*, PPNG).

Control

Strain variation and poor antigenicity have precluded an effective vaccine to date, so contact tracing and treatment of all infections, celibacy, monogamy, or safer sex must suffice.

N. MENINGITIDIS (MENINGOCOCCUS)

N. meningitidis is a Gram-negative capsulated diplococcus; it is aerobic, capnophilic, fragile and fastidious. It is divided into serogroups by polysaccharide capsular antigens, of which A, B, C, Y and W 135 are most common (these are further divided into serotypes by outer membrane

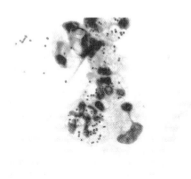

Fig. 1 *N. gonorrhoeae* **using Gram stain.**
Note the intracellular Gram-negative diplococci, i.e. within neutrophils.

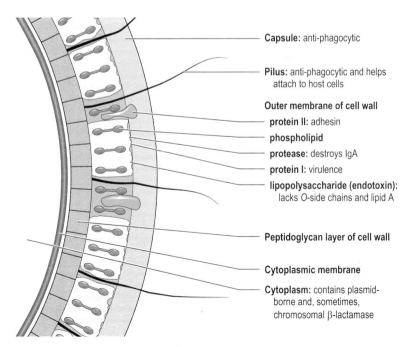

Capsule: anti-phagocytic

Pilus: anti-phagocytic and helps attach to host cells

Outer membrane of cell wall
protein II: adhesin
phospholipid
protease: destroys IgA
protein I: virulence
lipopolysaccharide (endotoxin): lacks O-side chains and lipid A

Peptidoglycan layer of cell wall

Cytoplasmic membrane

Cytoplasm: contains plasmid-borne and, sometimes, chromosomal β-lactamase

Fig. 2 *N. gonorrhoeae:* **structure and function.**

proteins, and immunotypes by lipopolysaccharides). The only known host is humans, and asymptomatic nasopharyngeal carriage occurs.

The meningococcus is very similar structurally to the gonococcus but its more prominent capsule is more protective against antibody-mediated phagocytosis so systemic blood spread is much greater. Continuous production of outer membrane fragments releases great quantities of endotoxin, so shock and haemorrhage are marked (pp. 86–7, 128–9).

Infection occurs in three stages, with progression to a further stage often not proceeding:

- growth in the nasopharynx: often asymptomatic
- invasion of blood: petechial rashes, bacteraemia, shock
- invasion of the meninges: meningitis.

Confirmatory tests
Microscopy of CSF samples is often positive for intracellular Gram-negative diplococci in meningococcal meningitis. Culture of blood and CSF is essential but may be negative because of prior antibiotic treatment or inhibitory substances in some media. Colonies are transparent and non-haemolytic; large capsules give mucoid colonies.

Biochemistry is also important. While **g**onococci produce acid from **g**lucose, **m**eningococci also utilise **m**altose (*N. lactamica* also uses **l**actose, and *N. sicca* also uses **s**ucrose, but not lactose).

Serology is useful, as latex agglutination (Fig. 3) detects soluble polysaccharide antigen, especially if empiric antibiotic treatment has made cultures negative before bacterial diagnosis.

Clinical syndromes
Acute meningitis and/or overwhelming septicaemia with haemorrhagic rash and shock are the two major diseases. The Waterhouse-Friderichsen syndrome is meningococcal septicaemia and shock with adrenal haemorrhage and adrenal failure. Rarely, pneumonia, arthritis or sub-acute low-grade septicaemia occur.

Chemotherapy
Urgent treatment with penicillin or a third-generation cephalosporin is essential, with appropriate supportive measures.

Control
A polyvalent vaccine is now available but is expensive and ineffective against the commonest type, B.

BRANHAMELLA

Branhamella are normal flora of the respiratory and genital tracts. The most important member is *B. catarrhalis* (formerly called *Moraxella catarrhalis*). Branhamella are small Gram-negative cocci that tend to grow in pairs end-to-end. Little is known of virulence factors, but about 80% of *B. catarrhalis* produce β-lactamase.

Confirmatory tests
B. catarrhalis is a capsulated Gram-negative coccus, oxidase and catalase positive, which grows on nutrient and blood agar, but does not produce acid from carbohydrates (unlike *Neisseria*).

Clinical syndromes and chemotherapy
B. catarrhalis causes bronchitis, sinusitis, otitis media, sometimes pneumonia and, rarely, osteomyelitis, meningitis or endocarditis. Other *Branhamella* spp. cause eye infections.

Amoxycillin plus clavulanic acid, or erythromycin, tetracycline or cotrimoxazole are all usually effective in treatment.

KINGELLA

Kingella are found in the human respiratory tract, where asymptomatic carriage occurs. They are small aerobic Gram-negative cocco-bacilli; some are encapsulated. They are oxidase positive but catalase negative, and β-haemolytic on blood agar. Microscopy and culture show these features, and biochemical tests confirm the identity. *K. kingii* is a rare cause of endocarditis and other infections. Penicillin is effective against many strains.

ACINETOBACTER

The classification of *Acinetobacter* spp. is confused. The most acceptable method uses one species, *A. calcoaceticus*, with two varieties, var. *anitratus* for acid-forming strains, and var. *lwoffi* for the remainder. All are small aerobic Gram-negative rods or cocco-bacilli, with some long forms, i.e. they are pleomorphic, ranging from cocco-bacilli to filaments. There are no spores or flagellae. They are oxidase negative, unlike all others described here. *Acinetobacter* spp. are normal flora in the respiratory tract and are identified by their physical characteristics and by biochemical tests. They sometimes cause hospital-acquired pneumonia, especially in ventilated patients. Community-acquired pneumonia is very rare. *Acinetobacter* spp. are relatively antibiotic resistant, though usually they respond to sulphonamides. Gentamicin and ticarcillin are often used in treatment.

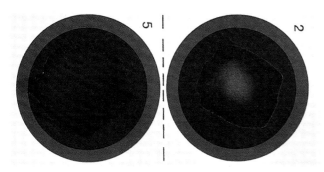

Fig. 3 *N. meningitidis:* latex agglutination.

Neisseria, Branhamella, Kingella and Acinetobacter

- *Neisseria* are Gram-negative aerobic cocci growing in pairs.
- *Neisseria* have no exotoxins but numerous virulence factors include capsule, pili, endotoxin and enzymes: gonococci also have proteins I and II.
- Gonococci are inhibited by serum and cause local urethral, cervical, tubal, pharyngeal and rectal gonorrhoea, and sometimes joint and disseminated disease.
- Meningococci are not inhibited by serum and cause meningitis and septicaemia with shock and haemorrhage.
- *Branhamella catarrhalis* is much less virulent and causes broncho-pulmonary infections.
- The related genera *Kingella* and *Acinetobacter* are of less importance medically.

HAEMOPHILUS, BORDETELLA AND LEGIONELLA

HAEMOPHILUS

H. INFLUENZAE

There are many similarities between the three encapsulated organisms *H. influenzae*, *N. meningitidis* and *S. pneumoniae* in virulence factors, pathogenesis and clinical syndromes, though *S. pneumoniae* being Gram positive lacks endotoxin.

H. influenzae is a pleomorphic aerobic Gram-negative coccobacillus which needs accessory growth factors X (haematin) and V (NAD) from blood. It is non-motile and has no spores. Virulent strains have large polysaccharide capsules, with type b the most important of six serotypes, a to f. Humans are the only known host, and non-encapsulated strains are normal respiratory flora in almost all people.

Structural features of importance are:

- the capsule, which is anti-phagocytic
- lipopolysaccharide lipid A of endotoxin which causes acute inflammation and shock
- a protease that destroys the Fc end of antibody.

Maternal or acquired antibody is strongly protective, so most infections are from age 3 months to 3 years.

Confirmatory tests

Gram stain is characteristic in CSF, pus or sputum (Fig. 1). Culture on chocolate agar (blood agar gently heated to liberate factors X and V and destroy inhibitors), or blood agar with a streak of *Staph. aureus* (for the same reasons) shows small opaque colonies growing best ('satellited') around the streak. Like the gonococcus, it utilises glucose only and is catalase positive.

Serology by latex agglutination confirms the serotype b.

Clinical syndromes

Non-encapsulated strains only cause local disease like bronchitis and otitis media, while capsulated strains, protected from phagocytes, cause meningitis and epiglottitis.

Chemotherapy

At least 20% of strains in most areas now produce β-lactamase, and some also have chromosomally mediated β-lactam resistance, so penicillin and ampicillin are now unreliable; hence a third-generation cephalosporin (or chloramphenicol in poorer communities) is used for serious infections, or ampicillin (if sensitive), cotrimoxazole or cefaclor for milder infections.

Control

Vaccines have been difficult to develop but are now widely available and effective,

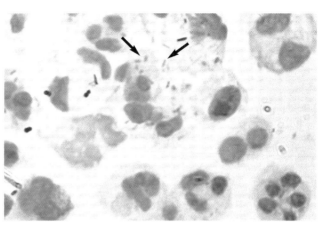

Fig. 1 *H. influenzae:* Gram stain of pink coccobacilli (few pneumococci also purple diplococci).

although antibody response is poorest in the 6–12-month-old child at greatest risk. Rifampicin is used for unvaccinated close contacts of children with meningitis or epiglottitis.

OTHER *HAEMOPHILUS* SPP.

Three other species from the respiratory tract are of some medical importance.

- *H. haemolyticus* because being β-haemolytic it may be mistaken for *S. pyogenes* if the Gram stain is not well decolorised.
- *H. parainfluenzae* (does not require X factor nor CO_2 and ferments sucrose as well as glucose) sometimes causes infective endocarditis and, rarely, other infections similar to those from *H. influenzae*.
- *H. aphrophilus* requires CO_2 but neither X nor V factors, and ferments glucose, sucrose and lactose. Colonies are slow-growing and adherent to chocolate agar, qualities reflected in its rare clinical presentation of subacute endocarditis.
- *H. ducreyi* is very different. It does not require V factor or CO_2, is catalase negative and ferments none of the usual three sugars. It causes the sexually transmitted disease **chancroid**. The synonym 'soft sore' is a reminder that the genital lesion caused is soft, ulcerated and painful, unlike the painless induration of a syphilitic chancre. Gram-stain from the edge of the ulcer or lymph node is more reliable than culture. **Treatment** of patient and partner(s) is usually with ceftriaxone, erythromycin or cotrimoxazole.

BORDETELLA

The genus Bordetella comprises tiny Gram-negative, aerobic, nutritionally fastidious rods.

B. PERTUSSIS

B. pertussis is non-motile and ferments no sugars. Humans are the only reservoir and source of infection. It causes pertussis, 'whooping cough', a severe but superficial infection of the epithelium of large airways.

Structural features of importance are:

- fimbriae, which strangely do not assist attachment
- filamentous haemagglutinin, which does assist attachment
- endotoxin, which causes fever and systemic symptoms
- exotoxins (Table 1).

Confirmatory tests

Gram-stain of nasopharyngeal swabs usually shows no organisms, but direct fluorescent antibody (DFA) stains may be positive. Culture needs bedside inoculation of special media, e.g. Bordet-Gengou is required. Minute colonies grow over 3–4 days. Specific antibody confirms identity.

Clinical syndrome and management

The distinctive disease of pertussis is described on page 109.

Erythromycin and other antibiotics do not alleviate symptoms but diminish infectivity to other humans. Vaccination is relatively effective, usually in triple antigen, but scare campaigns about side-effects have decreased usage and increased epidemics.

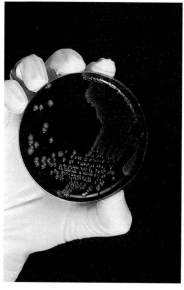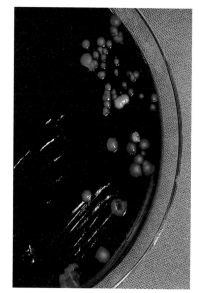

Fig. 2 **L. pneumophila** cultured on BCYE agar.

Table 1 **Virulence factors of *B. pertussis***

Virulence factor	Site of action	Actions
Filamentous haemagglutinin	Host epithelium	Attachment, also RBC agglutination
Pertussis toxin (numerous actions, hence many names)	Local and systemic (ADP-ribosylation)	Attachment, blocks cell signals, impairs phagocyte function, lymphocytosis
Tracheal cytotoxin	Ciliated epithelium	Initially ciliastasis, then cell death
Adenylate cyclase toxin	Local conversion of ATP to cAMP	Impairs phagocyte function, capillary leak, oedema
Dermonecrotic toxin	Local epithelial vasoconstriction	Ischaemia then necrosis
Endotoxin	Local and systemic	Fever, systemic symptoms

OTHER *BORDETELLA* SPP.

B. parapertussis causes a minority of cases of pertussis. It is less fastidious than *B. pertussis* and grows on ordinary media, which it may colour brown. It is more active metabolically, being citrate and urease positive.

B. bronchiseptica is an animal pathogen that rarely can cause respiratory infections in humans, including immunocompromised patients. It also grows well on ordinary media, and reduces nitrate as well as being citrate and urea positive.

LEGIONELLA

L. pneumophila was eventually isolated from the lungs of delegates to the 1976 American Legion convention in Philadelphia who had developed a new type of pneumonia, which can often be fatal. Other species less commonly cause less severe but similar infections.

Legionella are small cocco-bacilli, pleomorphic on culture. Virulent strains are serum resistant, and they survive in macrophages by inhibiting lysosomal fusion. They cause tissue damage, probably by their proteases, lipase, nuclease and phosphatase.

Legionella have a number of unusual features:

- they have a typical Gram-negative cell wall, yet do not stain with Gram's stain (Gimenez and Dieterle silver stain are effective)
- they do not grow on usual media, needing L-cysteine for growth, which is enhanced by iron and inhibited by sodium and aromatic compounds
- they are widely distributed in water and soil, in the sediment in water systems and survive 45°C but not 60°C.
- they are intracellular parasites, in macrophages.

Confirmatory tests
These are a combination of staining, culture and serology.

- *Microscopy.* Non-specific stains (Gimenez, Dieterle) are of limited use when other organisms are present in, for example, sputum, so the direct fluorescent antibody (DFA) test using labelled specific antibody is best on clinical specimens.
- *Culture.* Must be on special media (see above) such as buffered charcoal yeast extract (BCYE) agar, which is buffered to pH 6.9, provides nutrients (L-cysteine and iron), is low in sodium, and absorbs aromatic compounds (Fig. 2). Identification of the species is done in reference laboratories only.
- *Serology.* This is not a reliable identification method. Antibodies take 4–8 weeks to appear. Antigen detection is specialised, specific but insensitive though useful in urine.

Clinical syndromes and management
Infection may be asymptomatic or result in acute short-lived flu-like 'Pontiac fever', or cause Legionnaires' disease, a severe pneumonia with multi-system (brain, kidney, gut) involvement (p. 113). Erythromycin is the drug of choice.

Control. Water systems, particularly air-conditioning and evaporative cooling towers and stagnant areas of hot water systems, must be disinfected or flushed regularly to decrease the numbers of *Legionella*, as total sterilisation is usually impossible.

Haemophilus, Bordetella, Legionella

- *H. influenzae* needs X and V factors from blood for growth.
- Virulence factors include an antiphagocytic capsule and endotoxin.
- It causes meningitis, epiglottitis and other respiratory infections.
- Many strains now produce β-lactamase.
- *H. ducreyi* does not require V factor and causes chancroid.

- *B. pertussis* is a small Gram-negative rod needing special media for growth.
- It has at least four exotoxins as well as endotoxin.
- It causes pertussis (as does *B. parapertussis*, though less often).

- *L. pneumophila* does not stain with usual stains or grow on usual media.
- It lives in the water supplies of buildings, and it does not spread person to person.
- It is an intracellular pathogen causing two different diseases, mild Pontiac fever and severe Legionnaires' disease.
- Treatment is erythromycin.

ENTEROBACTERIACEAE (1)

Aerobic, gram-negative bacilli can be divided into three broad groups for practical purposes: firstly, the coliforms, i.e. intestinal bacteria of the family Enterobacteriaceae (e.g. *Escherichia, Shigella, Salmonella, Citrobacter, Klebsiella, Enterobacter, Serratia, Hafnia, Proteus, Providentia* and *Morganella*); secondly, the parvobacteria (e.g. *Haemophilus, Bordetella, Brucella, Yersinia* and *Pasteurella*) and thirdly, *Pseudomonads* and related bacteria (pp. 46–7). The Enterobacteriaceae are all aerobic (facultatively anaerobic) non-spore forming bacilli which live particularly in the intestine and, for convenience, are often called 'coliforms' or 'enteric bacteria'. They cause a range of gastrointestinal, intra-abdominal, urinary, wound and other infections.

Identification of enteric bacteria
Most species are morphologically indistinguishable: they are usually straight rods but considerable variation occurs.

- motility: shigellae, yersiniae and klebsiellae are non-motile
- culture: *Proteus* spp. 'swarm' and smell of ammonia; and *Klebsiella* spp. produce large mucoid colonies.
- biochemical tests: multiple tests distinguish species by differences in metabolism. These tests are usually performed with commercial kits or automated machines (Fig. 1)
- serology: important species are identified by detection of specific antigens by agglutination reactions.

Table 1 shows general features of the main distinguishing biochemical tests; most enteric bacteria are positive for these tests but the exceptions help in identification (Fig. 2). Lack of gas from glucose, non-lactose fermentation (NLF) and positive production of H$_2$S are all signs of serious pathogens. Other specific tests include ability to grow in KCN, phenylalanine utilisation, ornithine decarboxylation and fermentation of numerous sugars.

Cell structure and function
Figure 3 shows the major structures and their functions in virulence and pathogenesis. All Enterobacteriaceae have H, K and O antigens and endotoxin. Some will produce exotoxins (heat-labile proteins).

Antibiotic sensitivity
Resistance to antibiotics is becoming an increasing problem in dealing with infections particularly when hospital cross-infection or community spread is rapid. Table 2 gives figures for one particular hospital. General features of Enterobacteriaceae infections are:

- *E. coli, Klebsiella, Proteus* and *Enterobacter* spp. are the most common in hospitals
- *E. coli, Klebsiella* and *Proteus* spp. are in general more sensitive; *Enterobacter, Serratia* and *Citrobacter* spp. are in general more resistant to antibiotics
- order of likely efficacy is gentamicin/ciprofloxacin, third-generation cephalosporin, timentin, cotrimoxazole
- amoxycillin, cephalothin and ticarcillin should not be used alone without sensitivity results.

Table 1 **Biochemical reactions used in identification**

Test	Positive result	Exceptions
Acid production from glucose	All	
Gas production from glucose	Most	Shigellae, *Y. pestis, S. typhi*
Lactose fermentation	Only *E. coli, Klebs, Enterobacter*	Salmonellae, Shigellae, *Y. pestis*
Catalase positive	All	
Reduce nitrate	All	
Oxidase negative	All	
Production H$_2$S	Only salmonellae, *Proteus* and *Y. pestis* always +ve	K-E-S-H group always −ve
Urease	*Proteus, Morganella, Yersinia* spp. strong	K-E-S-H group −ve
Indole production	*E. coli, Proteus* spp. strong	*P. mirabilis*, Salmonellae, K-E-S-H group
Utilises citrate	Most	*E. coli*, Shigellae, morganellae, yersiniae
Gelatin liquefaction	Most	Salmonellae, shigellae
VP	K-E-S-H group	Most negative

K-E-S-H group: *Klebsiella, Enterobacter, Serratia, Hafnia* spp. −ve, negative; +ve, positive.

Table 2 **Incidence of Enterobacteriaceae isolated, and sensitivity to chosen antibiotics[a]**

Organism	No. patients	AMOX	CEF1	Sensitivity (%) GENT	TIM	SuTM	CIP	CTX
E. coli	453	47	82	98	78	82	100	100
Shigella	2	0	100	100	100	50	100	100
Salmonella	9	100	100	100	100	100	100	100
Citrobacter spp.	14	0	0	100	100	100	100	100
Klebsiella	99	0	72	86	83	91	87	97
Enterobacter	51	0	0	96	56	96	100	54
Serratia	15	0	0	100	100	100	100	100
Hafnia	4	25	0	100	100	100	100	100
Proteus	90	79	100	98	100	90	100	100
Providentia	2	0	0	100	0	NT	100	100
Morganella	13	0	0	100	100	100	100	100
Yersinia	0	–	–	–	–	–	–	–
Total	752							

AMOX, amoxycillin; CEF1, cephalothin; GEN, gentamicin; TIC, ticarcillin; SuTM, cotrimoxazole; CIP, ciprofloxacin; CTX, ceftriaxone (a third-generation cephalosporin). NT, not tested
Tested in 4 months at Alfred Hospital, Monash University, Melbourne, Australia, Jan–April, 1998.

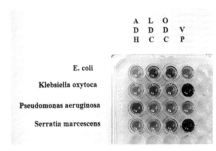

Fig. 1 **Commercial kit for biochemical reactions.**

ESCHERICHIA COLI

E. coli is a common component of the aerobic bowel flora and causes urinary, wound, lung, meningeal and septicaemic infections. Some strains are important causes of travellers' diarrhoea and the haemolytic–uraemic syndrome.

Identification of *E. coli* and its variants is important because:

- it is normal commensal flora that must be distinguished from intestinal pathogens
- its presence in water supplies is evidence of faecal contamination
- it can be a pathogen.

Pathogenesis and virulence

Entero-pathogenic *E. coli* strains (EPEC) are classified by their O antigens (and subdivided by H and K antigens). Important strains are:

- enteroadhesive (EAEC)
- enteroinvasive (EIEC)
- enterotoxigenic (ETEC): exotoxins
- enterohaemorrhagic (EHEC): cytotoxic verotoxin.

Confirmatory tests

Microscopy and staining shows a motile Gram-negative rod. Culture shows dry flat lactose-fermenting colonies on MacConkey agar (Fig. 2a), and often haemolytic colonies on blood agar.

Biochemical tests used include those in Table 1. Note that *E. coli* is citrate–H$_2$S–urea–VP negative, indole positive and ferments both glucose and lactose with gas. It is motile, unlike *Shigella* spp.

Clinical syndromes

There are four entirely different groups of clinical disease caused by *E. coli*:

- neonatal meningitis from maternal bowel flora and cross-infection (p. 86)
- organ system infections, from the patient's own normal flora or from crossinfection, causing UTI, wound, intraperitoneal, blood-stream or lung infections (pp. 128, 142 and 148)
- gastroenteritis or dysentery from the EPEC strains noted above (p. 138) (EAEC, EHEC, EIEC, ETEC)
- haemolytic uraemic syndrome (p. 138).

Chemotherapy

Ampicillin and cotrimoxazole resistance is now common. Gentamicin remains reliable. Newer penicillins and cephalosporins may be used.

Control

This depends on hygiene, hospital infection control and the availability of pure food and water.

SHIGELLA

There are four species, *S. flexneri, S. sonnei, S. boydii* and the most virulent, *S. dysenteriae*. They are non-motile, aerobic Gram-negative rods that are citrate, H$_2$S, urea, VP and often indole negative and do not produce gas from glucose, nor ferment lactose. As few as 200 organisms can cause bacillary dysentery. Spread is primarily direct faecal–oral (i.e. person to person), rarely via food or water.

Shigellae produce an enterotoxin (shiga toxin) which gives diarrhoea, and are invasive into colonic mucosa. *S. dysenteriae* has a neurotoxin, so headache, meningismus (meningeal symptoms without meningitis) and even fits occur.

Confirmatory tests

Microscopy of stool samples shows many white blood cells but few if any organisms. Culture and biochemistry show no motility, no gas from glucose and no lactose fermentation (Fig. 2b). Serology distinguishes serotypes of all except *S. sonnei*.

Clinical syndromes

Shigella spp. all cause bacillary dysentery, ranging from mild to fatal (pp. 138–9).

Chemotherapy

Only *S. dysenteriae* infections usually need antibiotics to control infection, though antibiotics decrease infectivity in all. Resistance to ampicillin, cotrimoxazole, tetracycline and even chloramphenicol is now widespread, so quinolones, cephalosporins or gentamicin are used.

Control

Good hygiene, especially hand-washing, is essential.

CITROBACTER

The most common *Citrobacter* spp. *C. freundii*, is similar to salmonellae in being H$_2$S positive, and indole and VP negative, but it is a late lactose fermenter. It is commonly present as normal bowel flora and has low virulence. Clinical syndromes are uncommon, but Citrobacter can cause hospital-acquired infections especially in immunocompromised patients. Gentamicin, ciprofloxacin or a third-generation cephalosporin are usually prescribed.

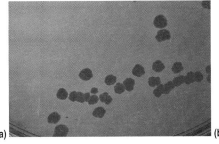
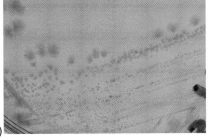

(a)   (b)

Fig. 2 **Lactose fermentation to distinguish species: (a)** *E. coli* **positive (pink), (b)** *Shigella flexneri* **negative (colourless).**

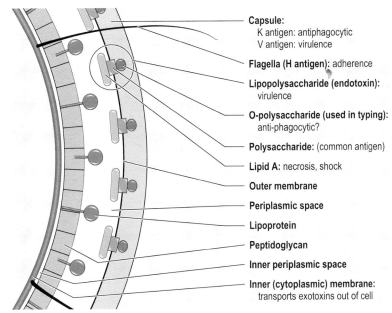

Capsule:
  K antigen: antiphagocytic
  V antigen: virulence

Flagella (H antigen): adherence

Lipopolysaccharide (endotoxin):
  virulence

O-polysaccharide (used in typing):
  anti-phagocytic?

Polysaccharide: (common antigen)

Lipid A: necrosis, shock

Outer membrane

Periplasmic space

Lipoprotein

Peptidoglycan

Inner periplasmic space

Inner (cytoplasmic) membrane:
  transports exotoxins out of cell

Fig. 3 **Structure of Enterobacteriaceae.**

ENTEROBACTERIACEAE (2)

SALMONELLA

Salmonella spp. are motile aerobic Gram-negative rods that infect humans, animals and birds. Classification is difficult, as the 2000+ serotypes based on O, H and K antigens (previously separate species) are very useful epidemiologically in tracing outbreaks and yet are not species by the usual rules. One classification has only three species: *S. cholerae-suis* (the type species, basic for nomenclature), *S. typhi* (the cause of typhoid fever, very different from all the others), and *S. enteritidis* ('of enteritis', all the others, causing gastroenteritis). This last group contains a number of serotypes called *Arizona*, formerly a separate related genus, then a sub-genus.

Salmonellae can be divided into two groups, epidemiologically and clinically:

- *S. typhi* and *paratyphi* serotypes: these infect humans only, producing severe illness with septicaemic as well as intestinal symptoms (enteric fever)
- all others: primarily animal pathogens, widespread in food animals, eggs and animal feed; these cause local gastroenteritis (salmonella food poisoning).

Structural features of importance are like *E. coli* except the K antigen of *S. typhi* is called Vi (for virulence, though other factors are important).

Confirmatory tests

Salmonellae are citrate positive, produce H_2S, gas from glucose (except *S. typhi*) and are non-lactose fermenters. Stool cultures are diagnostic in gastroenteritis, while typhoid fever often needs stool, urine and blood cultures plus the Widal agglutination test with specific antisera.

Clinical syndromes

Pathogenesis follows very different patterns.

- **Localised gastroenteritis** (p. 138) without systemic spread (caused by *S. enteritidis* serotypes) results from local invasion (by the bacteria) of the epithelial cells of the gut, then migration to the lamina propria layer where they multiply and stimulate active fluid secretion into the gut lumen.
- **Septicaemia** (p. 130) particularly from *S. typhi, S. paratyphi, S. cholerae-suis* and *S. dublin*. Focal abscesses are common.

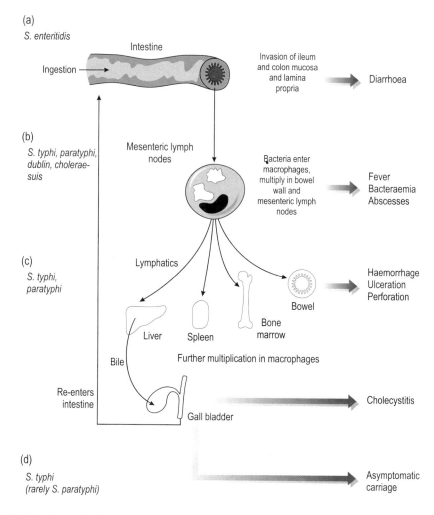

(a) *S. enteritidis*

(b) *S. typhi, paratyphi, dublin, cholerae-suis*

(c) *S. typhi, paratyphi*

(d) *S. typhi (rarely S. paratyphi)*

Fig. 1 **Pathogenesis of salmonella infections.**

- **Asymptomatic carriage**, especially in the gallbladder, with *S. typhi*.
- **Systemic 'enteric fever' (typhoid)** (p. 133) in which organisms invade the mucosa of ileum and colon, entering and multiplying in macrophages to be carried by them to liver, spleen and bone marrow where further multiplication occurs giving systemic illness. Liver infection spreads in the bile to the gall bladder, and thence again to the bowel. Colonic mucosal infection progresses to haemorrhage, ulceration and sometimes perforation. This sequence explains the 10-day incubation period and the prolonged illness (Fig. 1). Typhoid fever is due to *S. typhi* with milder paratyphoid fever from paratyphoid A, B and C serotypes.

Chemotherapy

Usually no chemotherapy is used for gastroenteritis. Owing to increasing antibiotic resistance, ciprofloxacin is replacing chloramphenicol, cotrimoxazole and ampicillin in treating typhoid. Ciprofloxacin or norfloxacin is used for carriers.

Control

Hygiene in food preparation and storage is essential, and vaccination is available in endemic areas or for travellers.

KLEBSIELLA

Klebsiella spp. are non-motile, capsulated aerobic Gram-negative rods which are citrate and VP positive, but indole, H_2S and urea negative, and produce gas by fermenting many sugars including glucose and lactose. Their classification has had many changes: *K. ozaenae* and *K. rhinoscleromatis* are clearly different biochemically and clinically, while *K. pneumoniae* (including the former *K. aerogenes*) is the commonest clinical isolate and *K. oxytoca* the next commonest. They are normal flora in the gut, and are found in soil, grain and water.

A prominent anti-phagocytic capsule causing large mucoid colonies is usual. The cell wall is typical of Gram-negative bacteria, and there are K, O and H antigens and endotoxin as in *E. coli* (Fig. 3, p. 41).

Laboratory identification depends on Gram stain, culture and biochemical tests. Serotyping is rarely used clinically.

Clinical syndromes

- *K. pneumoniae* causes a primary, cavitating, serious pneumonia (Fig. 1, p. 114).
- *K. pneumoniae* and *K. oxytoca* are important in hospital-acquired infections (p. 190) of urine, wounds and respiratory tract from the hospital environment, gut colonisation and other infected patients.
- *K. rhinoscleromatis* causes rhinoscleroma, an unusual chronic infection of the nose and adjacent tissues (p. 108).
- *K. ozaenae* is associated with ozaena, a foul-smelling purulent atrophic rhinitis.

Chemotherapy

Most strains produce penicillinases, so gentamicin, quinolones, third-generation cephalosporins or timentin are used.

Control

Cross-infection measures are necessary, and eradication of hospital reservoirs is desirable.

Fig. 2 *Serratia marcescens* with bright red pigment.

Fig. 3 *P. mirabilis* in urine: affected by antibiotic.

ENTEROBACTER, SERRATIA, AND HAFNIA

Enterobacter are very similar to *Klebsiella* except that they are motile, and ornithine positive. They are divided into several species by their biochemistry, especially by arginine and lysine decarboxylase (e.g. *E. cloacae, E. aerogenes, E. sakazakii*). They are found as faecal flora and in soil, sewage and water. *Serratia* have the same major reactions but usually are pigmented bright pink or red (Fig. 2). They, with *Hafnia*, are distinguished from *Enterobacter* by sorbitol fermentation and DNAase production. They are found in faeces, soil, sewage and water.

Their structure and function are unexceptional, and their virulence is relatively low. Laboratory identification is by Gram stain, culture, and their biochemical characteristics.

Clinical syndromes

They are seldom primary pathogens but as opportunists can cause hospital-acquired wound, blood, lung and urinary tract (UTI) infections, especially in immunocompromised patients (pp. 190–3).

Chemotherapy

For infections, gentamicin or quinolones are used; third-generation cephalosporins are becoming unreliable owing to the organisms' extended spectrum cephalosporinases.

Control

This depends on hygiene, handwashing and elimination of sources.

PROTEUS, PROVIDENTIA AND MORGANELLA

All are aerobic, motile, non-spore-forming Gram-negative rods (Fig. 3) and, alone among Enterobacteriaceae, all are phenylalanine deaminase positive. A commonly accepted division is into *P. mirabilis* (the commonest) and *P. vulgaris* (next commonest) as urease positive, H_2S and indole positive and VP negative with marked swarming over agar without discrete colony formation. *P. mirabilis* is indole negative. *Providentia* are similar in major reactions except that they are urease negative and do not swarm. *M. morganii* is like *P. vulgaris* in major reactions except that it is citrate negative. They are all found in human faeces and in soil.

Features of importance are:

- urease which splits urea in urine, raises urine pH and encourages renal stone formation
- pili may assist both attachment and phagocytosis
- flagellar action may assist spread.

Laboratory identification depends on Gram stain, culture and biochemical tests as above.

Clinical syndromes

Of this group, *P. mirabilis* is the commonest in community acquired UTI; the others are more important in hospital-acquired infections (UTI, wound, blood, and lung). The source is infected patients, not the environment or gut colonisation (contrast with *Klebsiella*) (pp. 190–3).

Chemotherapy

P. mirabilis has been more sensitive to antibiotics, but resistance is increasing and often gentamicin, quinolones or a third-generation cephalosporin is needed for any of these genera.

Control

This is particularly by hygiene and handwashing.

Enterobacteriaceae

- They are all aerobic, non-spore-forming Gram-negative rods usually found in the gut.
- They are distinguished by biochemical tests, especially citrate, H_2S, indole, VP, urea, sugar fermentation, and serology.
- Virulence factors include capsular K, flagellar H and somatic O antigens, endotoxin, and some have exotoxins.
- A few are aggressive primary pathogens causing bacillary dysentery and typhoid fever, gastroenteritis, meningitis, pneumonia and urinary infections, but more are opportunist pathogens infecting hospitalised and immunocompromised patients.
- Gentamicin, quinolones and later cephalosporins are often necessary because of antibiotic resistance.
- Control is rarely by vaccine, usually by hygiene and hand washing.

VIBRIO, CAMPYLOBACTER, HELICOBACTER, AEROMONAS, PLESIOMONAS

These organisms form, with *Neisseria* and *Pseudomonas*, one of only three groups of **oxidase positive** Gram-negative pathogens.

VIBRIO

Vibrios are short, curved, asporogenous, aerobic (facultatively anaerobic), Gram-negative rods. Special features are their comma shape, motility by a polar flagellum, oxidase positivity and salt-tolerance. Major pathogenic species are classified into *V. cholerae, V. parahaemolyticus, V. vulnificus* and *V. alginolyticus*. They live in fresh, brackish or salt water and infect humans directly or via shell-fish or raw fish.

They have typical Gram-negative cell structure. The somatic O antigens of *V. cholerae* give six serogroups. Most pathogenic strains are O1, further divided into two biotypes, 'cholerae' and 'El Tor', and three serologic subgroups, useful epidemiologically. All antigens are poorly immunogenic, so repeat infections occur, and vaccines are relatively ineffective.

Pathogenesis and virulence

Virulence of *V. cholerae* results from the potent cholera toxin which produces massive secretory diarrhoea (Fig. 1). In addition, cyclic AMP inhibits chemotaxis and phagocytosis. *V. cholerae* is thus an exceptional (and exceptionally effective) pathogen, able like *C. tetani*, to kill by up-regulating a normal biochemical host pathway without any cell damage or invasion.

Other vibrios are pathogenic by tissue invasion, though *V. parahaemolyticus* has a mild enterotoxin.

Confirmatory tests

Gram stain is characteristic. Culture of fresh specimens is best on special media such as thiosulphate-citrate-bile-sucrose (TCBS) agar, giving sucrose-positive (Fig. 2) yellow colonies for *V. cholerae* and *V. alginolyticus*, sucrose-negative blue-green colonies for *V. parahaemolyticus* and *V. vulnificus*. Further biochemistry and serology is performed.

Clinical syndromes and management

Syndromes depend on the infecting species:

- *V. cholerae* causes cholera (p. 139) (by toxin)

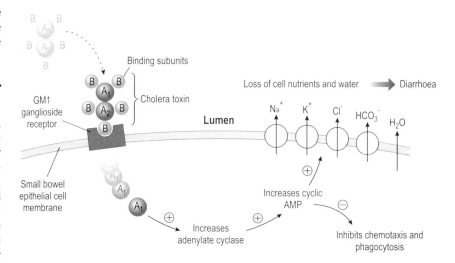

Fig. 1 **The action of cholera toxin.** A_1, active toxin subunit; A_2 binds A_1 to B; B binding subunit to GM1.

- *V. parahaemolyticus* causes gastroenteritis (p. 139) (by toxin and invasion)
- *V. vulnificus* causes severe spreading cellulitis, or septicaemia with unusual skin lesions (by local invasion and distant spread)
- *V. alginolyticus* causes superficial wound, eye and ear infections (by local invasion).

Fluid and electrolyte replacement is the more important treatment for cholera but tetracycline shortens the course. Ampicillin or cotrimoxazole are also effective.

Cholera epidemics can only occur under conditions of poor hygiene. A vaccine gives limited short-term protection. Control involves avoiding exposure and the provision of pure drinking water.

CAMPYLOBACTER

Campylobacters were long considered to be mainly animal pathogens, but once microaerophilic faecal cultures at 42°C were introduced, *C. jejuni* became recognised as the commonest cause of bacterial gastroenteritis.

Campylobacters are curved, asporogenous, microaerophilic or anaerobic Gram-negative rods. Special features are their curved shape, motility by a polar flagellum, oxidase positivity and need for only 3–15% oxygen. They are unusual metabolically as they use energy from TCA intermediates or amino acids and they neither ferment nor oxidise carbohydrates. Hence they are inactive in

many routine biochemical tests. Major pathogenic species are classified into *C. jejuni. C. coli, C. lari* (formerly *laridis*), and *C. foetus* (with two subspecies). The pathogenicity of other species is being established. As with salmonellae there is a large animal reservoir of campylobacters; infections result from consumption of contaminated food including meat, and milk from bottles with tops pecked by birds. Rarely, homosexual men have acquired infection as a STD.

Like salmonellae, they are not resistant to gastric HCl, and the infective dose is relatively high, usually 10 000 organisms or more. They infect the jejunum, ileum and colon. Probably both tissue invasion and a cytolytic endotoxin are important in pathogenesis. IgG, IgM and IgA antibodies develop and may be protective. *C. foetus* resists complement and antibody-mediated killing in serum much more than *C. jejuni*, so systemic spread occurs.

Confirmatory tests

Gram stain is characteristic ('campyle' means curved), but phase-contrast microscopy may be necessary to detect darting motility. Culture needs the special microaerophilic atmosphere, a temperature of 42°C, and selective media. Identification of species is by biochemical tests including nitrate, H_2S and hippurate hydrolysis, and by antibiotic sensitivity.

Clinical syndromes and management

C. jejuni and, rarely, other species cause enterocolitis (pp. 136–8).

Septicaemia, organ damage (pneumonia, meningitis) and vascular endothelial infections occur with *C. foetus*.

Fluid and electrolyte replacement are important but erythromycin is helpful if given early in enterocolitis. Gentamicin is used for serious infections.

Control. This depends on good food-handling practices and hand-washing.

HELICOBACTER PYLORI

Helicobacter is a recently recognised pathogen now accepted as the major causative factor in chronic gastritis and found in 95% of people with duodenal ulcer.

Initially classified as a campylobacter, it is a curved motile micro-aerophilic Gram-negative rod (Fig. 3a); is unable to grow at 42°C, is strongly urease positive (Fig. 3b) and is nitrate negative.

Motility assists the organism to move to the chemotactic factors haemin and urea in gastric pits. Urease generates NH_4^+ from urea, neutralising gastric acid. Haemin stimulates further growth of *H. pylori*, then enzyme and mediator release cause chronic inflammation. Little immunity develops and relapse is frequent.

Confirmatory tests
Special silver stain or modified Giemsa stain on gastric specimens for histology is distinctive. Biochemistry shows strong and very rapid urea production (Fig. 3b). Culture is on selective media at 37°C, in a microaerophilic atmosphere, but it is not done routinely as staining and urease are uniquely diagnostic on gastric specimens. Antibiotic sensitivity tests are not done routinely. Serology by ELISA is only helpful with individual patients if negative, as 30% of people in many communities are positive without active disease.

Clinical syndromes
These include chronic gastritis and duodenal ulcer.

Chemotherapy
The best therapy is not certain. Triple therapy with bismuth, amoxycillin (or tetracycline) and metronidazole (or tinidazole) probably gives most cures and fewest relapses.

Control
Until the reservoir and route of transmission are known, no logical control measures are possible.

AEROMONAS

Aeromonas spp. are short, asporogenous, aerobic (facultatively anaerobic), Gram-negative rods. Special features are their motility by a polar flagellum, oxidase positivity and production of gas by sugar fermentation. Major pathogenic species are usually classified into *A. hydrophila* and *A. liquefaciens*. Their habitat is water. Virulence is relatively low, and pathogenesis is usually by an enterotoxin.

Laboratory identification is by culture on blood agar and MacConkey's medium, Gram stain, oxidase reaction and biochemical tests including fermentation of sucrose, mannitol and inositol and ornithine-lysine-arginine to differentiate it from pseudomonads, *Plesiomonas*, vibrios and other enteric pathogens.

Aeromonas spp. cause gastroenteritis in normal hosts, and systemic opportunistic infections in compromised hosts (p. 169). Chemotherapy, if necessary, is usually by gentamicin, cotrimoxazole, tetracycline, or chloramphenicol. Control depends on avoidance of infected water.

PLESIOMONAS

P. shigelloides is the only species in the genus; it is a short, asporogenous, aerobic (facultatively anaerobic), Gram-negative rod. Special features are its motility using several polar flagella, oxidase positivity, and no production of gas by sugar fermentation. Its habitat is brackish water.

An O antigen is related to some *Shigella sonnei*. Virulence is relatively low, and pathogenesis is probably by localised invasion by organisms from infected shellfish.

Laboratory identification depends on the tests listed above under *Aeromonas*. Presentation is as gastroenteritis, or rarely cellulitis or systemic infection.

Chemotherapy, if necessary, is usually by cotrimoxazole, gentamicin, tetracycline or chloramphenicol. Control is by avoidance of infected shellfish.

Fig. 2 *Vibrio cholerae* culture on TCBS medium, yellow colonies.

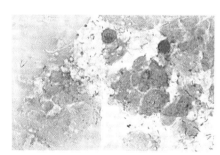

(a)

(b)

Fig. 3 *Helicobacter pylori*: (a) Gram stain and (b) urease reaction.

Vibrio, Campylobacter, Helicobacter, Aeromonas and Plesiomonas

- All are motile, oxidase-positive, aerobic or microaerophillic Gram-negative rods, often associated with water.
- They cause disease by enterotoxins or by local invasion, rarely by distant spread and invasion.
- The laboratory should be notified if they are suspected, to use special media or techniques, and oxidase and other biochemical tests.
- Gastrointestinal infection is most common, but systemic infection can occur, especially in compromised hosts.
- Chemotherapy is usually tetracycline, gentamicin or cotrimoxazole.

PSEUDOMONAS AND RARE GRAM-NEGATIVE RODS

PSEUDOMONAS

Pseudomonads are an entirely different family from the Enterobacteriaceae. They are Gram-negative rods but are:

- strict aerobes (a few can grow anaerobically using nitrate as electron acceptor)
- non-lactose fermenting
- motile
- oxidase positive, with oxidative metabolism, never fermentative
- able to survive with few nutrients, e.g. acetate, glucose
- widely distributed, therefore, in nature, in fluids, in hospitals
- normal bowel flora in only about 8% of healthy people
- opportunist pathogens infecting those with impaired defences (except for the primary pathogens *P. mallei* and *P. pseudomallei*) (Table 1).

Their structure is typical of Gram-negative bacteria. From a fluid reservoir by various routes they need to rupture our defences through a portal of entry such as burnt skin, intravenous drug abuse or medical use, instrumentation or disease. Then their numerous virulence factors (Table 2) overcome weakened host defences, including neutropenia or immune defects.

Confirmatory tests

Gram stain shows slender Gram-negative rods (Fig. 1). Culture is easy on blood or MacConkey agar. *P. aeruginosa* has flat, spreading matt or mucoid colonies, green pigmentation, and a typical grapelike sweetish smell.

Biochemical tests including gelatin, arginine, poly-β-hydroxy-butyrate and use of various carbon sources determine the species. Phage typing, pyocin typing and serotyping are research tools.

Clinical syndromes

- *P. aeruginosa* causes lung infections (especially in cystic fibrosis, p. 111), septicaemia (p. 128), wound, burn (p. 171), ear (p. 94) and other organ infections.
- *P. pseudomallei* causes melioidosis, a serious systemic infection (p. 120).
- *P. mallei* is a zoonosis causing glanders in horses and donkeys; it rarely affects humans and causes acute or chronic suppurative lesions, acute pneumonia, or acute fatal septicaemia.
- Other pseudomonads mainly cause opportunistic infections, with wound infections and spread in the blood to various organs, e.g. lungs.

Chemotherapy

Usually two antibiotics are used, as pseudomonads are intrinsically quite resistant, and host defences are often impaired. Gentamicin (or tobramycin) with either an anti-pseudomonal penicillin like ticarcillin or a third-generation cephalosporin, especially ceftazidime, are usual.

XANTHOMONAS AND STENOTROPHOMONAS

Like pseudomonads, these are strictly aerobic, asporogenous, motile, Gram-negative rods but are oxidase negative (or weakly positive) and have more complex minimal growth requirements. The most important is *Stenotrophomonas maltophilia* (formerly *Xanthomonas maltrophilia*), which produces acid with maltose,

Table 1 **Pathogenesis**

Primary pathogens	Opportunist pathogens		Accidental infection from animal flora
	Environmental	Normal Flora	
P. pseudomallei	*P. aeruginosa*	*A. actinomycetemcomitans*	*S. minor*
P. mallei	Other pseudomonads	*C. hominis*	*S. moniliformis*
C. granulomatis	*S. maltophilia*	*E. corrodens*	
	Flavobacterium spp.		
	C. violaceum		

Table 2 **Virulence factors of *Pseudomonas***

Virulence factor	Actions
Common to many Gram-negative bacteria	
Capsule, polysaccharide	Attachment, anti-phagocytic
Fimbriae	Attachment (respiratory)
Endotoxin	Fever, shock, DIC
Proteases	Tissue damage, antibody and neutrophil inhibition
Specific to pseudomonads	
Elastase	Vascular endothelial damage, neutrophil inhibition
Exotoxin A, exoenzyme S (stable)	Protein synthesis inhibition, cell damage, dermonecrosis
Leucocidin	Phagocyte inhibition
Phospholipase C	Haemolysis, tissue damage

Fig. 1 *Pseudomonas aeruginosa*: Gram stain.

not glucose. It also is found in fluids: water, milk and frozen foods. Characteristic virulence factors and pathogenesis are little studied.

Confirmation is by culture and biochemical tests including those above listed for pseudomonads.

Clinical presentation is as opportunistic infections from fluids and the environment, including bacteraemia, wound, lung, urinary and other organ system infections, often hospital acquired (p. 190).

Chemotherapy is often difficult, as resistance to aminoglycosides, imipenem and aztreonam is characteristic, and sensitivity to cephalosporins is unreliable. Most are surprisingly sensitive to cotrimoxazole.

RARE GRAM-NEGATIVE RODS

The following organisms are unrelated to the pseudomonads and, mostly, to each other. They are listed here for convenience, being uncommon Gram-negative organisms which at times cause important diseases. Table 1 shows the different patterns.

Flavobacterium spp.

These are non-motile, aerobic, oxidase-positive, weakly fermentative Gram-negative rods which are widespread in water in the environment and, hence, can contaminate hospital-fluids, causing hospital-acquired opportunist infections of blood stream, lungs and meninges. *F. meningosepticum* is the commonest. They grow easily on routine media and are identified by usual biochemical tests, including citrate, indole, urea, MR/VP, gelatin and sugars, and distinctive yellow ('flavo') pigment (Fig. 2).

Treatment is often difficult because of resistance to penicillins, cephalosporins and aminoglycosides. They may be sensitive to rifampicin, cotrimoxazole and (unusually for Gram-negative bacteria) erythromycin and vancomycin. Control is by using sterile, uncontaminated fluids where necessary.

Chromobacterium spp.

These bacteria are pigmented ('chromo' means colour), and biochemically similar to pseudomonads and *Flavobacterium* spp; they are differentiated by a series of biochemical tests.

They also are found widely in environmental water and cause similar opportunistic infections to those of flavobacteria. In addition, *C. violaceum* in hotter areas causes infection in swimmers, probably through skin wounds, resulting in systemic sepsis and liver abscesses, often fatal. Treatment may be with gentamicin, chloramphenicol or tetracycline.

Fig. 2 *Flavobacterium* **showing pigmented colonies.**

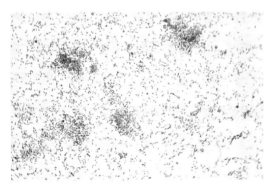

Fig. 3 *Actinobacillus actinomycetemcomitans.*

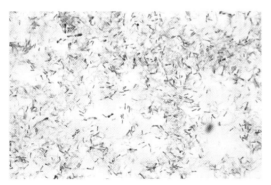

Fig. 4 *Cardiobacterium hominis.*

Actinobacillus actinomycetemcomitans

The extraordinary name means the 'ray-shaped bacillus accompanying actinomyces', as it was first found with actinomycosis. It is a small, capnophilic Gram-negative cocco-bacillus often taking 7 days to grow, even on special media (Fig. 3). It is normal oral flora and causes local rapidly destructive juvenile periodontitis. Bacteraemia can lead to endocarditis, endarteritis or organ infections and abscesses.

Abscesses may respond to tetracycline or chloramphenicol, while endocarditis needs bactericidal drugs in combination, usually a penicillin and an aminoglycoside.

Cardiobacterium hominis

This also is a small, capnophilic, slow-growing (1–2 weeks) Gram-negative bacillus (Fig. 4), part of normal oral flora, causing subacute endocarditis from dental disease or procedures.

It grows on blood or chocolate agar, not on MacConkey agar. It is identified biochemically.

Treatment with penicillin for 4–6 weeks is usually successful.

Eikenella corrodens

This is another small, capnophilic, relatively slow growing (2–3 days) Gram-negative rod; 45% of isolates pit ('corrode') the agar surface. It is oxidase and nitrate positive. Being normal mouth and respiratory flora, it (1) infects human bites; (2) causes endocarditis or disseminated polymicrobial infection after dental procedures, and (3) can cause opportunistic infections in patients with oral disease or immune deficiency. It also has unusual sensitivities for a Gram-negative aerobe: it is sensitive to penicillins and resistant to aminoglycosides. It is also resistant to clindamycin and metronidazole, often inappropriately used alone in treating bite infections.

Calymmatobacterium granulomatis

This unusual capsulated ('calymma' means sheathed), Gram-negative bacterium is difficult to grow and hence poorly characterised. It is diagnosed from tissue smears or sections, usually within macrophages in groups of 15–30 organisms called Donovan bodies.

It causes granuloma inguinale by sexual transmission or, occasionally, by trauma. It is usually treated by at least 3 weeks of tetracycline, cotrimoxazole, chloramphenicol or gentamicin.

Spirillum minor and Streptobacillus moniliformis

S. minor is a non-culturable spiral Gram-negative rod with characteristic darting motility from its polar flagella; *S. moniliformis* is a non-motile, pleomorphic, capnophilic Gram-negative coccobacillus in chains or filaments, growing best with 20% serum. They are normal oral flora in the rat and either one causes rat bite fever, which is treated with penicillin for 10–14 days.

Pseudomonas and rare Gram-negative rods

- Pseudomonads are strict aerobes widely distributed in nature.
- They are mainly opportunist pathogens infecting those with impaired defences.
- Virulence factors enable spread of infection from the site of entry, including septicaemia.
- Some rare GNRs are environmental organisms causing hospital-acquired infections from contaminated fluids or equipment.
- Other rare Gram-negative rods are normal or animal flora causing opportunistic infections including dental infections, infected bites and systemic infections.

BACTEROIDES, FUSOBACTERIA AND OTHER ANAEROBES

The family Bacteroidaceae are obligate anaerobic, non-sporing Gram-negative bacteria, of variable morphology. A major part of normal oropharyngeal, bowel and genital tract flora, they cause infections if introduced to sterile tissues.

BACTEROIDES AND *FUSOBACTERIA*

The obligate anaerobes of the family Bacteroidaceae are classified into over a dozen genera of which *Bacteroides, Fusobacterium, Prevotella* (Fig. 1) and *Porphyromonas* (Fig. 2) are clinically important. There have been several changes in classification from the original genera defined (Table 1). All are non-sporing, and growth requires reduced oxygen tension.

Bacteroides produce mixtures of acetic, formic, lactic and other acids from peptones or glucose; *Fusobacteria* produce butyric acid as a major product. Further biochemical tests determine the species and other genera. Their predominant locations as normal flora in mouth, bowel and vagina are also shown in Table 1.

Bacteroides have a typical Gram-negative cell structure except that the lipopolysaccharide (LPS) of *Bacteroides* does not have endotoxin activity because it lacks two unique carbohydrates. *B. fragilis*, although a minor member of normal flora, becomes the major pathogen because of its numerous virulence factors including a prominent capsule, and its relative aerotolerance (Table 2).

The pathogenesis of infection is multifactorial: usually the introduction of a mixture of aerobes and anaerobes into sterile tissues produces damage and anaerobic conditions, multiplication of anaerobes and abscess formation by enzymic action.

Confirmatory tests

Gram stain on pus showing thin pleomorphic pale-staining Gram-negative rods is highly suggestive (Fig. 3). Culture is in specific media, e.g. cooked meat broth, and on agar plates in an anaerobic jar or chamber (Fig. 4).

Biochemical tests provide definitive identification. Gas liquid chromatography (GLC) detects the acetic, butyric, propionic and other acids formed by metabolism and is a rapid test for the presence of many anaerobes. It helps to diagnose a particular species only in pure culture, not in clinical polymicrobic specimens.

Clinical syndromes

The hallmarks of anaerobic infection are abscess formation and foul pus from polymicrobial, endogenous infection. These include inhalation pneumonia, lung abscess, intra-abdominal and pelvic abscesses, all from adjacent normal flora.

Table 1 **Name changes of anaerobic bacteria, and usual habitat as normal flora**

Current name	Previous name(s)	Normal flora in			
		Mouth	Small bowel	Large bowel	Vagina
Fusobacterium nucleatum subspecies *nucleatum*	*Fusobacterium nucleatum*	+	+	+	+
Peptococcus spp.	All *Peptostreptococcus*,	+	+	+	+
Porphyromonas asaccharolytica	*Bacteroides asaccharolyticus*	+	–	–	–
Porphyromonas gingivalis	*Bacteroides gingivalis*	+	–	–	–
Prevotella bivia	*Bacteroides bivius*	–	–	–	+
Prevotella disiens	*Bacteroides disiens*	–	–	–	+
Prevotella melaninogenica	*Bacteroides melaninogenicus*	+	+	–	–
Prevotella intermedia	*Bacteroides intermedius*	+	–	–	–
Bacteroides fragilis group	As current name	–	–	+	–
B. fragilis group *B. distasonis, B. fragilis, B. ovalis, B. thetajotamicron, B. vulgaris*					

Table 2 **Virulence factors and pathogenesis**

Virulence factor	Action	*B. fragilis* group	*P. melaninogenica*	*Fusobacteria* spp.
Structural factors				
Capsule (polysaccharide)	Anti-phagocytic, adherence	+++	+/–	–
Fimbriae	Adherence	–	+	+
Lipopolysaccharide	Endotoxin	NO!	–	+
	WBC migration	+	+	+
	WBC chemotaxis	+	+	+
Enzymes				
Catalase, superoxide dysmutase	Aerotolerance	+	+	+
β-Lactamases	Anti-antibiotic	+++	++	+
Collagenase	Lyses collagen	+	+	–
DNAase	Liquefies pus	+	+	+
Fibrinolysin	Lyses fibrin	+	+	–
Neuraminidase	Hydrolyses protein	+	+	–
Ig proteases	Inactivate antibody	–	+	–
Chondroitin sulphatase, heparinase and hyaluronidase	Reduces cell adhesion	+	–	–

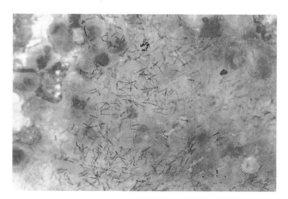

Fig. 1 *Prevotella disiens* Gram stain.

Chemotherapy

Metronidazole is effective against almost all clinically significant anaerobes. Resistance to the previously effective clindamycin and tetracyclines is now common in *B. fragilis*. Some cephalosporins (cefoxitin, cefotetan) are reasonably effective, but β-lactamases of varying types are now widespread.

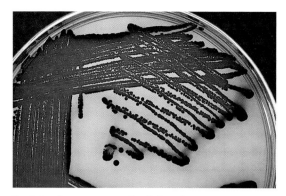

Fig. 2 *Porphyromonas asaccharolytica* culture.

Fig. 3 Mixed anaerobes including *Fusobacteria* sp. (Gram stain).

Control

Chemoprophylaxis including anti-anaerobic drugs before colonic and other high-risk surgery is essential and is now routine.

PEPTOCOCCUS AND PEPTOSTREPTOCOCCUS

These Gram-positive cocci are obligate anaerobes which can use peptones or amino acids as their sole energy source. Reclassification in 1983 left only one species in *Peptococcus* (*P. niger*, named from its black colonies), with all others transferred to join the peptostreptococci, of which *P. anaerobius* is commonest. All clinically important species are, like Bacteroidaceae, normal flora in the mouth and upper respiratory tract, the bowel and the vagina.

Peptococcus and *Peptostreptococcus* differ from *Bacteroides* in at least four important ways:

- they are Gram-positive
- they are less virulent, and usually only found in mixed infections with other bacteria, including *Bacteroides*
- they actually outnumber the anaerobic Gram-negative bacilli in the vagina, hence are prominent in pelvic infections

- they are normal skin flora, hence found in surgical and wound infections.

Laboratory identification is by Gram stain, anaerobic culture, biochemical tests and GLC.

Clinical syndromes

As with the Bacteroidaceae, the hallmarks of anaerobic infection are abscess formation and foul pus from polymicrobial, endogenous infection. Anaerobic cocci form about 25% of the isolates in such infections. These include inhalation pneumonia, lung abscess, intra-abdominal and pelvic abscesses, plus surgical and wound infections, all from adjacent normal flora.

Chemotherapy and control

This is usually with penicillins, cephalosporins or clindamycin. Metronidazole is less effective.

Control is by perioperative chemoprophylaxis, as for Bacteroidaceae.

VEILLONELLA

Veillonella parvula is an anaerobic Gram-negative coccus, but otherwise has many similarities with the peptostreptococci:

- normal flora in mouth, colon and vagina
- low virulence
- polymicrobial infections (but only 1% of anaerobic isolates)
- clinical laboratory methods similar
- clinical infections in normally sterile lungs, abdomen and pelvis by spread from adjacent normal flora
- chemotherapy by penicillins, cephalosporins or clindamycin (metronidazole is usually effective)
- control by chemoprophylaxis.

AEROCOCCUS

Aerococcus viridans is microaerophilic and Gram-positive but is listed here for convenience. It is slow growing, α-haemolytic, weakly catalase positive, and identification is by biochemical tests. Unlike the other bacteria described here, it is not part of our normal flora, but an environmental organism, found in the air, dust and surfaces. Therefore it is occasionally an opportunist pathogen in wounds but is usually only a contaminant; treatment with penicillin, erythromycin or a cephalosporin is therefore seldom necessary.

Bacteroides, Fusobacteria and other anaerobes

- *Bacteroides* and *Fusobacteria* spp. are obligate anaerobic non-sporing Gram-negative bacilli.
- They are normal flora in oropharynx, bowel and vagina and cause polymicrobic infections (with foul pus in abscesses) by numerous virulence factors when they are spread to adjacent tissues.
- Chemotherapy or prophylaxis is usually metronidazole or selected cephalosporins.

- Peptococci and peptostreptococci are obligate anaerobic non-sporing Gram-positive cocci.
- They are normal flora in oropharynx, bowel and especially vagina and skin, and are part of polymicrobic infections when spread to adjacent tissues.
- Chemotherapy or prophylaxis is usually penicillins or cephalosporins.

- Veillonellae are obligate anaerobic non-sporing Gram-negative cocci, similar to the above but less virulent and, hence, less important.

- Aerococci are microaerophilic Gram-positive cocci quite different from all the above, being an environmental airborne organism, usually found only as a contaminant.

Fig. 4 Anaerobic jar used for culture.

ZOONOTIC BACTERIA

Zoonoses are human diseases caused by a pathogen that has an animal reservoir. This excludes parasitic diseases such as malaria where humans are an essential part of the life cycle of the pathogen. Table 1 lists some important zoonoses caused by bacteria. Zoonoses are also caused by fungi, viruses, helminths and protozoa.

BRUCELLA

Brucellae are small, non-spore-forming, non-motile, non-capsulated strictly aerobic Gram-negative cocco-bacilli, needing several vitamins yet still slow to grow. *B. abortus* needs added CO_2. They are catalase, oxidase, nitrate and urea positive, citrate, indole and VP negative. There are four medically important species, differentiated biochemically and by growth or inhibition by two dyes, basic fuchsin and thionin.

Brucellae are animal pathogens which infect humans by direct contact (farmers, vets, abattoir workers), or through unpasteurised milk or cheese, or through inhalation (lab workers).

Virulence and pathogenesis
The marked difference in virulence of different species is not well understood:

- 'smooth' organisms are more virulent than 'rough' but have no detectable capsule
- they are intracellular pathogens, protected from many host defences
- they cause degeneration and lysis of neutrophils and macrophages by unknown mechanisms, blocked by antibody from previous infection.

The pathogenesis is better understood, for brucellae are phagocytosed into macrophages and carried to the reticuloendothelial system (liver, spleen and lymph nodes), where they multiply, form granulomata and spread by bacteraemia to cause destructive lesions in bones, joints and other tissue.

Confirmatory tests
Staining is of little use because of their small size. Culture is more useful, but the laboratory must be notified to use rich blood media and prolonged incubation (and CO_2 for *B. abortus*). They take great care to avoid laboratory infections!

Serology is helpful but easy to confuse. The standard (tube) agglutination test (SAT or STA) mainly measures IgM (not to *B. canis*) and so is positive in acute disease. False positives from cross-reactivity occur from cholera vaccination, cholera and tularaemia. The complement fixation test (CFT) mainly measures IgG so is positive in subacute and chronic disease, when SAT can have become negative.

Clinical syndromes
The course of infection may be subclinical, acute, subacute, relapsing or chronic. The intensity may be mild (*B. abortus*, *B. canis*), or severe (*B. melitensis*, *B. suis*).

The manifestations are extremely variable but combine:

- undulant fever (one name of the disease)
- systemic symptoms such as weakness and weight loss
- widespread or focal organ involvement.

If your joints give you a pang,
And your gum falls away from
   your fang,
If you shiver and shake,
And your testicles ache,
   You've *Brucella abortus* (Bang).

Chemotherapy
Usually tetracycline for 6 weeks and streptomycin for 2 weeks are used, but doxycycline plus rifampicin is an alternative.

Control
Eradication campaigns particularly in cattle have been very successful in numerous countries.

Table 1 Sources, spread and syndromes

Organism	Epidemiological source	Transmission from source	Pathogenic mechanism	Clinical syndrome
Brucella spp.	Cattle, goats, sheep, dogs, pigs	Direct contact, ingestion (inhalation)	Distant spread	Brucellosis
Yersinia enterocolitica	Animals, meat, milk	Ingestion	Local invasion	Enterocolitis
Y. pestis	Rat, flea (humans)	Flea bite, inhalation	Local spread Distant spread Aspiration	Bubonic plague Septicaemic plague Pneumonic plague
Pasteurella multocida	Cats, dogs	Animal bite, inhalation	Local spread Aspiration Distant spread	Cellulitis Pneumonia Opportunist
Francisella tularensis	Rabbits, ticks, rodents, water	Bites, ingestion, contact	Local spread Distant spread	Ulceroglandular, oculoglandular Typhoidal tularaemia
Bacillus anthracis[a]	Cattle, goats, horses, sheep, pigs	Contact, inhalation, ingestion	Local spread Distant spread	Anthrax, septicaemia
Salmonella spp.[b]	Many animals	Ingestion	Local invasion Distant spread	Gastroenteritis Typhoid fever
Leptospira interrogans[c]	Mammals	Infected animal urine	Distant spread	Leptospirosis
Borrelia burgdorferi[d]	Wild mammals	Tick bite	Local and distant spread	Lyme disease

See pp. 32, 170[a]; pp. 42, 135, 140[b]; pp. 53, 186[c]; pp. 52, 134[d]

YERSINIA SPP.

Y. enterocolitica
Y. enterocolitica is a non-lactose fermenting, aerobic, Gram-negative rod in the family Enterobacteriaceae. It is motile at 22°C but not at 37°C. It is found in numerous domestic animals, in water, milk and other foods. It is identified biochemically and seriologically. Characteristics of virulence and pathogenesis are not well understood but include V and W antigens, endotoxin and an enterotoxin, serum resistance and penetration of epithelium. After ingestion, it causes enterocolitis and mesenteric lymphadenitis.

Laboratory isolation is unlikely from faeces without cold 'enrichment' at 4°C for several weeks. Culture from tissues is easier but slow. Serology is only useful retrospectively.

Clinical syndromes include enterocolitis (p. 139), which is more usual but still not common. Septicaemia even in immunocompromised patients is rare.

Chemotherapy is of doubtful value in the self-limited enterocolitis. In septicaemia, gentamicin, a third-generation cephalosporin or chloramphenicol is given (mortality 50%). Control is by hygiene to prevent water- and food-borne infections.

Y. pseudotuberculosis
Y. pseudotuberculosis was first isolated from humans in 1953. It is similar to *Y. enterocolitica* but can be distinguished biochemically and serologically. It is found in many wild and domestic animals and

birds, water and milk. It causes pseudotuberculosis. Its virulence and pathogenesis result from an endotoxin and its ability to invade and to survive intracellularly.

Y. pseudotuberculosis causes mesenteric adenitis, mimicking acute appendicitis. Rarely, erythema nodosum or septicaemia occur. Chemotherapy for septicaemia may be ampicillin and gentamicin, and control is by hygiene to prevent infection.

Y. pestis

Plague is known from the 6th century pandemic. The organism was discovered in Hong Kong in the third pandemic, 1894.

Y. pestis (formerly *Pasteurella pestis*) is also an aerobic Gram-negative rod, but it is non-motile, and has bipolar staining. It has two epidemiologic cycles, sylvatic (forest) plague in wild mammals and their fleas, and urban plague in rats and their fleas. Humans are infected either from the animals (now more usual), from fleas (classically) or from humans with pneumonic plague (rare but terrible) (Fig. 1).

Laboratory identification is by Gram-stain of lymph node aspirate (positive in 85% of cases), and blood culture.

The organism is readily phagocytosed but survives, and multiplies within the cells (causing enlarged painful lymph nodes), is released and is then resistant to host defences. Endotoxaemia, shock, haemorrhage and death follow. Clinical presentation is as:

- bubonic plague: swollen nodes then systemic symptoms
- septicaemic plague: fever, prostration and death in 2 days
- pneumonic plague: fever, X-ray changes, pneumonia, death.

Chemotherapy is with streptomycin or tetracycline and control is best achieved by avoidance of infection and selective prophylaxis with tetracycline for close contacts.

PASTEURELLA

P. multocida is the major human pathogen in this genus, which is part of the same family as *Haemophilus*. Pasteurellae are aerobic, non-motile Gram-negative rods that are catalase and oxidase positive, ferment sugars with acid, not gas production. *P. multocida* are carried in the respiratory tract of and are pathogenic to many animals ('multocida = killing many').

Laboratory identification is by Gram stain, culture on blood or chocolate agar, and usual biochemical tests.

Clinical syndromes are:

- cellulitis and lymphadenopathy from an animal bite (p. 170)
- opportunist infections in immunocompromised patients, and, rarely
- respiratory infection from inhalation.

Chemotherapy is with penicillin or a cephalosporin, and control is by animal avoidance!

FRANCISELLA TULARENSIS

F. tularensis is a tiny fastidious aerobic Gram-negative pleomorphic cocco-bacillus occurring with several **biovars** (**bio**chemical **v**arieties). It causes tularaemia in rabbits, rodents and ticks, and in humans by four routes:

- from bites
- animal contact
- infected meat or water (10 million organisms needed)
- inhalation (10 to 50 organisms needed).

Virulence depends on an antiphagocytic capsule, on endotoxin and on protection from antibody while intracellular. Bacteraemia follows entry, then transport and multiplication in phagocytes is followed by focal necrosis and granulomata in infected organs.

For laboratory identification, Gram stain is usually unsuccessful, and culture is difficult, delayed and dangerous. Serology is, therefore, the usual method of diagnosis but does not distinguish recent from previous infection.

Clinical syndromes (p. 185) depend mainly on the route of infection:

- ulceroglandular or glandular from bite or skin contact
- typhoidal (febrile, systemic, pulmonary) from any route
- oculoglandular or oropharyngeal from local infection.

Chemotherapy is by streptomycin or gentamicin, and control is best achieved by avoiding the reservoirs and vectors.

Brucella
- Infection is by direct contact, usually occupational, or by ingestion.
- Diagnosis is by culture (dangerous) and serology.
- Brucellosis is often a long-continued systemic infection with organisms multiplying inside the cells of the reticuloendothelial system producing a granulomatous reaction, then spreading to many systems.
- Chemotherapy is with tetracycline and streptomycin, but is often ineffective with relapses occurring.

Yersinia
- Infect humans from animals and their products.
- *Y. enterocolitica* principally causes enterocolitis, *Y. pseudotuberculosis* causes pseudotuberculosis, principally mesenteric adenitis.
- *Y. pestis* causes bubonic, septicaemic and pneumonic plague.
- Chemotherapy is tetracycline, streptomycin or gentamicin.
- Control is avoidance of the animal reservoir, or vector.

Pasteurella
- *P. multocida* is pathogenic to many animals, usually infects humans by an animal bite, and is treated usually by penicillin.

Francisella
- *F. tularensis* is dangerous to culture and causes tularaemia in humans when caught from animals, e.g. from rabbits and ticks.
- Disease is ulcerative, or glandular, or systemic ('typhoidal') and is treated with streptomycin or gentamicin.

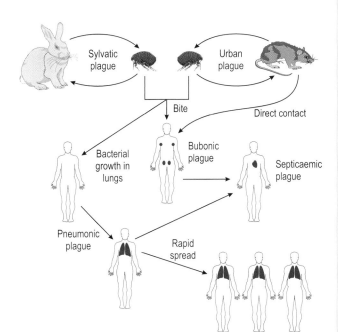

Fig. 1 **Spread of plague**.

SPIROCHAETES: TREPONEMES, BORRELIA, LEPTOSPIRES

The family Spirochaetaceae include three medically important genera: *Treponema, Borrelia* and *Leptospira*. These bacteria differ considerably from all those already considered:

- spiral shape: long, thin and poorly staining
- cell wall Gram-negative-like but has no lipopolysaccharide (LPS)
- culture difficult or impossible in the laboratory
- virulence and pathogenesis thus little understood
- diagnosis by special microscopy and serology, not by usual Gram stain and culture.

TREPONEMA PALLIDUM

Syphilis was probably brought to Europe by Colombus' sailors 500 years ago. The long, thin spiral is about 0.15 μm wide and 12 μm long, with internal axial fibrils inserted at each end (Fig. 1) resulting in rapid corkscrew movements and periodic angular bending of the cell body. It is actually classified and named *T. pallidum* subspecies *pallidum* but is usually called simply *T. pallidum*. It is very fragile, does not live in culture and dies quickly with drying or heat. It infects only humans and rabbits.

Virulence and pathogenesis

Study of virulence and pathogenesis is limited by failure of laboratory culture (except in rabbit testes). Treponemes enter through intact or abraded skin or mucous membrane, then multiply locally causing tissue destruction, i.e. an ulcer or the **primary chancre**. This heals while the organism spreads systemically, giving **the secondary stage** of fever, 'flu', rash (Fig. 2) and lymphadenopathy 2–6 weeks later. In 70% of patients this passes to **asymptomatic latent syphilis** for 3–30 years, then **tertiary syphilis** develops. This is vasculitis and chronic inflammation, both by direct spirochaetal action (gumma formation) and by cell-mediated hypersensitivity to spirochaetal antigens without live spirochaetes. Antibody is not protective, though useful in diagnosis.

Confirmatory tests

Diagnosis is usually by serology, though immediate dark-ground microscopy (DGM) of exudate from skin ulcers or genital mucous membrane is reliable if positive. (DGM is not suitable for mouth swabs, as non-*pallidum* spirochaetes are normal flora.)

Serology diagnoses spirochaetal disease, not syphilis uniquely. The tests are (1) non-specific, such as the VDRL (Venereal Disease Research Laboratory) and RPR (rapid plasma reagin) for IgG and IgM 'reagin' antibodies to lipids from damaged host cells (Fig. 3); (2) specific, using antitreponemal antibody in the FTA-ABS (fluorescent treponemal antibody absorption) test, the TPI (*Treponema pallidum* immobilisation) test or other tests.

All tests may be negative in primary syphilis, become positive during the secondary stage, and, by definition, one must be positive in latent syphilis. The FTA becomes negative in 5% and the VDRL in up to 50% of patients with tertiary syphilis. The non-specific, but not the specific, tests become negative with treatment. The specific tests are, therefore, used to diagnose if a patient ever had syphilis, while the non-specific are used to assess the success of treatment.

Clinical syndrome

Syphilis is described on pages 88–9, 100, 163 and 175. Endemic syphilis spread non-venereally by direct contact is called Bejel (p. 175).

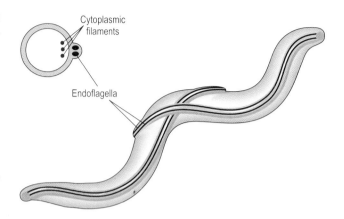

Fig. 1 *T. pallidum* **structure.** Organisms are long and helical shaped with their flagella between the cell wall and the outer membrane. The flagella originate at each end of the cell and wrap around it in a spiral to overlap at the middle of the cell. From the flagellar basal bodies, cytoplasmic filaments originate.

Chemotherapy

Fortunately penicillin remains very effective in primary, secondary, untreated latent, and tertiary syphilis with gumma formation and live spirochaetes. Tertiary syphilis with few or no live spirochaetes, such as tabes dorsalis or meningovascular syphilis (pp. 88–9), of course, does not respond as well.

Control

Congenital syphilis is controlled by routine antenatal testing and treatment of infected mothers. Venereal syphilis is partly controlled by treatment of infectious patients and their partners, by diminished promiscuity and by condom use; no vaccine is available.

OTHER TREPONEMATA

T. pertenue

T. pertenue is almost indistinguishable from *T. pallidum* and is correctly named *T. pallidum* subsp. *pertenue*. Why it causes **yaws** rather than syphilis is a mystery. Laboratory identification is by DGM and serology. Clinical presentation is yaws, occurring in three stages, with widespread skin nodules at early stages, and bony and other lesions late if untreated (p. 175).

Chemotherapy is by penicillin, with remarkable response, and control is by mass penicillin campaigns in endemic areas.

T. carateum

T. carateum is similar to *T. pallidum* structurally and in its methods of identification. Its virulence factors are unknown and it causes **pinta**, a depigmenting skin disease in South and Central America (p. 175). Chemotherapy is by penicillin, and control is by treating infected patients and contacts.

BORRELIA SPP.

The borreliae of the relapsing fevers have been gradually recognised over the last 120 years, while in 1975 two alert mothers in Lyme and Old Lyme, USA, recognised a new arthritis, now called Lyme disease.

Borreliae are microaerophilic, fastidious, slow-growing, weakly Gram-negative spiral bacilli, rather bigger than

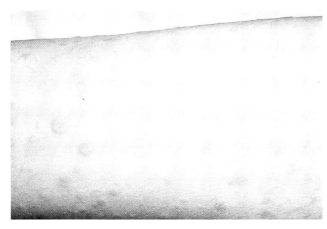

Fig. 2 **Rash of secondary syphilis.**

treponemes, with multiple internal periplasmic flagellae. *B. burgdorferi* infects hard ticks (*Ixodes* spp.), *B. recurrentis* infects the human body louse (*Pediculus humanus*) while the others infect soft ticks of *Ornithodorus* spp.

Being difficult to culture and study, little is known of virulence. No toxins are known. Antigenic variation of waves of borreliae emerging from tissues protects them from host antibody, hence the recurrent fevers. The late organ damage in Lyme disease may be caused by direct borrelial action or by immunologic cross-reactivity to host antigens shared with *B. burgdorferi*.

Confirmatory tests

Relapsing fever is often diagnosed by seeing borreliae in Giemsa stains of blood films taken during fever. Culture is slow, specialised and seldom useful. Serology by ELISA for Lyme disease is useful in the later stages.

Clinical syndromes

Clinical syndromes (p. 132) vary with infecting species:

- *B. burgdorferi*: Lyme disease
- *B. recurrentis*: epidemic louse-borne relapsing fever
- other *Borrelia* spp. (about 15): tick-borne relapsing fever.

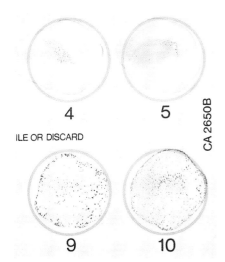

4   5

ILE OR DISCARD

CA 2650B

9   10

Fig. 3 **RPR serology test for syphilis.**

Chemotherapy

Tetracycline is the usual therapy, though chloramphenicol can be used in relapsing fever, and penicillin in Lyme disease.

Control

Vaccines are not available and the only effective control is by avoiding tick and louse bites, and by rodent and lice control.

LEPTOSPIRA INTERROGANS

The genus *Leptospira* now has only two species, the pathogenic *L. interrogans* ('shaped like a question mark') with over 170 serotypes, and the saprophytic *L. biflexa*.

Leptospires are helical, motile, obligate anaerobic organisms, staining poorly with Gram stain. Virulence is not well understood. Infection occurs from the urine of rodents, cattle, dogs etc. After entry through intact or abraded skin, they spread via the blood to many tissues, damaging the vascular endothelium, causing haemorrhage, meningitis, conjunctivitis and renal and hepatic impairment. Phagocytosis occurs, and antibodies clear leptospires from the blood. The kidneys of many animals including rats and cattle are infected, often lifelong and asymptomatically.

Laboratory identification is sometimes by dark ground microscopy (DGM), rarely by culture and usually by serology. However agglutinating antibodies only appear 1–2 weeks after infection.

Infection may be asymptomatic, or a flu-like illness, or progress to meningitis, haemorrhage and hepatic or renal failure (p. 184). Chemotherapy with penicillin is only helpful early, when diagnosis is hardest. Control is by rat control and avoidance of water contaminated with animal urine, especially by farmers, vets and abattoir workers.

Treponemes

- *T. pallidum* is a non-cultivable spirochaete which causes syphilis.
- It can be diagnosed by dark ground microscopy, or more usually by serology, which can also be used to monitor disease progress and treatment.
- Untreated syphilis progresses in stages: primary chancre; secondary fever and rash; an asymptomatic latent stage (3–30 years); and a tertiary stage with vasculitis and chronic inflammation damaging brain, heart and other organs in four distinct clinical patterns.
- Syphilis is treated with penicillin.
- *T. pertenue* causes yaws with skin nodules progressing to bony lesions if untreated.
- *T. carateum* causes pinta, with skin depigmentation.

Borreliae

- These spiral organisms are very difficult to culture, thus diagnosed by microscopy or serology.
- They cause Lyme disease (*B. burgdorferi*), epidemic louse-borne relapsing fever (*B. recurrentis*) and endemic tick-borne relapsing fever (other *Borrelia* spp.).

Leptospirae

- *L. interrogans* are long, thin, spiral bacteria, difficult to culture, hence diagnosed by serology and microscopy. There are over 170 serotypes.
- They cause leptospirosis by contact with infected animals or their urine.
- Human infection is often asymptomatic, may be flu-like with myalgia, or may be serious with haemorrhage and organ failure.
- Early penicillin treatment may help.

MYCOBACTERIA

Mycobacteria are non-motile, non-sporing, strictly aerobic rods. The cell wall differs from the other genera considered in having an outer lipid layer (25% by weight) which makes it resistant to acid decolourisation (i.e. *acid-fast*) (Fig. 1).

This cell wall gives mycobacteria great resistance to:

- environmental stresses like heat, cold, and drying, resulting in infectivity over months or years
- disinfectants, hence dangerous in hospitals
- usual stains and acid, hence identification requires special acid-fast stains
- nutrient uptake, hence slow growth and special culture media
- host defences, hence high infectivity, initial inapparent disease, intracellular bacteria, late reactivation, chronic inflammation, delayed hypersensitivity, host damage and chronic disease
- antibiotics, hence special antimycobacterials.

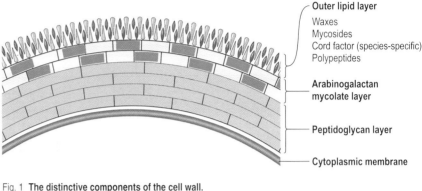

Outer lipid layer
Waxes
Mycosides
Cord factor (species-specific)
Polypeptides

Arabinogalactan mycolate layer

Peptidoglycan layer

Cytoplasmic membrane

Fig. 1 **The distinctive components of the cell wall.**

M. TUBERCULOSIS (MTB)

Tuberculosis (TB) is found in the bones of Egyptian mummies 4000 years old. Growth on special media takes 2–6 weeks. Biochemical tests are used to classify the species. Only humans are naturally infected, by person-to-person spread, usually through aerosols. The closely-related *M. bovis* from cattle infects through unpasteurised milk.

Virulence and pathogenesis
No toxins are known, and virulence depends on its cell wall, its intracellular nature and resistance to host defences; pathogenesis is the same in all tissues, though, of course, the clinical disease varies with the organs infected. Once inhaled, ingested or implanted, the tubercle bacillus ('MTB'), enters macrophages, multiplies and is disseminated to regional lymph nodes; tissue damage is by **tubercles** which are small granulomas of epithelioid cells and macrophages as multinucleated giant (Langhans) cells (Fig. 2). Central cheesy necrosis is called **caseation**; macroscopic damage causes **cavitation**. Damage may be local, or MTB may be disseminated by the bloodstream to organs including kidneys, meninges, bones and joints, where the multiplication and damage continues. Cell-mediated immunity is important, but antibodies are not protective.

Confirmatory tests
The diagnosis of TB is suggested by clinical signs and supported by chest X-ray changes and a positive skin reaction in the Mantoux (tuberculin) skin test (by PPD). These tests are confirmed by microscopy using acid-fast techniques such as Ziehl–Neelsen (Fig. 3) on sputum and tissues (less reliable on urine). Culture is slow (2–6 weeks) and requires special media. Further biochemical tests define the species, and antibiotic sensitivity tests assist treatment.

Clinical syndromes
Pulmonary TB (p. 118) is the commonest form of infection; extra-pulmonary TB occurs mainly in lymph nodes (25%)

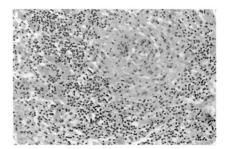

Fig. 2 **Miliary tuberculosis showing small granuloma in lung with peripheral epithelioid cells and two central giant cells.**

(p. 123) and pleura (20%), genitourinary tract (15%), bone and joint (10%) miliary (10%), meningeal (5%), and elsewhere.

Chemotherapy
This is with three or four special drugs initially, reduced in number after 1 or 2 months or when sensitivity results are available. 'Short' courses of 6 or 9 months are now more used (p. 204).

Control
This is by treatment of infectious patients, examination of contacts, chemoprophylaxis with isoniazid for recent Mantoux converters, and improvement of socioeconomic conditions and housing. In some countries, mass radiography and mass BCG immunisation are used.

M. LEPRAE

The leprosy bacillus was the first ever recognised as a cause of human disease, seen in tissues by Hansen in 1874. *M. leprae* is an obligate intracellular pathogen of humans. The strongly acid-fast parallel-sided rods 0.4 µm by 1–8 µm do not grow in artificial media but can be grown in the armadillo and mouse foot-pads, i.e. at temperatures lower than 37°C.

Little is known of virulence factors. Pathogenesis is likewise difficult to study but is much influenced by the host response, with active delayed-type hypersensitivity (DTH) in tuberculoid leprosy, but profound anergy (loss of immune function) in lepromatous leprosy (see pp. 92, 174).

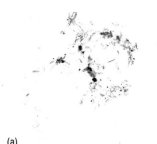
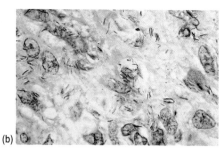
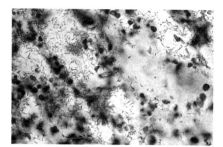

Fig. 3 **Acid-fast stain: (a)** *M. tuberculosis*; **(b)** *M. leprae*.

Fig. 4 *M. ulcerans*: acid-fast stain.

Confirmatory tests
Nasal mucosal smears, skin snips from eyebrows and other sites (and nerve or other tissue biopsies from lepromatous leprosy, but not other forms) are stained with acid-fast stains (Fig. 3b).

Culture and serology are useless, but histology is used for non-lepromatous leprosy.

Clinical syndromes
There is a spectrum of disease from lepromatous leprosy (LL) with chronic inflammation particularly of the skin and mucous membranes, through borderline disease to tuberculoid disease (TT) with infiltration of peripheral nerves leading to anaesthesia and secondary trophic changes (Fig. 2, p. 92).

Chemotherapy and control
Chemotherapy is usually with rifampicin, dapsone and clofazimine. Control depends more on chemotherapy than isolation, as

Fig. 5 **Mycobacterial blood cultures.**

leprosy is not highly infectious; the nasal discharge is infectious in LL patients.

ATYPICAL MYCOBACTERIA

The atypical mycobacteria are acid-fast rods, distinguished into species by special biochemical tests. The Runyon classification divides them into four groups depending on pigmentation and rate of growth (Table 1). They are environmental organisms of variable pathogenicity which is usually by local invasion, followed by dissemination in immunocompromised patients. Little is known to explain their lower virulence compared with *M. tuberculosis* and *M. leprae*.

Confirmatory tests
Microscopy and acid-fast stains are important for rapid diagnosis (Fig. 4); culture is slow but shortened to 8–14 days by the 'Bactec' system (Fig. 5). A specific nuclear probe for MAC is now available.

Clinical syndromes
Infection by atypical mycobacteria has become more common with the increase in immunocompromised patients. They cause pulmonary infection (*M. kansasii*), skin and soft tissue ulcers (*M. ulcerans, marinum*) and systemic infections (*M. avium-intracellulare* complex (MAC)) in AIDS.

Chemotherapy and control
Chemotherapy is difficult, as many strains are multiresistant. Control is also difficult, as mycobacteria are widespread in the environment and particularly infect the immunocompromised.

Table 1 **Runyon classification of some atypical Mycobacteria**

Group and examples	Pigmentation	Growth rate[1]	Optimal temperature for growth (°C)
Group I			
M. kansasii	Photochromogens[2]	Slow	37
M. marinum	Photochromogens[2]	Moderate	32
Group II			
M. scrofulaceum	Scotochromogens[3]	Slow	37
Group III			
M. avium-intracellulare	None	Slow	37
M. ulcerans	None	Slow	32
Group IV			
M. cheloneae	None	Rapid	37
M. fortuitum	None	Rapid	37

[1]Rapid growth is about 2 weeks by conventional methods; slow growth is 4–6 weeks.
[2]Photochromogens produce vivid yellow caratenoids only with light; [3]Scotochromogens produce pigment in the dark as well.

Mycobacteria
- Mycobacteria have a special cell wall rich in lipids, making them very resistant to environmental stresses, host defences, disinfectants, stains and antibiotics.
- They are slow-growing and need special laboratory techniques including 'acid-fast' stains and specific media.
- They are intracellular bacteria, particularly in macrophages, where they multiply, disseminate and cause tissue damage as granulomas, with central caseation in tuberculosis.
- *M. tuberculosis* causes the various forms of tuberculosis.
- *M. leprae* causes the various forms of leprosy.
- The atypical mycobacteria are environmental species that can colonise or contaminate or cause pulmonary and systemic infections and skin and soft tissue ulcers. They are prominent infections of the immunocompromised.

ACTINOMYCES, NOCARDIA AND RARE GRAM-POSITIVE BACILLI

ACTINOMYCES

Actinomyces are Gram-positive anaerobic bacteria, which require rich media, are capnophilic, grow slowly and need temperatures around 37°C for growth. These features correlate with their habitat, the human oral cavity, and contrast with *Nocardia*.

Actinomyces means 'ray fungus' and refers to their apparently fungal form with branching filaments in tissues (Fig. 1). However, these break into bacilli and coccoid forms, and their cell structure is clearly procaryotic, with no nuclear membrane and a Gram-positive cell wall. Disease begins when these normal flora enter adjacent sterile tissues by implantation or inhalation.

Confirmatory tests. This depends on Gram stain of granules in the pus (for organisms are elsewhere scanty), on anaerobic culture with 10% CO_2 for at least 2 weeks, and biochemical differentiation of species. Histopathology or culture from tissues is advisable because isolates from surfaces may be only normal flora.

Clinical syndromes. Actinomycosis occurs worldwide in five major forms:

- cervico-facial ⎫
- abdominal ⎬ by invasion locally
- mycetoma (by implantation)
- thoracic (by inhalation)
- disseminated disease (including brain).

The pathological features are swelling, slow progression, sinus formation, sclerosis, scarring and sulphur yellow granules in the pus.

Chemotherapy and control. Penicillin is the drug of choice. Control is by dental care and appropriate chemoprophylaxis.

NOCARDIA

Nocardia are Gram-positive and weakly acid-fast aerobic bacteria (Fig. 2), which grow on simple media over a wide temperature range. Filaments are better seen in liquid media.

Infection occurs by inhalation or by implantation from soil.

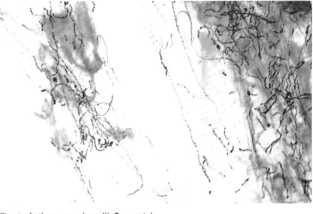

Fig. 1 *Actinomyces israelii:* Gram stain.

Confirmatory tests. Identification depends on Gram stain, acid-fast stains (using weak acid only), and aerobic culture for at least a week. Species differentiation is by biochemical tests.

Clinical syndromes. Pulmonary pseudotuberculosis and brain abscesses occur especially in the immunocompromised, while mycetoma follows skin implantation (pp. 67, 174). Unlike actinomycosis, sinuses are not a feature and sulphur granules do not occur. Suppuration and scarring with granulation tissue occur, but granulomas as in tuberculosis do not.

Chemotherapy and control. Chemotherapy is usually with sulphonamides, and control involves care of soil-contaminated wounds and of immunocompromised patients.

RARE GRAM-POSITIVE BACILLI

Lactobacillus, Propionibacterium, Eubacterium and *Bifidobacterium* spp. are all Gram-positive bacilli, at times bifid or branching; most are obligate anaerobes and form part of the normal flora of the skin (propionibacteria), bowel and vagina. They have minimal virulence.

Pathogenesis is only understood for *P. acnes* in acne, where the production of propionic acid deep in the follicle leads to irritation and formation of the comedone.

Confirmatory tests. These are usually routine Gram stain (Fig. 3), anaerobic culture and biochemical testing. Propionic acid from *Propionibacteria* is detected by GLC (p. 209).

Clinical syndromes. These rare Gram-positive bacilli are found with other pathogens in mixed infections; rarely they are found alone, in opportunist infections such as endocarditis or device-related infections. *P. acnes* is a common contaminant from skin when taking blood cultures or swabs, and is a major factor in acne.

Chemotherapy is usually unnecessary, but penicillins, clindamycin and vancomycin are usually active.

Actinomyces

- Anaerobic, Gram-positive, branching, pleomorphic bacteria, slow growing and fastidious.
- Part of normal oral flora, they cause cervico-facial infections or mycetoma if implanted, or thoracic infections if inhaled.
- Penicillin is the drug of choice.

Nocardia

- Aerobic, Gram-positive, weakly acid-fast, soil bacteria.
- Slow growing, they cause mycetoma if implanted, pulmonary cavitation if inhaled, and abscesses if disseminated.
- Sulphonamides are drugs of choice.

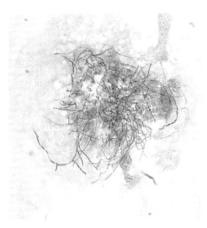

Fig. 2 *Nocardia asteroides:* acid-fast stain.

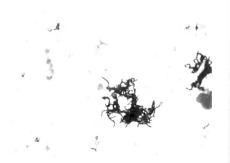

Fig. 3 Gram stain of propionibacteria.

MYCOPLASMA AND UREAPLASMA

Mycoplasmataceae is a family of small bacteria with two genera, *Mycoplasma* and *Ureaplasma*. They have at least four unique properties:

- the smallest free-living organisms known (diameter only 0.2–0.7 μm)
- no cell wall, so do not stain with Gram stain and are resistant to cell-wall-active antibiotics like penicillin
- cell membrane is unusual, containing sterols which must be provided for growth in media
- generation time about 6 hours.

Not surprisingly they were originally thought to be viruses. However, they are bacteria with prokaryotic cell structure and divide by binary fission.

Mycoplasma are ubiquitous, frequently contaminating tissue cultures. *M. pneumoniae* is probably found only in those infected, but about one in six sexually active adults are colonised with *M. hominis*, and between 50% and 75% with *U. urealyticum*. Many other human mycoplasmas have been found, but their role in disease is questionable. Virulence varies between species: *M. pneumoniae* has a special terminal protein called P1 which adheres (Fig. 1) to a specific glycoprotein receptor in respiratory epithelium near the base of the cilia and microvilli. Cilial movement then stops, and mucosal (not alveolar) damage occurs through the action of ill-understood tissue toxins including H_2O_2. Inflammatory infiltrate is lymphocytic, not neutrophilic, and the mycoplasma remain extracellular. The pathogenesis is thus very different from pneumococcal or other pyogenic pneumonias.

Confirmatory tests
Mycoplasma do not stain so microscopy of specimens cannot be used. Culture requires special media and is slow; colonies on specific soft agar have a 'fried egg' appearance (Fig. 2).

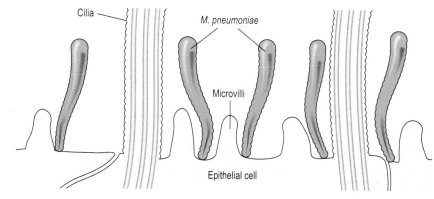

Fig. 1 *M. pneumoniae* attachment to respiratory epithelium.

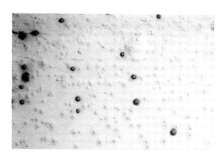

Fig. 2 **Mixed *Mycoplasma* and *Ureaplasma* spp. on culture.**

Fig. 3 *M. pneumoniae* antibody detection by particle agglutination.

Serology (Fig. 3) is the major diagnostic test, either for non-specific cold agglutinins (an early IgM antibody which agglutinates red cells at 4°C) or for specific complement-fixing IgG, which peaks at 4–8 weeks, has numerous false negatives (positive only in two-thirds of patients) and has some false positives also. Specific nucleic acid probes are being developed.

Clinical syndromes
- *M. pneumoniae* is an important cause of atypical pneumonia in young people (pp. 112–13); it is considered atypical compared with classical pneumococcal pneumonia because it is less acute (slower to develop), less serious, not lobar, not responsive to penicillin and slower to recover. Erythema multiforme rash (Fig. 4) may occur.
- *M. hominis* causes urinary infections, pelvic inflammatory disease and puerperal infections (p. 158).
- *U. urealyticum* probably causes half of non-chlamydial non-gonococcal urethritis (p. 154).

Chemotherapy
This is usually with erythromycin, though tetracycline is useful in adults. Prevention of infection is by avoiding close contact with respiratory infections, and by condom use or continence sexually.

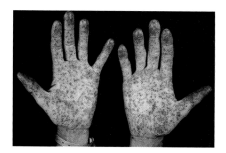

Fig. 4 *M. pneumoniae* erythema multiforme skin rash.

Mycoplasma and Ureaplasma

- *Mycoplasma* and *Ureaplasma* are unique, very small bacteria with no cell wall, a sterol-containing cell membrane and very slow growth.
- They are therefore difficult to stain and culture, and are resistant to cell-wall-active antibiotics like penicillin.
- *M. pneumoniae* causes atypical pneumonia, while *M. hominis* causes genitourinary infections, and *U. urealyticum* causes urethritis: all mucosal surface infections.
- Treatment is by erythromycin or tetracycline.

CHLAMYDIA

Chlamydiae are small, obligate intra-cellular bacteria which like most bacteria:

- have a cell wall with inner and outer membranes
- have procaryotic ribosomes, RNA and DNA
- synthesise their own macromolecules
- are susceptible to numerous anti-bacterial antibiotics.

Unlike other bacteria,

- they have no peptidoglycan in their cell wall
- they are unable to synthesise ATP
- they lack oxidative enzymes such as flavoproteins and cytochromes so are 'energy parasites' and cannot replicate extracellularly.

As a result of this they have a di-morphic life cycle. The first form is the **elementary body** (EB), which is small (0.3–0.4 μm), spherical and stable to environmental stresses. It exists extra-cellularly as the infective form, rather like clostridial spores or amoebic cysts. The second form is the **reticulate body**, (RB), which is the fragile intracellular replicating form with an internal net-like structure.

Three species are now recognised (Table 1): *C. trachomatis*, which causes four groups of infections; *C. pneumoniae*, recently found to cause respiratory infections; and *C. psittaci*, the cause of psittacosis, a pneumonia caught from birds.

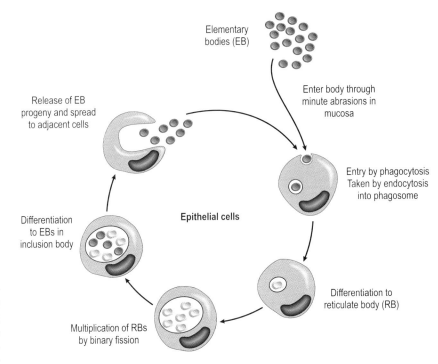

Fig. 1 **The life cycle of *Chlamydia*.**

Table 1 **Chlamydial infections**

Organism source	Epidemiological source	Pathogenic mechanism	Clinical syndrome
C. trachomatis Serovars A-C	Infected eyes	Local invasion	Trachoma
B, D-K	Urogenital infection	Local invasion Local invasion Mother to child	Urethritis Genital infections Perinatal infections
L_1, L_2, L_3	Genital infection	Lymphatic spread	Lymphogranuloma venereum
C. pneumoniae	Respiratory infections	Local invasion	Respiratory tract infections, pneumonia
C. psittaci	Birds	Local invasion, distant spread	Psittacosis with pneumonia

CHLAMYDIA TRACHOMATIS

Trachoma was known over 2000 years ago. *C. trachomatis* can be divided into 15 antigenically different serotypes (serovars):

- 12 of the trachoma biovar, including A, B, Ba and C causing trachoma itself, and D to K causing genital and neonatal disease
- three (L_1, L_2, L_3) of the LGV biovar (Table 1).

All are human pathogens, spreading by direct contact, with no animal host, though insects are sometimes a vector for trachoma.

Pathogenesis

The life cycle of chlamydiae is shown in Figure 1. Unless coated with antibody, the surface components of the EB prevent the usual fusion of the phagosome with lysosomes and so they are not destroyed. In lymphogranuloma venereum (LGV), the same cycle occurs, but in macrophages that are carried to regional lymph nodes, thus explaining the clinical picture of enlarged, infected nodes.

Confirmatory tests

Microscopy on urethral, cervical and eye swabs (Fig. 2) is more effective using direct immunofluorescence (DIF, Fig. 3) than Gram or Giemsa stains. Culture is in tissue monolayers e.g. treated McCoy cells, which show glycogen-positive inclusion bodies in 2–4 days. In LGV, lymph node aspirate is used.

Serology is useless, as only genus-specific antigen is usually available and it does not distinguish present from past disease. Antigen detection by ELISA is only reliable in specimens from symptomatic women.

Clinical syndromes

C. trachomatis causes four groups of infections:

- trachoma (p. 99)
- numerous 'non-specific' genital tract infections (pp. 154–60)
- infections from the genital tract during childbirth (p. 187)
- a specific STD, lymphogranuloma venereum (p. 163).

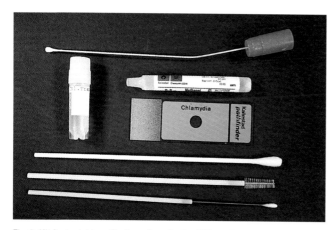

Fig. 2 **Kit for bedside collection of swabs for** *Chlamydia.*

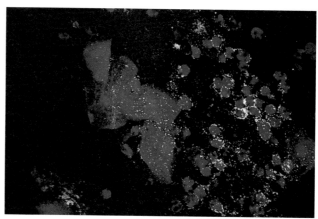

Fig. 3 **Direct immunofluorescent microscopy of chlamydiae.**

Chemotherapy

Chemotherapy uses tetracycline or erythromycin, which are effective but slow. Control depends on the treatment of infections, safer sex practices and improved socioeconomic conditions.

C. PNEUMONIAE

C. pneumoniae was first described in 1986 as the TWAR agent in Taiwan conjunctival and acute respiratory specimens.

C. pneumoniae is by DNA homology a separate species from the other two chlamydiae, but shares the unique qualities of the genus. It is a worldwide human pathogen. No animal host is known.

Pathogenesis is similar to *C. trachomatis*, but affecting respiratory rather than genital epithelium!

Confirmatory tests

Microscopy by direct immunofluorescence is available in some centres and is quick and specific. Culture is not yet useful routinely. Serology by microimmunofluorescence is specific if available.

Clinical syndromes

C. pneumoniae is a common cause of 'atypical' pneumonia, clinically very like mycoplasmal pneumonia. It also causes bronchitis and upper respiratory infections.

Chemotherapy

Chemotherapy with tetracycline or erythromycin is effective. There are no specific control measures at present.

C. PSITTACI

C. psittaci is a typical chlamydia but produces a glycogen-negative inclusion body. The epidemiology is also entirely different, as it infects many wild and pet birds, which may be ill or asymptomatic. The organism is on their feathers and in dried bird faeces. The infective dose to humans by inhalation is small, as infection has occurred after short exposure in an area previously occupied by an infected bird. Kissing love-birds can also infect humans!

Virulence and pathogenesis

While structure and function of the bacterium are typical of the genus, the pathogenesis of psittacosis is different, for this virulent pathogen causes a systemic disease, not simply a mucosal or epithelial one. Infection occurs by inhalation, but organisms are carried from the lungs to the reticuloendothelial cells in liver and spleen. Replication occurs there with focal necrosis, then blood-borne spread occurs back to lungs and to other organs including meninges and brain, heart, gut and adrenals. Inflammation is lymphocytic with macrophages containing glycogen-negative cytoplasmic inclusions, surrounded by oedema, haemorrhage and necrosis.

Confirmatory tests

Microscopy is unhelpful and culture is dangerous. Serology gives the diagnosis after 2 or 3 weeks, by a rise in titre in paired acute and convalescent sera.

Clinical syndrome

The pathogenesis explains the long incubation period, systemic symptoms of malaise and myalgia, marked headache and mentation changes, macular rash and multiple organomegaly (especially splenomegaly), mucoid sputum and marked pneumonia.

Chemotherapy

Tetracycline is the usual treatment, but erythromycin is also effective.

Control. Psittacosis is only a risk disease for those in contact with birds, particularly pet psittacines (birds with hooked beaks, e.g. parrots).

Chlamydia

- Chlamydia are unique obligate intracellular bacteria, unable to synthesise ATP and hence dependent on host cells.
- They exist in two forms: a robust infective extracellular elementary body and a fragile intracellular replicating reticular body.
- *C. trachomatis* causes trachoma, non-gonococcal urethritis, cervicitis and other genital infections, neonatal infections from an infected mother, and lymphogranuloma venereum.
- *C. pneumoniae* causes respiratory infections including 'atypical' pneumonia.
- *C. psittaci* causes psittacosis, a systemic disease with pneumonia, from infected birds.
- Chlamydial infections are treated with tetracycline or erythromycin.

RICKETTSIA, COXIELLA, ROCHALIMAEA AND EHRLICHIA

In the family Rickettsiaceae there are five medically important genera: *Rickettsia, Orientia, Coxiella, Rochalimaea* and *Ehrlichia*, which cause a number of important diseases including the typhus group and Rocky Mountain spotted fever (RMSF). Rickettsia are small aerobic bacteria which like other bacteria:

- have a cell wall with inner and outer membranes
- have procaryotic ribosomes
- have RNA and DNA
- synthesize their own macromolecules
- have peptidoglycan in the cell wall.

However, they lack enzymes for energy metabolism, and so are obligate intra-cellular parasites, and cannot replicate extracellularly (except for *Rochalimaea*). They require ATP, coenzyme A and NAD from the host cell and can exchange some of their ADP for host ATP. Their dimorphic life cycle involves an animal or human reservoir and an arthropod vector (Fig. 1), so rickettsial diseases each have a special geographic distribution.

Virulence and pathogenesis
Virulence varies between species, and possible factors are:

- lipopolysaccharide, like other typical Gram-negative rods
- toxin, ill-defined and of uncertain importance
- phospholipase A to dissolve the phagosomal membrane (Fig. 1).
- cell destruction and death after intracellular replication.

Pathogenesis is similar in all but *Coxiella* infections (Fig. 1). Doubling time is slow, about 8 hours, but in 40–48 hours the host cell dies, ruptures and liberates large amounts of rickettsiae to infect further cells. Vasculitis (damage to small blood vessels) is the most striking feature (Fig. 2b).

Antibody, either naturally developing or from vaccine, prevents escape from the phagosome so the rickettsiae are destroyed by lysosomal enzymes.

Confirmatory tests
Microscopy is generally not useful except for *R. rickettsii* where direct immuno-fluorescent (DIF) microscopy of skin biopsies is quick and specific.

Culture is difficult and dangerous so is rarely attempted outside specialist laboratories. Serology is the main identi-fying method. The Weil–Felix agglutina-tion test uses patterns of cross-reactivity

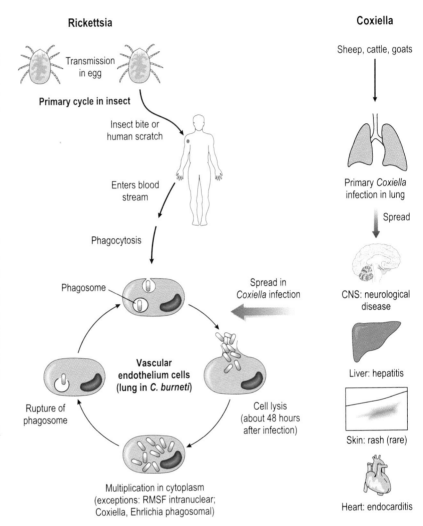

Fig. 1 Typical events in *Rickettsia* and *Coxiella* infection.

Table 1 Weil–Felix test patterns

Disease	Organism	P. vulgaris antigens		
		OX19	OX2	OXK
Epidemic typhus	R. prowazeckii	+	+/–	–
Brill-Zinsser disease	R. prowazeckii	+/–	–	–
Murine typhus	R. typhi	+	+/–	–
Scrub typhus	O. tsutsugamushi	–	–	+
RMSF and 'regional' rickettsioses	R. rickettsii	+	+	–
	R. australis, conorii etc			
Rickettsial pox, Q fever and Ehrlichiosis are negative to all.				

of antibodies to rickettsiae with antigens in *Proteus vulgaris* strains (Table 1). Complement fixation tests (CFT) will also differentiate some species. IF, Latex and EIA are now available commercially.

R. PROWAZECKII

R. prowazeckii causes epidemic louseborne typhus; a mild endogenous recrudes-cence 20–40 years later is called Brill-Zinsser disease.

Unlike other rickettsioses, humans are the principal host and the vector is the human louse, *Pediculus humanus* var. *corporis*, which dies from the infection in 2–3 weeks. All major epidemics of typhus have been associated with wars and famine, where conditions predispose to louse infestation.

Confirmatory tests
Diagnosis is primarily clinical, but serol-ogy can be used (Table 1).

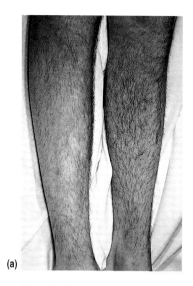

Fig. 2 **Rocky Mountain spotted fever. (a)** Rash on legs; **(b)** Vasculitis in skin biopsy.

Clinical syndromes

Epidemic typhus is a serious disease, and mortality in malnourished populations has been over 60%. It is characterised by fever, headache, prostration and a rash beginning on the trunk and spreading to the extremities (the pattern in all rickettsial rashes except RMSF).

Brill–Zinsser disease is a mild recrudescence many years later.

Chemotherapy and control

Tetracycline or chloramphenicol are effective. Louse control with DDT, lindane or malathion is essential. A vaccine is available for special risk groups.

R. TYPHI

R. typhi, formerly *R. mooseri*, causes endemic murine typhus. The natural reservoir is the rat, and transmission is by the rat flea, *Xenopsylla cheopis*, neither of which are killed by the rickettsia. The flea defecates on feeding, and rickettsia in the faeces enter through the bite wound or scratches.

Confirmatory tests

Diagnosis is usually clinical, but Weil–Felix serology distinguishes murine typhus from RMSF, but not from epidemic typhus, which needs specific specialised indirect immunofluorescence.

Clinical syndrome

Murine typhus is similar to epidemic typhus, but milder. It is distinguished from RMSF by (1) being an urban disease of late summer and autumn with no tick exposure, while RMSF occurs in spring-summer after tick exposure; (2) having a rash which spreads from trunk to extremities, whereas RMSF does the reverse; and (3) by showing a positive CFT, which usually differentiates.

Chemotherapy and control

Tetracycline or chloramphenicol are effective chemotherapy, and control involves rat and flea control.

R. RICKETTSII

R. rickettsii, the cause of Rocky Mountain spotted fever (RMSF), is present in many animals, including mammals, birds and dogs in endemic areas in the USA. It is transmitted by ticks, especially *Dermacentor andersoni* and *D. variabilis*, the dog tick.

Confirmatory tests

Initial diagnosis must be clinical; DIF microscopy of skin biopsy and serology are used.

Clinical syndromes

Rash which begins on the extremities (Fig. 2) and spreads centrally, headache, fever, myalgia, and prostration occur acutely, and mental clouding, splenomegaly, pulmonary oedema, disseminated intravascular coagulation (DIC), shock and death follow in patients diagnosed or treated late.

Chemotherapy and control

Tetracycline or chloramphenicol are used as chemotherapy, with supportive care. Control depends on tick avoidance or their early removal.

COXIELLA BURNETII

Q fever (for query, as the cause was unknown) was first described in Queensland, Australia and later attributed to *C. burnetii* infection.

C. burnetii is in a separate genus because it differs in being very robust, surviving in the environment for months or more, hence can infect humans directly by inhalation, without an arthropod vector. A further difference is its existence in two phases: (I) when isolated from animals and in Q fever endocarditis, and (II) when grown in eggs.

Spore-like structures unique among the Rickettsiaceae may explain its robustness. It is virulent, with a low infective dose. The **pathogenesis** (Fig. 1) is different from all other Rickettsiaceae. It is inhaled, and replicates slowly in the lungs causing, in 20 days, an 'atypical' interstitial pneumonia with lymphocytic infiltrate. It also disseminates by the blood stream and causes granulomas, in the liver particularly. It also infects abnormal 'native' and prosthetic heart valves and perivalvular tissues.

Confirmatory tests

Antibodies to *C. burnetti* do not react with *Proteus* OX antigens, so the Weil–Felix test is negative. Phase II antibodies develop, first IgM in 2–3 weeks, then IgG after 12 weeks, persisting often for 12 months. Titres of phase I antibody higher than phase II are found in chronic disease, particularly endocarditis.

Clinical syndromes

Infection is usually occupational, with abattoir workers the major group infected. The commonest syndrome is atypical pneumonia with systemic disease including hepatitis, while endocarditis is fortunately rare.

Chemotherapy

Tetracycline, chloramphenicol and erythromycin are rickettsiostatic and may shorten the acute illness. Endocarditis is extremely difficult to treat even with combined surgery and antibiotics: tetracycline with a second drug such as rifampicin or cotrimoxazole for 2 to 5 years may be needed.

Control

A vaccine is now available in some countries.

Rickettsiaceae

- Rickettsiae are small obligate intracellular bacteria.
- Most have an animal host and an arthropod vector and so have a geographical distribution.
- Pathogenesis is similar with vasculitis, except in Q fever with pneumonia and hepatitis (*Coxiella burnetii*).
- Diagnosis is clinical and usually serological.
- Tetracycline is the usual treatment.

ASPERGILLUS AND CANDIDA

Aspergillus and *Candida*, two common fungal genera, are similar in two respects:

- they cause opportunistic infections in immunocompromised hosts
- they do not cause invasive infections in normal hosts.

However, they also differ in two important respects:

- *Aspergillus* spp. are filamentous fungi, while *Candida* are yeasts.
- *Aspergillus* are environmental fungi, while *Candida* are normal flora.

ASPERGILLUS

Although *Aspergillus* infection has been known for a century, opportunistic invasive disease is now more common because of the great numbers of patients immuno-compromised by disease, drugs or defects in the skin–mucous membrane barrier.

Aspergillus spp. are branching, septate, filamentous fungi, 7–10 μm in diameter, and several hundred micrometres long. They are classified by their sexual structures, but distinguished in clinical medicine by their asexual structures including the shape of the conidiophore, the supporting vesicle and phialides (sterigmata), and the conidiospores (Fig. 1). The most important species causing human disease are *A. fumigatus* (Latin for 'smoky', referring to the smoky blue-grey mycelium on white colonies) (Fig. 2), and *A. flavus,* meaning yellow (colonies). They tolerate temperatures up to 50°C, and, therefore, grow widely in the environment: in soil and (warm) decaying vegetation including haystacks and compost heaps, resulting in easy dispersal into the air.

They have typical fungal eucaryotic cell structure with a nucleus and varied cytoplasmic structures (p. 6). Different species differ in virulence and produce disease by all three fungal mechanisms:

mycotoxins, hypersensitivity and invasion, particularly of blood vessel walls (see pp. 8, 9).

Confirmatory tests

Both false-positive and false-negative results are frequent for three reasons:

- *Aspergillus* species are common in the environment and, therefore, are common contaminants of clinical specimens in the absence of disease
- serious infections are usually enclosed in deep tissues
- they grow poorly in blood cultures and some other media.

Definitive diagnosis therefore depends on histopathology showing septate hyphae with dichotomous branching (bifurcating, into two at acute angles), as well as Gram stain and repeated cultural isolation.

Clinical syndromes

The three pathogenic mechanisms lead to (Table 1):

- mycotoxicosis from coumarin-like aflatoxins, causing bleeding
- hypersensitivity causing allergic rhinitis, asthma or hypersensitivity pneumonia by inhalation
- local infection such as bronchopulmonary aspergillosis
- invasive opportunistic infection in the immunocompromised host, with a fungus ball in a pulmonary cavity, severe pneumonia, vascular infection

or disseminated abscesses in organs such as brain, kidneys, liver, skin or bone.

Chemotherapy

This is unsatisfactory: amphotericin B is still the treatment of choice in invasive disease but is not always effective. Itraconazole is proving useful in some patients.

Control

Usually impossible, except by removal from the cause in allergic disease, or filtered air and exclusion of plants and soil for highly immunosuppressed patients.

CANDIDA

Superficial candidial infections are less troublesome now because of better control of predisposing factors, better diagnosis and better treatment, but systemic invasive candidiasis is more common, for the reasons given for invasive aspergillosis.

Candida are small round yeasts which multiply by budding (blastocondidia formation), forming either pseudo-hyphae ('germ tubes', Fig. 3) or septate hyphae. They are classified into species by carbohydrate assimilation and fermentations. *C. albicans* is the commonest and with *C. tropicalis, C. glabrata* and *C. parapsilosis* is part of the normal oropharyngeal, gut and vaginal flora, but they are uncommon on the skin unless it is macerated or damaged.

Table 1 **Sources, spread and syndrome**

Organism	Epidemiological source	Pathological mechanism	Clinical syndrome
A. fumigatus and *Aspergillus* spp.	Air, soil, mouldy vegetation	Toxin production Hypersensitivity Opportunistic direct invasion Systemic spread	Bleeding Asthma, pneumonia Pneumonia, fungus ball Organ abscesses
C. albicans and other spp.	Own normal flora	Opportunistic local infection Systemic spread	Skin, mucosal infection Organ infections

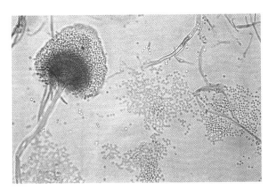

Fig. 1 *Aspergillus.* Hyphae and conidiophore.

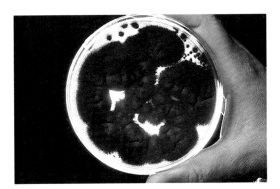

Fig. 2 *A. fumigatus.* Culture.

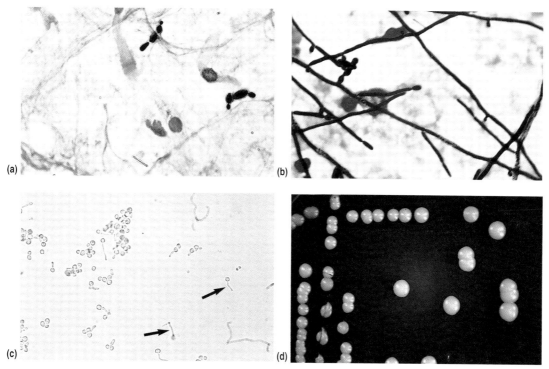

Fig. 3 *C. albicans.* (a) Gram stain of sputum. (b) Blood culture showing enormous pseudo-hyphae. (c) Germ tubes (no stain). (d) Colonies cultured on Sabouraud's medium.

Virulence and pathogenesis

Candida spp. have typical fungal cell structure (p. 6). They have minimal virulence for normal hosts, and pathogenesis depends on one or more defects in host defence, including:

- ruptured normal skin-mucous membrane defences caused by mechanical breaches including catheters, by changes in normal flora owing to disease or drugs, or by maceration, trauma or disease
- impaired neutrophil or eosinophil function
- impaired monocyte or macrophage function
- impaired lymphocyte function
- impaired alternative complement pathway.

Antibody is usually produced but is probably not protective. The known adherence of *Candida* spp. to many cells may be important.

Confirmatory tests (Fig. 3)

Typical Gram stain, culture on usual laboratory media such as blood agar, and germ tube formation in serum in 2–3 hours are the usual tests for *C. albicans,* with carbohydrate assimilation and fermentation tests for other species. Although *C. albicans* can grow in vented blood culture bottles, disseminated candidiasis often has negative blood cultures initially. CT scan of liver and spleen often shows characteristic small, regular lesions, so histopathology of biopsies is necessary less often.

Clinical syndromes

These include (Table 1):

- local skin or nail disease including napkin rash, intertrigo or paronychia, occurring in hosts with only locally impaired defences
- local mucous membrane disease such as oral or vaginal thrush with locally impaired defences
- local but chronic and severe chronic mucocutaneous candidiasis (CMC) in children with specific T-cell defects

- organ system infection, including CNS, pulmonary, cardiovascular, urinary, eye, bone or joint, or disseminated infection, in the immunocompromised patient.

Chemotherapy

Local nystatin or imidazoles are used for local infections, keto-conazole for CMC; deep, systemic or disseminated infections need specialised treatment with amphotericin B, supplemented at times with an imidazole or flucytosine.

Control

This depends on treating or removing the predisposing defects in host defences. Chemoprophylaxis is used in vulnerable patients.

Aspergillus

- Heat-tolerant, environmental, septate filamentous fungi.
- *A. fumigatus* is the commonest pathogen.
- *Aspergillus* spp. cause disease by mycotoxins, by hypersensitivity and by opportunistic invasion of the immunocompromised patient.
- Amphotericin B is the standard treatment, itraconazole is promising; systemic infection is often fatal.

Candida

- Round yeasts which multiply by budding.
- *C. albicans* is the commonest yeast in the normal flora of the oropharynx, gut and vagina but is not found on normal skin.
- Local infection of skin, nails or mucous membrane occurs when local, first-line defences are impaired.
- Systemic opportunistic infection occurs in the immunocompromised patient.
- Nystatin or imidazoles are used for local infections and for prophylaxis, while systemic infections need amphotericin B.

CRYPTOCOCCUS AND HISTOPLASMA

In contrast to *Aspergillus* and *Candida*, both *Cryptococcus neoformans* and *Histoplasma capsulatum* cause disease (systemic mycoses) in normal hosts, as well as severe or fatal disease in immunocompromised patients.

CRYPTOCOCCUS NEOFORMANS

C. neoformans has two varieties, var *neoformans* and var *gattii*, which differ genetically, biochemically and ecologically. Both however are monomorphic, being yeasts in both culture and tissues, (unlike the other four fungi causing systemic mycoses, see below and pp. 66–67). The sexual forms are varieties of *Filobasidiella neoformans*. *Cryptococcus* is found world-wide in pigeon excreta and nests (the organism's urease allows the use of nitrogen sources) and in certain trees. It infects humans (but not the pigeons!) by inhalation (Fig. 1).

Cryptococci have a typical round yeast structure, with a large characteristic mucopolysaccharide capsule, important in pathogenicity and helpful in diagnosis (Fig. 2). About 50% of patients have no known predisposing factor, while the remainder have immune defects, usually of T-cell and macrophage function, less often of neutrophil, B cell or complement. CNS infections are prominent because soluble serum anti-cryptococcal factors are absent in CSF, CSF is a good growth medium, and the CNS inflammatory response is minimal, partly because complement also is absent in the CSF.

Confirmatory tests
Microscopy of the CSF is often positive when Indian ink is used as a negative stain to show the clear capsulated cryptococci against a black background (Fig. 3). Gram stain and microscopy of sputum are less reliable. Culture should always be done to confirm the microscopy.

Biochemical tests including urease confirm the identification. Antigen detection by latex agglutination is now sensitive and 90% reliable on serum and CSF.

CT scan is often diagnostic, especially in immunocompromised patients (e.g. in AIDS) with cerebral lesions. Histopathology may be needed for unusual pulmonary lesions.

Clinical syndromes
Inhalation may lead to an asymptomatic pulmonary lesion, rarely to a primary pneumonia, or, most frequently, to meningitis or cerebral infection with single or

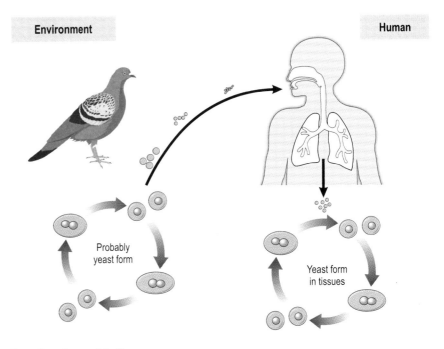

Fig. 1 *C. neoformans* **infection.**

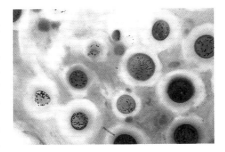

Fig. 2 **Encapsulated** *C. neoformans.*

multiple scattered lesions. Skin, bone or visceral lesions also occur.

Chemotherapy
The usual treatment has been amphotericin B plus flucytosine, but fluconazole is also effective in treatment. Pulmonary lesions may need excision for diagnosis.

Control
Fluconazole is used for long-term prophylaxis in the immunocompromised patient, particularly in AIDS.

HISTOPLASMA CAPSULATUM

'Tutankhamen's curse', the disease affecting those who opened his tomb, was probably histoplasmosis from still viable spores!

H. capsulatum is a dimorphic fungus, with a filamentous spore-forming mould form in the environment and in culture at 25°C, and the pathogenic yeast form

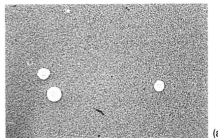

(a)

(b)

Fig. 3 *C. neoformans:* **Indian ink stain. (a)** Low power; **(b)** high power.

in tissues and macrophages (Fig. 4). The sexual stage is an ascomycete, *Ajellomyces capsulata. H. capsulatum* var *capsulatum* is found in most countries, but highly endemic areas include the Ohio–Mississippi valleys in the USA. *H. capsulatum* var *duboisii* is found in Africa. The mould grows in soil, nitrogen-enriched from bird or bat droppings, and humans are infected by spore inhalation, especially when first exposed during work or leisure.

The filaments are thin, branching and septate, the spore being either tuberculate macroconidia or round microconidia.

Probably yeast form

Yeast form in tissues

Environment

Human

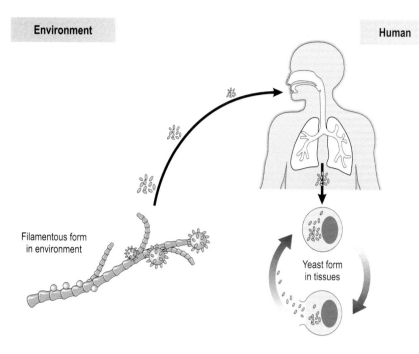

Filamentous form
in environment

Yeast form
in tissues

Fig. 4 *H. capsulatum* infection.

Table 1 **The interaction between organism and host in *H. capsulatum* infection**

Organism	Infecting dose	Host tissue	Host defence	Clinical result
Var *capsulatum*: usual organism, lesser virulence	Usual	Normal (lung)	Immune / Non-immune	Asymptomatic infection, 90% / Flu-like illness / Latent infection
	High	Normal	Non-immune	Acute pulmonary histoplasmosis
	Usual or high	Normal	Excessive response	Peri-pulmonary fibrosis
	Usual	Normal	Impaired response	Disseminated histoplasmosis
	Usual	Normal	First normal then impaired response	Re-activation and dissemination
	Usual	Chronic lung disease	Local defect	Chronic pulmonary histoplasmosis
Var *duboisii*: unusual infection, greater virulence	Usual	Normal	Normal	Bone and subcutaneous histoplasmosis

Confirmatory tests

The diagnosis is initially clinical in those exposed, especially in an endemic area, and chest X-ray showing small scattered infiltrates plus hilar lymphadenopathy is characteristic. Calcification is common on healing (Fig. 5).

Microscopy of sputum is not useful, and culture is slow and only useful in chronic disease.

Histopathology with silver stains is usually diagnostic but seldom necessary.

Serology is useful; the complement fixation test is standard on paired sera. Immunodiffusion in agar gel is more specific but less sensitive, and rapid antigen detection tests are available.

Clinical syndromes

These provide a wonderful illustration of the interaction between: the organism's forces (infective dose, varying virulence and tissue trophism) and the host first-line defences (healthy or diseased tissues) and subsequent immune response (normal, or impaired, or initially normal then impaired, or excessive) (Table 1).

Chemotherapy

Amphotericin B is the classical treatment for severe or prolonged disease, while ketoconazole or itraconazole are less effective but useful orally.

Control

This depends on avoiding exposure, or using ketoconazole as long-term suppression in chronic disease.

The yeast form is ovoid, about 2 μm by 3 μm. No toxins are known. Inhaled spores germinate to the yeast form, which is ingested by macrophages and transported to mediastinal lymph nodes, spleen and liver. Granulomata form and heal, giving rise to diagnostic calcification. Immunity is cell mediated and wanes with time, though repeated attacks are less severe.

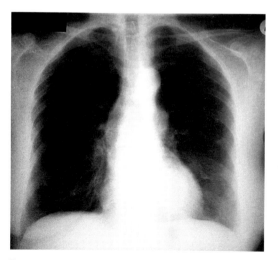

Fig. 5 **Histoplasmosis healed in lung showing scattered small calcified spots.**

Cryptococcus

- *C. neoformans* is a monomorphic yeast with a characteristic capsule, found particularly in pigeon droppings.
- It can infect humans with a normal immune system, by inhalation, though 50% of those infected have detectable immune deficiency.
- Infection causes pulmonary masses (often asymptomatic), chronic meningitis or cerebral masses.
- Classical treatment is amphotericin plus flucytosine, though fluconazole is now being used.

Histoplasma

- *H. capsulatum* is a dimorphic fungus: the filamentous mould form is found in soil with bird or bat droppings, and highly endemic areas are notorious.
- Humans are infected by inhalation of spores; these germinate to the yeast form, which is ingested by macrophages.
- The clinical disease depends on the dose inhaled, the immune state and local lung disease.
- Amphotericin B is the usual treatment, but ketoconazole has an increasing role.

BLASTOMYCES, COCCIDIOIDES, PARACOCCIDIOIDES

BLASTOMYCES DERMATITIDIS

B. dermatitidis (like *H. capsulatum*) is a dimorphic fungus with a filamentous spore-bearing mould form in the environment and on culture at 25°C, and a pathogenic yeast form in tissues (Fig. 1). The sexual form is an ascomycete called *Ajellomyces dermatitidis*. The organism, and hence blastomycosis, is endemic only in parts of North America and Africa. The mould is difficult to find in soil and infects humans, dogs and horses by inhalation (Table 1).

The pyriform or pear-shaped macroconidia are 2–4 μm in diameter, the yeast is bigger (8–15 μm), and divides by broadly based buds. No toxins are known and the pathogenesis is initially like histoplasmosis, with yeasts inhaled and spread by macrophages, followed by granuloma formation. However, calcification does not follow, and chronic suppuration is common.

Confirmatory tests

Microscopy of abscess pus shows broad-based budding yeasts, and culture is relatively easy. Serology is relatively unhelpful (unlike histoplasmosis), but histopathology, if necessary, is usually diagnostic.

Clinical syndromes and management

Asymptomatic pulmonary infection is probably common. Treatment is essential if skin or bone infection, or the less common symptomatic pulmonary or visceral disease, develop.

Chemotherapy

Ketoconazole or itraconazole are used for uncomplicated pulmonary disease, but amphotericin is necessary for systemic infection.

COCCIDIOIDES IMMITIS

C. immitis is a dimorphic fungus, with a filamentous spore-forming mould form in the environment and on culture at 25°C, and a pathogenic yeast form in tissues (Fig. 2). The sexual form is unknown.

The scattered endemic foci are in southwest USA, Central and S. America. It is a soil organism, infecting by inhalation (Table 1).

Alternate cells in the filaments develop into barrel-shaped arthrospores (arthroconidia), while the second characteristic structure is the large (20–70 μm) spherule in the infected tissues. The spherule is filled with endospores which develop into further spherules, so the budding typical of the other systemic mycoses is absent. Granuloma formation resembling tuberculosis is usual.

Confirmatory tests

Microscopy of sputum may show spherules with endospores. Culture is

Table 1 **The systemic mycoses**

Organism	Epidemiological source, route, area	Pathogenic mechanism	Clinical syndrome
Blastomyces dermatitidis	Soil and old buildings By inhalation N. America, Africa, Middle East	Local (inapparent) Distant spread	(Pulmonary) Bone and skin
Coccidioides immitis	Soil, dust storms By inhalation USA, Central and S. America	Local infection Distant spread	Pulmonary or none CNS, bone, skin
Paracoccidioides brasiliensis	Soil (probably) By inhalation, (?implantation) S. and Central America	Local (inapparent) Distant spread	(Pulmonary) Nose and mouth
Cryptococcus neoformans	Pigeon droppings By inhalation Worldwide	Local invasion Distant spread	Pulmonary CNS infection
Histoplasma capsulatum	Soil from birds and bats By inhalation Worldwide, mainly Americas and Africa	Local invasion Distant spread	Pulmonary Disseminated disease

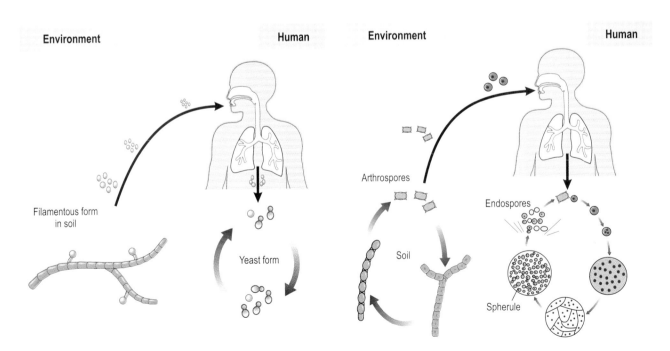

Fig. 1 **Environmental filamentous and pathogenic yeast forms of** *B. dermatitidis*.

Fig. 2 **Environmental filamentous and pathogenic yeast forms of** *C. immitis*.

only done with special facilities because of the risk of laboratory-acquired infections.

Serology by latex agglutination or agar immunodiffusion is useful, and CFT may be used to follow progress; histopathology is diagnostic if necessary (Fig. 3).

Clinical syndromes
Probably 60% of infections are asymptomatic, while the remainder result in an acute pulmonary infection. Most heal, but some will progress to chronic pulmonary disease or disseminated CNS, bone or skin disease.

Chemotherapy
Amphotericin B is the standard treatment. Ketoconazole or intraconazole are usually only suppressive.

PARACOCCIDIOIDES BRASILIENSIS

P. brasiliensis (like *H. capsulatum*) is a dimorphic fungus with a filamentous spore-bearing mould form on culture at 25°C, and a pathogenic yeast form in tissues (Fig. 4). The sexual form is unknown. The organism, and hence paracoccidioidomycosis

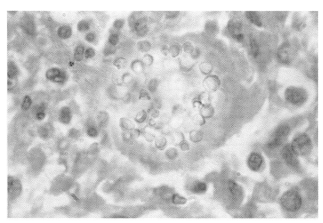

Fig. 3 **Tissue section showing *C. immitis* spherules.**

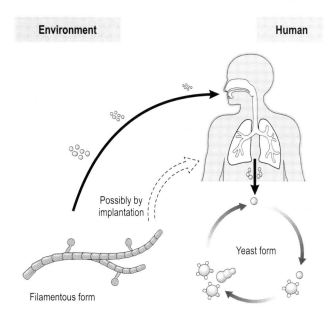

Fig. 4 **Environmental filamentous and pathogenic yeast forms of**
P. brasiliensis.

(South American blastomycosis), is endemic only in parts of Central and South America, especially Brazil. The mould is extremely difficult to find in soil and probably infects humans by inhalation, though local implantation is possible.

The mycelium and sporulation have no special characteristics. The yeast size varies from 4–40 μm, and budding is multiple and circumferential, giving a 'pilot's wheel' formation.

The mycelium–yeast transformation is inhibited by 17β-oestradiol, the probable reason why adult females are rarely infected. No toxins are known, and infection leads to granuloma formation, chronic suppuration and hyperplasia. Cellular immunity is depressed by infection and improves with recovery.

Confirmatory tests
Microscopy of abscess pus shows 'pilot's wheel' yeasts and culture is on Sabouraud–dextrose agar at 25°C.

Serology is by immunodiffusion or CFT. Histopathology, if necessary, is usually diagnostic. Skin testing is often negative on diagnosis.

Clinical syndromes and management
This is a progressive chronic disease, mainly in adult males. Primary pulmonary disease is often asymptomatic, and years later chronic 'mulberry' mucous membrane or warty skin ulcers develop. Lymph nodes, adrenals and other viscera can be infected.

Chemotherapy
Ketoconazole or intraconazole are replacing amphotericin and sulphonamides for chemotherapy, which are still used if the azole fails.

Blastomyces dermatitidis

- A dimorphic fungus endemic in parts of N. America and Africa.
- A soil organism causing blastomycosis by inhalation.
- Symptomatic pulmonary infection is rare, and chronic bone and skin disease is usual, with broad-based budding yeasts and granuloma formation.
- Ketoconazole or intraconazole are used for mild infection, amphotericin for systemic infection.

Coccidioides immitis

- A dimorphic fungus endemic in areas of the Americas.
- A soil organism causing coccidioidomycosis by inhalation.
- Asymptomatic pulmonary infection is commonest, followed by symptomatic pulmonary disease; progressive or systemic disease is rare.
- Arthroconidia and tissue spherules are characteristic, with granulomata.
- Amphotericin is needed for systemic or severe disease.

Paracoccidioides brasiliensis

- A dimorphic fungus endemic in S. America.
- A soil organism, it probably infects by inhalation, or possibly implantation.
- Chronic progressive skin and mucous membrane ulcers are usual, pulmonary and systemic disease occur rarely.
- Multiple budding gives a 'pilot's wheel' appearance microscopically, and granulomata occur.
- Ketoconazole or intraconazole are replacing amphotericin and sulphonamides in treatment.

MYCOSES OF SKIN AND ADJACENT TISSUES

SUPERFICIAL MYCOSES

The superficial mycoses are fungal infections of the outermost layers of skin and hair.

Tinea versicolor (pityriasis versicolor)

This is a result of skin infection by *Malassezia furfur (Pityrosporum orbiculare)*, a lipophilic yeast. Budding yeasts up to 8 μm across and short mycelial fragments ('spaghetti and meat balls') are seen in KOH-treated skin scrapings from the scaly hypopigmented infected skin. Culture requires a medium rich in fatty acids and is not undertaken routinely.

Tinea versicolor should be differentiated from **erythrasma**, which is caused by *Corynebacterium minutissimum* (p. 174). The latter fluoresces pink under UV light.

Tinea nigra of the skin

Exophiala (previously *Cladosporium*) *werneckii*, a dimorphic fungus containing melanin gives the green-black colonies and causes brown-black skin lesions. Initially double-celled oval yeasts grow (seen in KOH mounts), then hyphae and conidia as the culture ages. The black skin lesions are described on page 172.

Black piedra of the hair

Piedraia hortai, an ascomycete, has its sexual stage on the hair, showing 2–8 slender ascospores in each ascus. Culture gives green-black or red-black colonies with chlamydospores. The hard black hair nodules and treatment by clipping are described on page 172.

White piedra of the hair

White piedra is caused by *Trichosporon beigelii*, a dimorphic fungus. The creamy-white mycelial collar on the hair shaft is of septate hyphae, which break to form arthroconidia. Culture (on media without inhibitory cycloheximide) gives soft white colonies, becoming wrinkled and yellow-grey with age. The arthroconidia become rounded blastoconidia, showing the dimorphic nature of the fungus. Full identification is by biochemical tests. Clinical features are given on page 173.

CUTANEOUS MYCOSES

Fungi known (wrongly) as dermatophytes ('skin plants') infect the keratinised surface of the body producing conditions

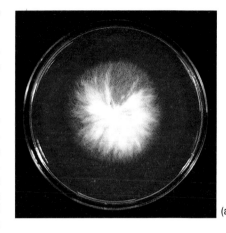

(a)

(b)

Fig. 1 *Microsporum canis.*
(a) Surface view of colony.
(b) Macroconidia.

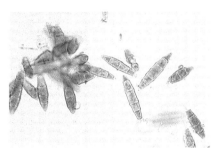

Fig. 2 **Macroconidia of *M. gypseum.***

Fig. 3 **Microconidia of *T. mentagrophytes.***

known collectively as tinea or ringworm. The infections are named after the area of the body affected:

- tinea capitis: scalp
- tinea barbae: beard
- tinea corporis: body
- tinea cruris: groin
- tinea pedis: feet (athlete's foot)
- tinea unguium (nails).

About 40 species from three genera – *Trichophyton, Microsporum* and *Epidermophyton* – are involved (Table 1); many contain keratinases and infect only keratin-containing tissue. The sexual stage of *Microsporom* and *Trichophyton* are ascomycetes, genus *Arthroderma*.

Certain generalisations may be drawn from Table 1:

- Each species of *Trichophyton* and *Microsporum* can cause infection of skin, hair or nails.
- *T. rubrum, T. mentagrophytes* and *M. canis* are most common, but frequency varies widely between tropical and temperate areas.
- *T. schoenleinii* uniquely causes **favus** at the hair follicle.
- *Epidermophyton* does not cause tinea capitis or barbae and very rarely causes tinea unguium.

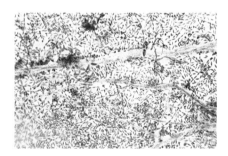

Fig. 4 *Paecilomyces lilacius*: a rare cause of phaeohyphomycosis.

- Some zoophilic species may be recognised by their names; these particularly infect the skin and hair of children, and the beard areas of rural men, owing to greater contact. They usually provoke greater host inflammatory response in humans.
- Geophilic species particularly infect the head of children, again related to their greater contact with soil.

Confirmatory tests

Scrapings from the lesions are treated with KOH to dissolve the keratin and expose the fungal bodies on direct microscopy (Figs 1–3):

- *Trichophyton*: abundant microconidia; rare, smooth, thin-walled macroconidia
- *Microsporum*: many, rough thick-walled macroconidia; rarely any microconidia
- *Epidermophyton*: many smooth-walled macroconidia; no microconidia.

Identification to species level requires culture on selective media, e.g. Sabouraud-cycloheximide agar.

Clinical syndromes

Tinea of the skin, hair and nails is described on page 172. Chemotherapy involves the use of topical anti-fungals including the imidazoles such as clotrimazole. Oral griseofulvin is used for hair, nail and severe skin disease.

FUNGI CAUSING SUBCUTANEOUS MYCOSES AND MYCETOMA

These uncommon diseases unfortunately have numerous causative fungi, many of which have several synonyms (Table 2). These mycoses are caused by the implantation of soil fungi into the subcutaneous tissues by thorns or other penetrating trauma, particularly in the tropics. They usually remain localised, except for sporotrichosis and mycetoma.

Subcutaneous mycoses and mycetoma are classified into:

- chromoblastomycosis
- lobomycosis
- phaeohyphomycosis (phaeomycotic cyst)
- rhinosporidiosis
- zygomycosis (subcutaneous)
- sporotrichosis (lymphocutaneous)
- eumycotic mycetoma.

The common causes and diagnostic mycology are summarised in Table 2.

Confirmatory tests

Culture of grains or pus from sinuses or of excised tissue, and histopathology are specialised but usually diagnostic. The clinical syndrome and geographic area are also helpful.

Clinical syndromes

All cause subcutaneous lesions which often progress to involve the overlying skin. Sporotrichosis and mycetoma spread as noted above (see page 174, 176).

Chemotherapy

Potassium iodide is used for treatment of subcutaneous sporotrichosis, and amphotericin B for extracutaneous sporotrichosis. Flucytosine is sometimes useful in chromoblastomycosis, but excision is often needed, as in the other subcutaneous mycoses. Mycetoma needs precise microbiologic diagnosis, and specialised medical and surgical treatment.

Table 1 **Cutaneous mycoses**

Fungus	Ecology	Clinical disease
Trichophyton		
T. concentricum	A	Tinea corporis
T. equinum, T. mentagrophytes var mentagrophytes, T. verrucosum	Z	Barbae–capitis
T. mentagrophytes var interdigitale T. rubrum	A	Pedis-manuum, corporis-cruris
T. schoenleinii, T. tonsurans	A	Tinea capitis
T. violaceum	A	Barbae-capitis
Microsporum		
M. audouinii	A	Tinea capitis
M. canis	Z	Barbae-capitis, corporis
M. equinum, M. gallinae	Z	Tinea capitis
M. ferrugineum	A	Tinea capitis
M. fulvum, M. gypseum, M. nanum	G	Tinea capitis
Epidermophyton		
E. floccosum	A	Pedis-manuum, corporis-cruris

A, anthropophilic (human source); G, geophilic (soil source); Z, zoophilic (animal source).

Table 2 **Subcutaneous mycoses and mycetoma**

Disease	Causative fungi	Diagnostic mycology
Chromoblastomycosis	*Cladosporium carrionii, Fonsecaea pedrosoi, F. compacta, Phialophora verrucosa, Wangiella dermatitidis*	Dematiaceous melanin-pigmented fungi: identified by type of sporulation; Histopathology shows pseudo-epitheliomatous hyperplasia, and golden-brown spherical 'sclerotic' bodies
Lobomycosis	*Loboa loboi*	Lemon-shaped yeasts in pairs or short chains; never cultured
Phaeohyphomycosis (phaeomycotic cyst)	*Alternaria alternata, Curvularia geniculata, Wangiella dermatitidis* and others	Dematiaceous fungi; short hyphal fragments, without hyperplasia on histology (Fig. 4)
Rhinosporidiosis	*Rhinosporidium seeberi*	Thick-walled cysts (10–200 µm) called spherules, filled with spores; never cultured
Zygomycosis (subcutaneous)	*Conidiobolus coronatus, Basidiobolus ranarum*	Culture after excision shows typical conidia
Sporotrichosis (lymphocutaneous): may spread along lymphatics	*Sporothrix schenckii*	Dimorphic: mould at 25°C, and yeast at 37°C and in tissues; conidia form typical rosettes
Eumycotic mycetoma: penetrates deeply to muscle, tendon and even bone	*Madurella grisea, M. mycetommatis Pseudallescheria boydii*	Culture and structure of fungi grown from grains from sinuses; distinguish from bacterial mycetoma

Note also non-fungal actinomycotic (bacterial) mycetoma p. 56

Superficial fungal infections

- Superficial mycoses are caused by four specific fungi and result in hypopigmentation or black skin, or black or white hairs.
- Diagnosis is clinical and by microscopy.
- Infected skin is treated topically, and infected hair clipped or shaved.

Cutaneous fungal infections

- Cutaneous mycoses result from infection by *Microsporum*, *Trichophyton* or *Epidermophyton* spp.
- Sources are humans, animals or the soil.
- Diagnosis is by microscopy and culture.
- Treatment is topical for mild skin disease, or oral griseofulvin for hair, nail or severe skin disease.

Subcutaneous fungal infections and mycetoma

- Subcutaneous mycoses and mycetoma are caused by penetrating trauma.
- Mostly localised but sporotrichosis and mycetoma can spread.
- Chemotherapy and often excision is required.

FUNGI CAUSING ZYGOMYCOSIS

Zygomycosis is infection with fungi of the class Zygomycetes, which includes the orders Mucorales and Entomophthorales. Mucormycosis is infection with fungi of the order Mucorales, including the genera *Mucor, Rhizopus, Rhizomucor* and *Absidia*. Entomophthoromycosis is infection with fungi of the other order, including *Basidiobolus* and *Conidiobolus*.

Life cycle and pathogenesis

The Mucorales, like *Aspergillus*, are monomorphic fungi, with only the mycelial form existing in all environments. The hyphae (Fig. 1) are non-septate, very big (10–15 µm across) and branch at right angles (unlike *Aspergillus*, which are septate, smaller and branch at acute angles).

Mucorales are soil organisms which grow easily on bread and fruit, and infection occurs usually by inhalation, rarely by implantation in wounds or burns.

Human infections are usually with *Rhizopus arrhizus* and *Absidia corymbifera*. The fungi invade tissues, particularly blood vessels, causing infarction and necrosis.

Virulence is increased by acidosis and corticosteroids, so diabetics and immunosuppressed patients are most at risk.

Confirmatory tests

As they are common surface contaminants, diagnosis of infection needs repeated isolation from non-sterile sites, culture from deep tissues, or **histopathology** (Fig. 2). These fungi cannot be distinguished from each other without culture. Rhizoids (roots) form from stolons (runners). The sporangiophores carry sporangia, containing spores (sporangiospores) (Fig. 1).

- *Rhizopus*: stalks of sporangiophores are opposite the rhizoids
- *Absidia*: stalks of sporangiophores are not opposite the rhizoids
- *Mucor*: no rhizoids.

Clinical syndromes

In recent years there has been an increase in generalised infections caused by fungi. Mucormycosis affects brain and lung most often. **Rhinocerebral zygomycosis** or mucormycosis is the commonest and most feared form of this rare infection (p. 91). Other rare syndromes are pulmonary (rather like aspergillosis), disseminated, gastrointestinal, wound, skin or visceral. Infected patients are usually diabetic, on steroids, or immunosuppressed by disease or drugs.

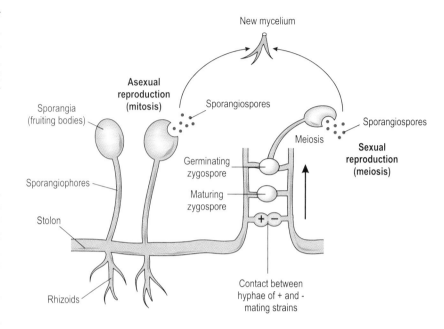

Fig. 1 **The structure and life cycle of a zygomycete.**

(a)

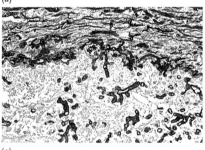

(c)

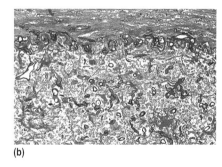

(b)

Fig. 2 **Mucormycosis. (a)** Skin biopsy.
(b) Histology, staining with H & E.
(c) Histology, staining with silver iodide.

Chemotherapy and control

Amphotericin B is essential. Control of the predisposing factors helps to prevent infection; early excisional surgery is required if infection is uncontrolled.

Fungi causing zygomycosis

- Zygomycosis is infection with fungi of the class Zygomycetes, including mucormycosis from *Rhizopus, Mucor, Absidia* and *Rhizomucor* of the Order Mucorales.
- Infection from these environmental organisms is by inhalation usually, implantation rarely. Acidosis and corticosteroids enhance their growth.
- Rhinocerebral mucormycosis is the commonest, often fatal form, but pulmonary, systemic and local infections can occur.
- Treatment is by amphotericin B, excision, and correction of predisposing factors, including diabetes and immunosuppression.

ARTHROPODS

The arthropods are invertebrates not microorganisms but they play an important role in human infections in several ways.

- **mechanical vectors** are involved in disease transmission but are not essential for the survival of the pathogen, e.g. house flies with bacterial enteric pathogens
- **biological vectors** transmit the disease pathogen but also provide a host for an essential part of the life cycle of the pathogen, e.g. the mosquito host for *Plasmodium* spp., which cause malaria
- **reservoirs**: some arthropods act as reservoirs for maintaining microorganisms between hosts, e.g. lice and *R. prowazekii*, causing typhus.
- **true infections** of the skin, e.g lice.

Table 1 lists some infections, covered elsewhere in this book, where an arthropod vector is involved.

ARTHROPOD INFECTIONS

PEDICULUS SPP.

Lice are blood-sucking insects that live on human skin: three main types are *Pediculus humanus* var. *corporis*, the human body louse; *Pediculus humanus* var. *capitis* the human head louse; and *Phthirus pubis* the human pubic louse.

The lice are about 3 mm long and have a 3-week reproductory cycle: 1 week for eggs to hatch and two more weeks to maturity.

They carry three major diseases:

- epidemic typhus fever (*Rickettsia prowazeckii*): body louse vector
- trench fever (*Bartonella quintana*): body louse host
- epidemic relapsing fever (*Borrelia recurrentis*): body louse vector.

Three predisposing factors are poor hygiene, close contact and shared possessions or partners. Head lice pass rapidly among children in schools and nurseries. Lice infestation can be indicated by bites, nits (the whitish eggs on hairs, Fig. 1) and lice (especially the pubic type).

Three common clinical manifestations are:

- pruritus: intense, prolonged and wide-spread
- haemorrhagic bites: especially near folds of clothes
- secondary bacterial infection.

Table 1 **Some diseases transmitted by arthropod vectors**

Disease	Pathogen	Vector
Tularaemia	*Francisella tularensis*	Ticks, deer flies
Plague	*Yersinia pestis*	Flea
Lyme disease	*Borrelia burgdorferi*	Tick
Epidemic typhus	*Rickettsia prowazekii*	Body louse
Rocky Mountain spotted fever	*R. rickettsii*	Tick
Malaria	*Plasmodium* spp.	Mosquito (Anopheles)
Sleeping sickness	*Trypanosoma brucei /gambiense*	Tsetse fly
Chagas' disease	*T. cruzi*	Reduviid bug
Filariasis	*Wuchereria bancrofti*	Mosquito
Loaiasis	*Loa loa*	Mango Fly

Rarer clinical manifestations are:

- blepharitis, from pubic lice and bacterial infection
- maculae caerulae (literally, sky-blue spots) on the abdomen and thighs from pubic lice, possibly from a haemolysin
- vagabond's disease, marked by chronicity, hyperpigmentation and lichenification of the skin.

Control measures range from prevention of overcrowding, through improved hygiene (washing clothing and bedding) to the use of insecticides: malathion, BHC (gamma benzene hexachloride), or DDT in past epidemic situations. The nits are removed by special fine-toothed combs.

SARCOPTES SCABIEI

The mite *S. scabiei* causes scabies, a parasitic skin infection. The mite lives in the epidermis of human skin and the females lay eggs in the stratum corneum. After a few days the eggs hatch and the larvae mature to spread across the skin (Fig. 2). The preferred areas are the wrists and between the fingers, and infection causes severe itching particularly at night. Secondary bacterial infections can occur where the skin is broken by scratching. The mites themselves do not cause itch directly; hypersensitivity reactions to the mites and their faeces is the powerful mechanism.

Norwegian scabies appears very different clinically with grossly thickened skin and thin scales in the immunosuppressed. Thick nails, pyoderma and pigmentation follow.

Scabies is transferred person to person by close contact, and from fomites such as clothes and bedding. Lindane (gamma-benzene hexachloride) kills the mites, and hygiene and improved living conditions reduce spread.

Fig. 1 **Nit on hair**.

Fig. 2 **Scabies**.

Arthropods

- May be mechanical vectors, biological vectors, reservoirs or cause infections themselves.
- *Pediculus* spp. cause lice infections, epidemic typhus, trench fever and relapsing fever.
- *Sarcoptes scabiei* cause scabies with severe nocturnal itch.

SPOROZOA: PLASMODIA, TOXOPLASMA, CRYPTOSPORIDIUM

PLASMODIUM

Plasmodia are sporozoan protozoan parasites of humans and animals, and cause malaria in humans. Four species infect humans:

- *P. falciparum*: causes malignant tertian malaria, the most severe type and the predominant type in the tropics.
- *P. vivax*: causes benign tertian malaria, the most widely distributed form and the predominant type in temperate areas.
- *P. malariae*: causes quartan malaria.
- *P. ovale*: causes a type of tertian malaria in West Africa and is relatively uncommon.

Life cycle and pathogenesis

The female *Anopheles* mosquitoes transmit the infection. The life cycle of plasmodia is complex with a sexual cycle occurring in the mosquito and two asexual (schizogony) phases in the human (Fig. 1), initially in the liver and then in the blood.

The liver cycle in the human involves the production of merozoites that can either enter new liver cells to repeat the cycle of production, or can enter red blood cells (RBC). The liver phase may persist for long periods, except in *P. falciparum*, and relapses triggered from these forms may occur although the parasite has been eliminated from peripheral blood.

The asexual stage in red blood cells consists of several distinct forms that are detected in blood films in diagnosis. Episodes of fever occur when further merozoites are released from the red blood cells. The intervals between the bouts of fever depend on the length of time the plasmodia take to complete the asexual cycle in the blood.

Some merozoites in RBC develop into male and female gametocytes which are ingested by mosquitoes to complete the life cycle.

Partial immunity from previous infections, and the sickle cell trait, decrease the severity of infection.

Confirmatory tests

Thick and thin films of blood are stained by Leishman or Giemsa stains and the red cells examined. Thick and thin blood films show the malarial parasites (MP) in various forms:

- ring forms, commonly seen as small round structures with a clear centre and a chromatin dot at one side
- rosettes, which are schizonts containing daughter merozoites
- gametocytes, rarely seen.

Each species has specific morphology, e.g. the gametocytes of *P. falciparum* are crescentic, the trophozoites of *P. vivax* are amoeboid.

Clinical syndromes

Acute malaria is a medical emergency (see pp. 92, 130).

Chemotherapy

This varies with the plasmodial species, the severity of illness and the state of immunity and chemoprophylaxis. Chloroquin, mefloquin and quinine are used, and recent WHO or Health Department recommendations must be followed. Primaquin is active against the liver cycles for eradication after acute treatment in non-endemic areas.

Control

This depends on mosquito control, protection from biting, and chemoprophylaxis in the non-immune during and after visits to endemic areas.

TOXOPLASMA GONDII

T. gondii is a coccidian parasite of cats. The sexual cycle takes place in the cat (Fig. 2a) producing **oocytes** in the faeces that can infect intermediate hosts: humans, many mammals and some reptiles. Human toxoplasmosis is absent on islands without cats but appears if they are introduced.

In the human or animal host, the asexual cycle produces slender **tachyzoites**, which infect cells (often macrophages and endothelial cells), or shorter broader dormant **bradyzoites**, which form large tissue cysts (Fig. 2b) with hundreds of organisms in tissues. These account for latency and are infectious to other animals if the tissue is eaten.

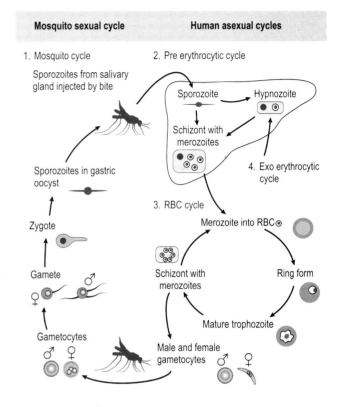

Mosquito sexual cycle	Human asexual cycles

Fig. 1 ***Plasmodium* life cycle.**

The asexual cycle also occurs in cats, perpetuating infection.

Humans are infected either by eating raw or undercooked meat (e.g. lamb, mutton) containing **tissue cysts**, by ingesting **oocysts** in food soiled with cat faeces, or by transplacental infection (congenital infection).

All strains appear of equal virulence. They survive initially in macrophages by preventing phagosome–lysosome activity and fusion, but are killed by activated macrophages.

Confirmatory tests

Serology is the major diagnostic method: ELISA for IgM and IgG is replacing IFA, IHAT, CFT and the Sabin–Feldman dye test.

CT brain scans in AIDS and other immunosuppressed patients are often characteristic. Histopathology on biopsies is diagnostic, if necessary.

Clinical syndromes

Most infections are asymptomatic: about 40% of adults have positive serology but recall no infection. A mononucleosis-like syndrome can occur with acute infection. Recrudescence of latent infection occurs with immunosuppression, e.g. in transplant patients and AIDS.

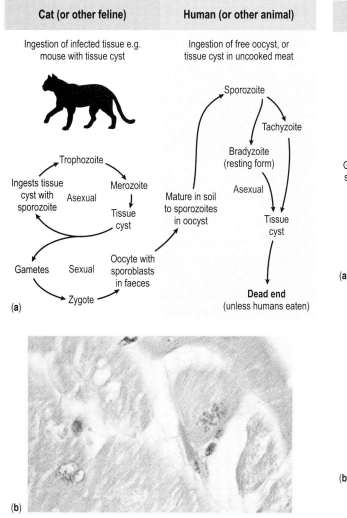

Fig. 2 *Toxoplasma*. (a) Life cycle; (b) tissue cyst.

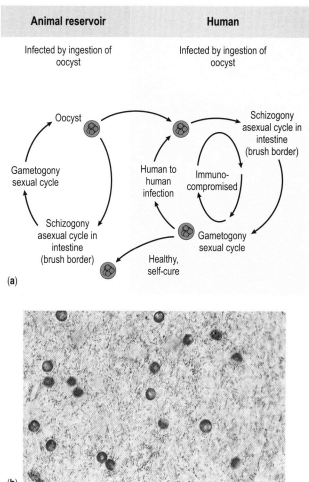

Fig. 3 *Cryptosporidum*. (a) Life cycle; (b) oocyst in faeces stained by acid-fast stain (*Cryptosporidum parvum*).

Congenital infection acquired in utero from an infected mother causes chorido-retinitis, convulsions, cerebral atrophy and calcification. The outcome is often fatal.

Chemotherapy

A combination of pyrimethamine and sulphadiazine is the usual treatment, while spiramycin, clindamycin and cotrimoxazole are possible alternatives. Treatment is suppressive as tissue cysts are resistant.

Control

Meat should be properly cooked, and hands washed after handling cats or their litter, after preparing raw meat and before eating. It is believed that humans are usually infected by eating food contaminated with oocysts. Cats are a health hazard particularly to women of child-bearing age.

CRYPTOSPORIDIUM SPP.

Cryptosporidium spp. are coccidial protozoa, found worldwide. They are animal parasites which infect humans, probably from calves and contaminated water.

Virulence and pathogenesis

Cryptosporidia have a simple life cycle (Fig. 3a). Only one or two cycles occur in normal hosts, but many cycles in the immunocompromised patient give prolonged severe watery diarrhoea.

The fertilised oocyst in faeces is diagnostic; it stains, unexpectedly, with acid-fast stains (Fig. 3b).

Clinical syndromes

Clinically, patients usually present with diarrhoea that lasts 1–2 weeks; in immunocompromised patients prolonged severe diarrhoea is caused by the multiple replication cycles.

Chemotherapy and control

No treatment is really effective, but spiramycin, azithromycin or paromomycin are sometimes useful. Handwashing and safer sexual practices offer some protection.

Sporozoan parasites

- Sporozoan parasites have life cycles with distinct sexual and asexual phases; the sexual phase occurs in the primary host (often non-human) and the asexual proliferation phase gives rise to the clinical symptoms.
- Chemotherapy is not always effective; control is by avoiding infection, reducing the insect carrier and careful handling of food and water.
- Malaria is caused by *Plasmodium* spp. which infect humans via the bite of the female *Anopheles* mosquito.
- Treatment for malaria includes drug regimens that must vary with parasite drug resistance. Prophylaxis is used for travellers.
- *T. gondii* is a cat parasite that causes toxoplasmosis in humans; congenital infection and that in the immunocompromised have severe consequences.
- *Cryptosporidium* spp. cause diarrhoeal symptoms that are usually severe in the immunocompromised.

AMOEBAE: ENTAMOEBA, NAEGLERIA, ACANTHAMOEBA

ENTAMOEBA HISTOLYTICA

E. histolytica is a pathogenic amoeba that causes amoebic dysentery, a disease mainly occurring in tropical and subtropical countries. Initial attacks are virulent: local colonic invasion causes acute dysentery (blood plus mucus with diarrhoea). Some immunity develops and, in endemic countries, mild chronic or intermittent diarrhoea, or asymptomatic cyst passing is more common.

E. histolytica lives in the mucosa of the large intestine producing local necrosis and ulcers. Local and distant spread can occur, with tissue lysis ('histolytica') causing abscesses particularly in liver, lung and brain. Infection is usually faecal–oral, sometimes oro–anal, rarely ano–penile in homosexuals.

The normal motile erythrocytophagic ('able to eat red blood cells') **trophozoite** comes from resistant non-motile **cysts** which survive outside the body for years.

Confirmatory tests

Microscopy of fresh stool for the fragile trophozoites or cysts (Fig. 1) is usual, but careful and expert microscopy on three or more specimens may be needed to find amoebae and distinguish normal *Entamoeba coli* from pathogenic *E. histolytica* (Table 1). Finding cysts (Fig. 1) does not prove active disease.

Serology by ELISA, IHA or CIE is most useful in non-endemic countries and for extraintestinal amoebiasis, using IgM detection for acute disease.

CT scans of liver and brain, and chest X-rays are often helpful.

Clinical syndromes

The symptoms of amoebic infection range from asymptomatic cyst passage to acute dysentery to chronic, intermittent dysentery, and to extraintestinal liver, lung and brain abscesses, or cutaneous infections (pp. 90, 140, 142–3, 176).

Chemotherapy

Metronidazole has displaced emetine as the 'gold standard', but intestinal antiseptics such as diloxanide furoate may be needed to complement metronidazole (or tinidazole). Abscesses seldom need drainage now.

Control

Hygiene including handwashing after defecation and careful preparation of food. Safer sex practices are also partly protective.

Other intestinal amoeba

Entamoeba coli, Entamoeba hartmanni, Endolimax nana, Iodamoeba butschlii are all considered non-pathogenic but must be carefully differentiated microscopically from *E. histolytica*. *Entamoeba polecki* from pigs may cause brief, mild diarrhoea. *Dientamoeba fragilis*, in spite of its name, is now classified as a flagellate (p. 75).

NAEGLERIA FOWLERI

N. fowleri is a free-living amoeba widely distributed in fresh water, especially where bacterial contamination occurs. The trophozoite is slow moving, but a mobile flagellate form occurs in culture and a cyst form in unfavourable conditions.

N. fowleri is highly virulent and causes amoebic meningoencephalitis (Fig. 2) by inhalation and penetration through the nasal mucosa and the cribriform plate.

Confirmatory tests

Microscopy of CSF shows the causative amoebae with many polymorphs. Trophozoites are 10–20 µm across, move slowly and have a large central karyosome in a round nucleus, features distinguishing them from monocytes in CSF. Protein is elevated, glucose low.

Clinical syndrome

Acute amoebic meningoencephalitis is rapidly progressive and fatal in 3–4 days unless treated early and vigorously.

Chemotherapy

Intravenous high-dose amphotericin B is essential and may be supplemented intrathecally, plus oral rifampicin and an imidazole.

ACANTHAMOEBA SPP.

These are also free-living amoebae, rather smaller than *Naegleria* and with foamy cytoplasm. They are found in contaminated water, soil and the human oral cavity. They cause either a more chronic granulomatous meningoencephalitis in immunocompromised patients, or corneal infection after trauma or contact lens contamination.

Amphotericin B is surprisingly ineffective; flucytosine and imidazoles are used.

Table 1 **Distinguishing *Entamoeba* spp.**

	E. histolytica	E. coli	E. nana
Trophozoite size (µm)	15–50	10–30	6–12
Nuclear karyosome	Central	Eccentric	Massive
Nuclear chromatin	Peripheral, fine	Clumped, coarse	Indistinct
Ingested RBC	Yes	No	No
Cyst size (µm)	8–20	10–30	8–10
Cyst nuclei	1–4	1–8	1–4
Chromatoid bars	Rounded	Split-enz	None

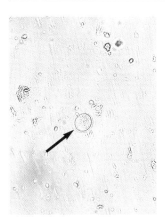

Fig. 1 *Entamoeba histolytica* cyst in stool.

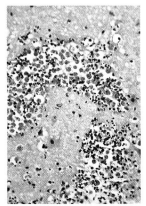

Fig. 2 **Amoebic encephalitis.**

Amoebae

- Pathogenic amoebae occur as amoeboid trophozoites or as resistant long-living cysts.
- Humans carry many non-pathogenic intestinal amoebae; *E. histolytica* causes amoebic dysentery and can spread to distant organs causing abscesses.
- Free-living amoebae *N. fowleri* and *Acanthamoeba* spp. are common in fresh water and cause severe meningoencephalitis.
- *Acanthamoeba* spp. also cause keratitis after trauma or contact lens contamination.

INTESTINAL AND VAGINAL FLAGELLATES AND CILIATES

GIARDIA LAMBLIA

This flagellate protozoan (previously *G. intestinalis*) is distinctive both in morphology and in infecting the duodenum and small intestine. Like amoebae, it has a fragile **trophozoite** (Fig. 1), a tougher infective **cyst** (Fig. 2) and divides by binary fission. Ingested cysts liberate trophozoites that adhere to the intestinal mucosa by a ventral sucking disc causing local damage, and malabsorption in heavy infection.

The reservoirs of infection are water, humans and animals including beavers. Infection is by water or food, or person-to-person by faecal–oral or oro–anal spread. Large outbreaks of giardiasis have occurred. It varies from asymptomatic infection to acute watery diarrhoea to chronic intermittent diarrhoea with malabsorption.

Diagnosis is by stool microscopy, which shows the distinctive cysts more often than trophozoites. The distinctive 'face with moustache' structure of the trophozoite is shown in Figure 1. Repeated specimens are often necessary because of intermittent excretion. If negative, a duodenal aspirate or biopsy with smear may be done.

Giardiasis is treated with metronidazole or quinacrin; repeat courses may be needed. Control depends on good hygiene including handwashing. Cysts resist usual water chlorination levels.

TRICHOMONAS VAGINALIS

T. vaginalis is a urogenital flagellate. It is very common in the vagina of sexually active women, and transmission is usually sexual, occasionally by fomites.

The trophozoite reproduces by binary fission, and no cyst form is known. Some strains have greater virulence, and pathogenesis involves epithelial adhesion, damage and microulceration. Asymptomatic infection is common in both women (vaginal) and men (urethral). Vaginitis with frothy watery yellow discharge is the most common symptomatic disease, while symptomatic urethritis, prostatitis and epididymitis are rare.

Direct microscopy of vaginal or urethral discharge by the wet mount technique shows the distinctive motile trophozoite.

Metronidazole is given to the patient and all sexual partners. During pregnancy, clotrimazole may be safer but is less effective.

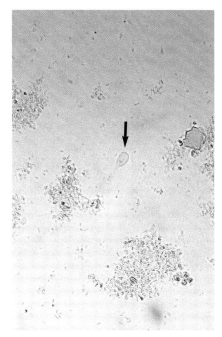

Fig. 1 *G. lamblia* trophozoite.

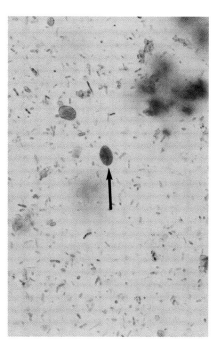

Fig. 2 *G. lamblia* cyst.

Safer sexual practices and care with shared personal articles help control transmission.

DIENTAMOEBA FRAGILIS

This parasite has several unusual characteristics:

- it is a highly motile flagellate, not an amoeba, despite its name
- no cyst form is known
- the delicate trophozoite is probably transmitted person-to person protected inside eggs of the pinworm, *Enterobius vermicularis*!
- It is bi-nucleate, with 4–6 chromatin granules in each.

D. fragilis causes mild local damage to the proximal colon and rarely causes abdominal and diarrhoeal symptoms.

Diagnosis is by stool microscopy, and treatment is iodoquin, tetracycline or paromomycin. Control is by hygienic measures and pinworm treatment.

BALANTIDIUM COLI

B. coli is the largest protozoan infecting mammals; the trophozoite is 50–200 µm in diameter, structurally complex and forms a resistant cyst that transmits disease. It is a parasite of pigs, infecting humans by the faecal–oral route, then from person to person. It invades the colonic mucosa, causing dysentery with blood and mucus in the stools.

Diagnosis is by stool microscopy, treatment is by tetracycline, or metronidazole, paromomycin or iodoquin, and control is by hygiene including handwashing.

Intestinal and vaginal flagellates and ciliates

- *G. lamblia* occurs in trophozoite and cyst form; it is found in water and it spreads rapidly person to person causing large outbreaks of giardiasis.
- Giardiasis usually causes diarrhoea and may result in malabsorption; several courses of metronidazole or quinacrin may be needed.
- *D. fragilis* rarely causes diarrhoea but *B. coli* causes dysentery.
- *T. vaginalis* is a sexually transmitted urogenital flagellate causing vaginitis and sometimes urethritis; treatment should include sexual partners.

BLOOD AND TISSUE FLAGELLATES

Haemoflagellates are protozoa with flagella which infect blood and tissues and are transmitted by biting flies or bugs, usually from animal reservoirs (see pp. 18, 71).

LEISHMANIA

Leishmania includes species with variants not considered here, hence the term 'complex' in *Leishmania donovani* complex, *L. tropica* complex, *L. brasiliensis* complex and *L. mexicana* complex. All are intracellular parasites of humans, **transmitted by sandfly bites**.

Life cycle and pathogenesis

Leishmaniae have a simple life cycle (Fig. 1) with only two forms:

- a rounded (2–3 μm), replicating **amastigote** with nucleus, parabasal body and kinetoplast, but no flagellum, in the mammalian host cells. Replication is by binary fission
- an elongated (2 × 20 μm) flagellated **promastigote** in the insect (sandfly) host.

All are virulent, causing severe disease in normal hosts. Pathogenesis begins with variable local destruction and invasion of macrophages; dissemination to spleen, liver and other lymphoid tissues occurs in the visceral form, and to mucosa in the mucocutaneous type. Activation of T-cells and cell-mediated immunity is important in host resistance.

Confirmatory tests

While clinical diagnosis is often possible in endemic areas, definite diagnosis depends on microscopy and specific staining (Giemsa, Wright) for amastigotes:

- in bone marrow, called Leishman–Donovan (LD) bodies, (Fig. 2) or other aspirate in kala-azar
- in ulcer scrapings, aspirates or biopsies in 'Old World' cutaneous disease (Fig. 3).
- in punch biopsies (smear plus histopathology) in 'New World' cutaneous and mucocutaneous disease.

Clinical syndromes

Four types of leishmaniasis occur (see page 179):

- *L. donovani* complex causes kala-azar (visceral) and post-kala-azar dermal leishmaniasis
- *L. tropica* complex causes oriental sore (cutaneous leishmaniasis)

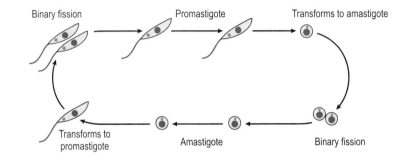

Sandfly vector	Human and animal reservoirs
Amastigote to promastigote	Promastigote to amastigote
Ingests amastigote	Injected with promastigote

Fig. 1 ***Leishmania* life cycle.**

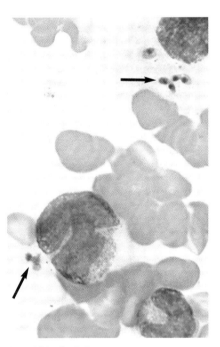

Fig. 2 **LD bodies in bone marrow.**

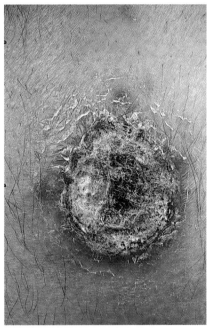

Fig. 3 **Cutaneous leishmaniasis sore.**

- *L. mexicana* complex causes American (New World) cutaneous leishmaniasis
- *L. braziliensis* complex causes mucocutaneous leishmaniasis.

Visceral disease is characterised by Sever and anaemia and has a high mortality. The cutaneous forms involve granulomatous and ulcerative sores.

Chemotherapy

Stibogluconate (used in the Old World) or the equivalent meglumine antimoniate (used in the New World) are the drugs of choice for all leishmaniasis. Alternatives are pentamidine in visceral disease, local heat in Oriental disease, and amphotericin B in American and Ethiopian disease.

Control

This depends on sandfly control, particularly around houses, and protection from biting.

TRYPANOSOMA

Human trypanosomes include two variants of *Trypanosoma brucei*, *T. brucei gambiense* and *T. brucei rhodesiense*, plus *T. cruzi*.

T. brucei causes the cattle disease, nagana, and human sleeping sickness. T. brucei live and multiply in the human blood stream and invade the CNS to cause sleeping sickness. T. cruzi is an intracellular parasite of humans, causing acute and chronic Chagas' disease, with cardiac and digestive tract involvement. **Both are spread by insect vectors, the tsetse fly for T. brucei and the reduviid ('kissing') bugs for T. cruzi.** The reservoir hosts are wild animals and cattle in some areas; in other areas the animal host is unknown.

Life cycle and pathogenesis

T. brucei have a simple life cycle (Fig. 4), with only two stages:

- the **trypomastigote** in human blood. Usually as a long slender (3 × 30 μm)

form with nucleus, parabasal body and kinetoplast, flagellum and full-length undulating flagellar membrane. It is somewhat polymorphic, so short broad forms without a flagellum, and intermediate forms occur. Replication is by binary fission
- the **epimastigote** in the tsetse fly with only a partial undulating membrane

In T. cruzi (Fig. 5) there is a third stage, the **amastigote**, which is the invasive, dividing, destructive tissue stage in humans. The pathogenic trypanosomes are virulent, causing severe disease in normal hosts. Pathogenesis begins with variable local destruction (the trypanoma) and invasion of the blood stream, followed by invasion and destruction of

cerebral tissue (T. brucei) or cardiac, oesophageal or colonic tissue (T. cruzi). T. brucei largely avoids host response by changing its surface antigens (glycoproteins) during bouts of parasitaemia.

Confirmatory tests

While clinical diagnosis is often possible in endemic areas, definite diagnosis depends on microscopy and specific staining (Giemsa) for amastigotes in anticoagulated blood, blood films or lymph node aspirates. Culture is often necessary in Chagas' disease where parasitaemia is less or absent. Serology is of some use.

Clinical syndromes

Details of the trypanoma, acute disease and progressive encephalitis of sleeping sickness from T. brucei, and of Chagas' disease from T. cruzi are covered on pages 131, 124.

Chemotherapy

Suramin is the drug of choice for early T. brucei infections, the alternative being pentamidine. Toxic organic arsenicals such as melarsoprol are needed for CNS infection. Nifurtimox is the drug of choice in T. cruzi infections.

Control

This depends on tsetse fly and reduviid bug control, particularly around houses, protection from biting, and treatment of infected humans.

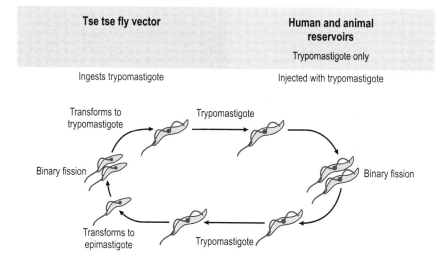

Fig. 4 *Trypanosoma brucei* life cycle.

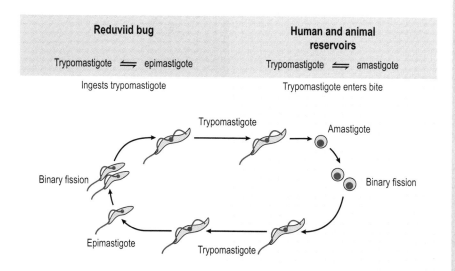

Fig. 5 *Trypanosoma cruzi* life cycle.

Blood and tissue flagellates

- Leishmaniasis can be visceral (kala-azar), cutaneous (oriental sore) or muco-cutaneous depending on the *Leishmania* spp. involved. All are transmitted by sandfly bites, and infections are diagnosed by microscopy of bone marrow, splenic, skin or mucosal ulcer specimens.
- *T. brucei* variants cause sleeping sickness in humans, spread by tsetse flies. Initial local trypanoma is followed by encephalitis. Parasites are detected in blood films.
- *T. cruzi* causes Chagas' disease in Latin America and is spread by reduviid ('kissing') bugs. Diagnosis is by microscopy of blood in early disease, or by culture in chronic disease with cardiac involvement, mega-oesophagus or mega-colon.

INTESTINAL NEMATODES

Helminths form three groups: nematodes, cestodes and trematodes. Some of the intestinal nematodes have an obligatory soil phase, others can have direct person-to-person spread or auto-infection. Infection is either by ingestion or skin penetration.

ENTEROBIUS VERMICULARIS (PINWORM OR THREADWORM)

The pinworm, a very common intestinal helminth, is small (male 2.5 mm, female 10 mm) like a piece of cotton thread. Unlike most other worms, it is more common in temperate climates because it spreads person-to-person by direct contact or fomites and, occasionally, by inhalation of egg-contaminated dust.

Life cycle and pathogenesis. Unusually, pinworms have no obligatory soil cycle, nor do they multiply in the body! The gravid female lays 10 000 to 15 000 bean-shaped eggs (50 × 25 μm) in a few minutes on the peri-anal skin and dies. Eggs are deposited on underclothes, bedding or under scratching finger-nails and ingested (Fig. 1a) or, occasionally, inhaled. The eggs can also carry *Dientamoeba fragilis* (p. 75).

Clinical syndrome. Light infections are asymptomatic; perianal female worms cause itching (pruritis ani) (p. 141) or vaginal irritation.

Confirmatory tests. Microscopy of adhesive tape previously pressed to the anus shows the typical egg (Fig. 2a). The adult worms may be seen macroscopically.

Chemotherapy. Treatment with pyrantel pamoate or mebendazole is given to the whole family.

Control involves family treatment, hand and nail hygiene, washing of bedding and night attire, and vacuuming to remove any eggs in dust.

TRICHURIS TRICHIURA (WHIPWORM)

Shaped like a whip (the tail) with a handle (the body), this worm is 3–5 cm long. Found worldwide, it is most common where human faeces enrich soil directly. Spread is indirectly faecal–oral, with no animal reservoir.

Life cycle and pathogenesis. The life-cycle is simple: ingested eggs hatch in the small bowel, the larvae migrate

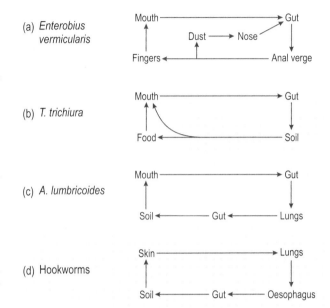

Fig. 1 **Life cycle of five helminths (including two hookworms).**

(Fig. 1b) to the caecum, the adult fertilised female lays 5 000 to 10 000 eggs daily and lives up to 7 years! Eggs passed in faeces must develop in the soil for 3 weeks to be infectious when ingested. Secondary bacterial infection can occur where the adult worm head penetrates caecal mucosa.

Clinical syndromes. Symptoms range from none to mild abdominal discomfort to bloody diarrhoea to rectal prolapse (p. 141). Luminal worms in appendicitis may not be causal.

Confirmatory test. Microscopy of stool shows the typical 'tea-tray'-shaped eggs, elongated ovals with bi-terminal plugs (Fig. 2b).

Chemotherapy. This is by mebendazole, which is less effective but better tolerated than thiabendazole.

Control. The main aims of control should be better sanitation and personal hygiene.

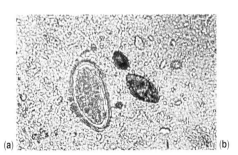

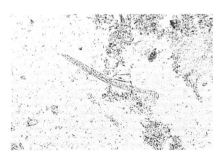

Fig. 3 **Strongyloides larva.**

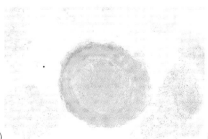

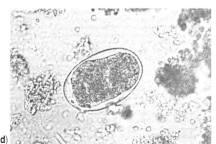

Fig. 2 **Helminth eggs in faeces, wet preparations.**
(a) *E. vermicularis*; (b) *T. trichiura*;
(c) *A. lumbricoides*; (d) hookworm.

ASCARIS LUMBRICOIDES (ROUNDWORM)

Shaped like a big earthworm 25–30 cm long, the roundworm is most common where human faeces enrich soil directly. Maturation of the egg in soil is essential, so spread is indirectly faecal–oral. There is no animal reservoir.

Life cycle and pathogenesis. The life-cycle involves an additional stage: pulmonary migration (Fig. 1c). The adult fertilised female lays up to 200 000 eggs daily for up to 1 year! Pathogenesis can include partial or complete mechanical obstruction to bowel, bile ducts or appendix, pneumonitis during pulmonary migration, or secondary bacterial infection when the adult worm head penetrates the small bowel (pp. 140–141).

Clinical syndromes. Symptoms range from none to abdominal discomfort to bowel obstruction, obstructive jaundice, appendicitis or peritonitis.

Confirmatory tests. Microscopy of stool shows the typical brownish oval eggs, about 45×70 µm, either fertilised or unfertilised, with or without (decorticated) their rough outer shell (Fig. 2c).

Chemotherapy. Mebendazole or pyrantel pamoate are used; the older, more toxic piperazine is a reserve drug.

Control involves education and better sanitation and personal hygiene, especially in food handlers.

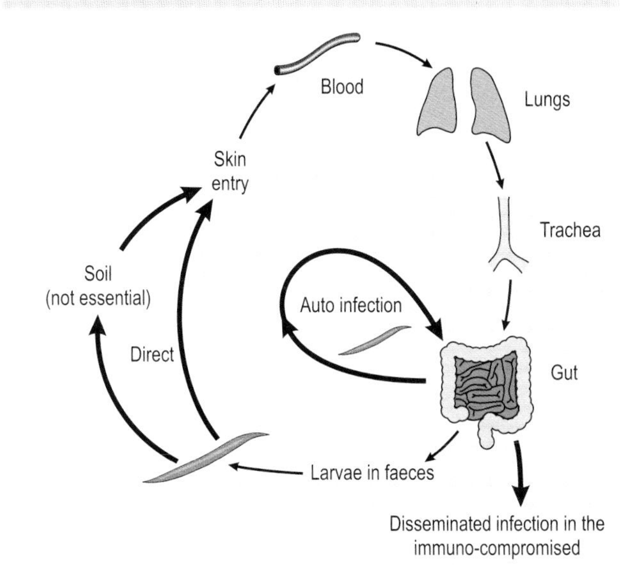

Fig. 4 **Life cycle of *S. stercoralis*.**

ANCYLOSTOMA DUODENALE AND *NECATOR AMERICANUS* (HOOKWORMS)

A. duodenale, the Old World hookworm is a little larger (1.2×0.6 mm) than *N. americanus*, the New World hookworm (1×0.4 mm) and has a different mouth. Both are most common where human faeces enrich soil directly. Maturation of the egg in soil is essential, so spread is indirectly faecal–skin. There is no animal reservoir.

Life cycle and pathogenesis. The life-cycle includes both skin penetration and pulmonary migration (Fig. 1d). The adult worms are about 1 cm long and live attached to intestinal mucous membranes by four hooked teeth. The adult fertilised female lays about 10 000 (*Necator*) to 30 000 (*Ancylostoma*) eggs daily for about 5 years. Pathogenesis is by the ingestion of blood from the human bowel mucosa, from 0.03 (*Necator*) to 0.3 ml (*Ancylostoma*) per worm per day: over 100 ml daily for a worm burden of 500 adults!

Clinical syndromes. Symptoms are itch at skin entry sites, allergic pneumonitis and, particularly, those of iron-deficiency anaemia (p. 142).

Confirmatory tests. Microscopy of stool shows the typical unstained oval eggs, about 40×70 µm, containing a developing larva (Fig. 2d).

Chemotherapy and control. Mebendazole or pyrantel pamoate, plus iron, are used in chemotherapy. Education, better sanitation, personal hygiene and the wearing of shoes should improve control.

STRONGYLOIDES STERCORALIS

This little skin-penetrating nematode (female 2.5 mm long, male only 0.7 mm) can cause severe disease. It is most common where human faeces enrich soil directly, for infective larvae pass in the faeces (Fig. 3). Maturation of the larvae in soil is not essential, so spread can be directly faecal–skin, or by direct anal contact (including sexual contact), or by autoinfection (see below). There is no animal reservoir.

Life cycle and pathogenesis. The life-cycle (Fig. 4) adds three further features to the cycle seen in the hookworm.

- The female may dispense with the male and lay fertile eggs parthenogenetically in the small bowel.
- The rhabditiform larvae hatch in the human bowel and are passed in the faeces to become infective filariform larvae or free-living adults in the soil.
- Rhabditiform larvae may become filiform larvae while still in the bowel and cause **autoinfection** through the small bowel mucosa or perianal skin.

Pathogenesis is by allergy in skin and lungs, by local bowel damage and invasion, and by disseminated infection in the immunocompromised patient (p. 193). At times, this is complicated by septicaemia from accompanying intestinal bacterial pathogens.

Clinical syndromes. Symptoms are itch at skin entry sites, allergic pneumonitis, abdominal symptoms, and disseminated infection in immunocompromised patients.

Confirmatory tests. Microscopy of repeated stool specimens or duodenal aspirate shows the filariform larvae, about 200–300 µm long (Fig. 3), with a longer oesophagus than hookworm larvae.

Chemotherapy. Thiabendazole provides more effective chemotherapy but is more toxic than mebendazole.

Control is enhanced by education, better sanitation, personal hygiene and the wearing of shoes. In endemic areas, patients for immunosuppressive therapy should be screened by at least three stool microscopies.

Intestinal nematodes

- Worm infections are common worldwide, with differing geographic distribution depending on the life cycle.
- Four have obligatory soil cycles, while pinworm and *S. stercoralis* do not, so spread directly person-to-person.
- They cause disease variously by allergy, mechanical obstruction, local bowel damage, blood loss, secondary bacterial infection and dissemination.
- Diagnosis is by stool microscopy, or anal adhesive tape microscopy for pinworm.
- Anthelminthics include mebendazole, pyrantel pamoate and thiabendazole; piperazine is a reserve drug.
- Control is by education, better sanitation, personal hygiene and therapy of infected people, and wearing shoes.

TISSUE NEMATODES

The nematodes (p. 12) infecting the blood-stream, lymphatics and tissues are considered in four groups

- filarial worms
- Guinea worm
- worms causing larva migrans
- *Trichinella spiralis*.

FILARIAL WORMS

Wuchereria bancrofti and *Brugia malayi*

These are sheathed filariae having many similarities and are considered together. They occur in tropical Africa and Asia, the Pacific and South America, are transmitted by mosquito bites and cause classical Bancroftian filariasis. They are coiled thread-like worms, 4–10 cm long. The life cycle is shown in Figure 1.

Infection is shown by detecting microfilariae in blood films obtained at night, except for the diurnal Pacific infection.

Clinically, hypersensitivity reactions cause lymphangitis and fever, and obstruction of the lymphatic circulation may give rise to elephantiasis (Fig. 2). Diethylcarbamazine can be used cautiously but there is no cure. Control is by decreasing mosquito bites.

Loa loa (African eye worm)

Loa loa is found in West and Central Africa. It is spread from infected humans by the bite of *Chrysops* spp. mango flies. The life cycle is similar to that of *W. bancrofti* (Fig. 1) but with several differences:

- insect host bites in the day
- larvae burrow through skin when deposited on it rather than being directly injected by the insect
- microfilariae (*Microfilaria diurna*) appear late (3–4 years) and are diurnal (present during the daytime)
- adult worms move in subcutaneous tissues near tendons (and across the front of the eye) for up to 20 years after infection (p. 99)
- allergic reactions to filarial toxins cause 'Calabar swellings' lasting 3 days on the hands, forearms or elsewhere.

Blood films show diurnal microfilaria, in adults more than children. Eosinophilia is usual, and serology is usually positive.

Chemotherapy is diethylcarbamazine, with or without surgery, and control is by education and avoiding fly bites.

Onchocerca volvulus (River blindness worm)

Onchocerciasis, leading to 'river blindness', affects about 25 million people in specific areas in West and Central Africa and Central and South America. The disease is transmitted by the bite of blood-sucking black flies, especially *Simulium damnosum*.

The life cycle, with stages in fly larvae and human, follows the familiar pattern. The adults live mainly in subcutaneous nodules, and the microfilariae migrate to various parts of the body. They rarely enter the blood. In heavy infections, they invade the eye causing blindness. The diagnosis is made by finding microfilariae in skin snips or in the eye, or adult worms in biopsies. Eosinophilia is common.

The classical clinical traid is dermatitis, nodules and keratitis (p. 99). Ivermectin is replacing older forms of chemotherapy. Control is by fly control programmes and treatment.

DRACUNCULUS MEDINENSIS (GUINEA WORM)

The guinea worm is found in Africa, the Middle East and South Asia and infection is by drinking water containing infected crustacea. It invades soft tissues, usually in the leg. The female is 60–90 cm long (rarely 120 cm), only 2–3 mm thick and is mainly a huge uterus packed with up to 3 million embryos!

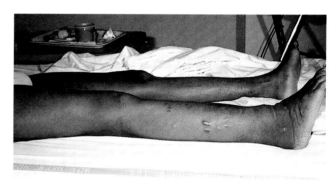

Fig. 2 **Filarial elephantiasis.**

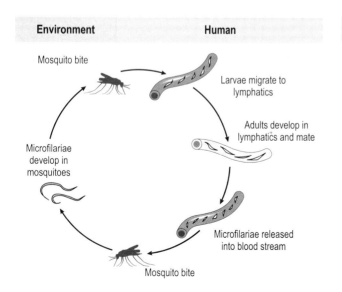

Fig. 1 **Life cycle of the filariasis worms.**

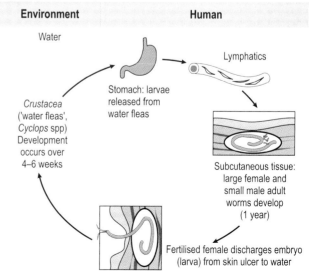

Fig. 3 **Life cycle of *Dracunculus medinensis* (Guinea worm).**

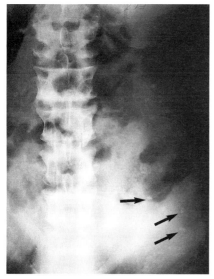

Fig. 4 **Dracunculiasis infection (calcification).**

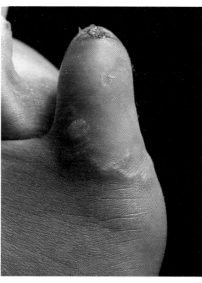

Fig. 5 **Cutaneous larva migrans.**

Ancylostoma braziliense (dog/cat hookworm)

This hookworm (and other animal hookworms) sometimes infects humans, particularly children. The hookworm eggs hatch in soil or sand and the filiform larvae can penetrate skin. The larvae migrate subcutaneously for weeks or months, causing cutaneous larva migrans (p. 176), with pruritis on infection ('ground itch'), then erythema and vesicles moving 1–2 cm daily, ('creeping eruption'). Diagnosis is usually clinical, rarely by biopsy, and treatment is thiabendazole.

Toxocara canis and *T. cati* (dog and cat ascaris)

Infection is by ingestion of ascarid eggs from faecally contaminated soil, especially by children. The eggs hatch in the human gut producing larvae that penetrate the gut wall to the blood stream and migrate into various tissues to produce visceral larva migrans (p. 133). Pathology includes haemorrhage, necrosis and granulomata, especially in lungs, liver and eyes, producing pneumonitis, hepatitis and retinitis.

Diagnosis is clinical and confirmed serologically. Infected pets are found by stool tests. Treatment is by mebendazole, thiabendazole or diethylcarbamazine. Control is by treatment of pets and disposal of animal faeces.

The life cycle is shown in Figure 3. Diagnosis is made by detecting embryos in washings of the ulcer, and X-rays (Fig. 4) with or without radio-opaque injection can show the adult. Eosinophilia is frequent. A skin ulcer with a worm visible at the base is very distinctive (pp. 176–7).

Chemotherapy by niridazole reduces inflammation and assists surgical removal. Classically the ulcer is wetted daily and the worm extracted by winding it gradually round a stick. However, septic complications are common. Control depends on education and clean water.

LARVA MIGRANS (CUTANEOUS OR VISCERAL)

Some nematodes, primarily of animals, can infect humans 'accidentally'; because the human is an abnormal host, the normal life cycle cannot be completed and disease is caused by the migrating larvae (Fig. 5). *Dirofilaria immitis*, the mosquito-borne dog heart worm, is one example; others are *Angiostrongylus, Gnathostoma* or *Anisakis* spp. Three others follow:

TRICHINELLA WORM

Trichinella spiralis (pig threadworm) infects humans when they eat incompletely cooked pork or bear meat containing encysted larvae (Fig. 6). Again, the life cycle is incomplete in the human 'dead-end' host.

Diagnosis is initially clinical but may be confirmed by finding encysted larvae in muscle biopsies or infected pork. Antibody rises after 3–4 weeks are detected by serology in some countries.

Clinical syndromes include abdominal symptoms, fever, periorbital oedema, splinter haemorrhages, myalgia and life-threatening cerebral and cardiac involvement (p. 185).

Chemotherapy is with thiabendazole or mebendazole with or without steroids. Control prevents pig infections from infected food, and cooking pork kills the encysted larvae.

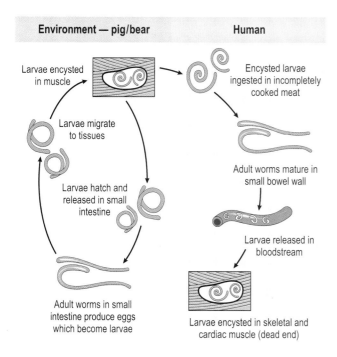

Fig. 6 **Life cycle of *Trichinella spiralis* (pig threadworm).**

> **Tissue nematodes**
> - Filarial infections are diagnosed by detecting microfilariae in blood films (taken at the right time of day), or larvae, or adult worms in ulcers or tissue biopsies.
> - Symptoms arise from hypersensitivity reactions and secondary blockage of lymphatics or tissue damage.
> - Treatment is often unsatisfactory, whether by surgery or chemotherapy.
> - As filarial infections are spread via intermediate hosts (often biting insects) control can be attempted at this step.
> - Dead-end infections result from accidental infection of humans with larvae or eggs from soil or larvae encysted in meat. Clinical symptoms result from cutaneous or visceral larva migrans or encysting larvae in skeletal and cardiac muscle.

CESTODES

Cestodes are flat segmented worms (tape-worms), and humans can be infected in two ways.

- As **definitive** hosts of the adult worm in the gut after ingesting encysted larvae in incompletely cooked meat; this produces mild or moderate intestinal disease in all but *Echinococcus*.
- As **intermediate** hosts of larvae in tissues after ingesting eggs in faecally contaminated food; this produces tissue disease from encysted larvae (e.g. cysticercosis) with space-occupying lesions (SOL).

TAENIA SAGINATA (BEEF TAPEWORM)

T. saginata is a common intestinal tape-worm found wherever raw or under-cooked infected beef is eaten and human faeces contaminate pastures. It does not occur in Hindus or vegetarians.

The life cycle is illustrated in Figure 1. The egg, indistinguishable from *T. solium* eggs, is round, 40 μm across and has three pairs of internal hooklets (Fig. 2). The worm can exceed 10 metres in length, and the segments (proglottids) each contain nervous, muscular, excre-tory and male and female genital sys-tems, the uterus having 15–30 lateral branches (*T. solium* has 7–12).

Clinical syndromes. Distress is either abdominal from the worm, or mental from seeing the proglottids in faeces or the bed!

Confirmatory tests. Stool microscopy for the egg confirms a tapeworm infec-tion. The tiny scolex (head) or the 1 × 1.5 cm proglottid confirms the species.

Chemotherapy. Praziquantel or niclosamide are effective orally. After treatment to expel the worm the faeces should be examined to ensure that the head is expelled, otherwise the worm will regenerate.

Control is by education, sanitation, meat inspection and proper cooking of beef (or not eating beef).

TAENIA SOLIUM (PORK TAPEWORM)

Taenia solium is a large intestinal tape-worm, common wherever raw or under-cooked infected pork is eaten, and human faeces contaminate pastures. It does not occur in Muslims or vegetarians.

The egg is indistinguishable from *T. saginata* eggs (Fig. 2). The worm

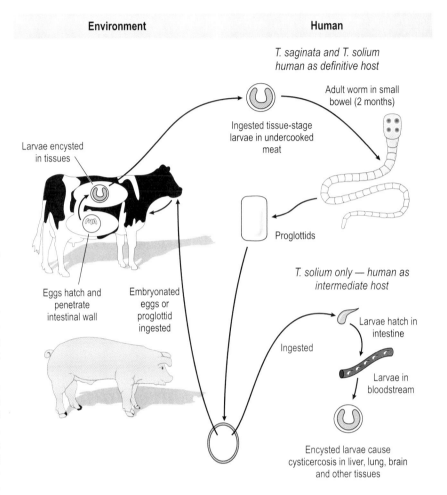

Environment

Larvae encysted in tissues

Eggs hatch and penetrate intestinal wall

Embryonated eggs or proglottid ingested

Human

T. saginata and T. solium human as definitive host

Adult worm in small bowel (2 months)

Ingested tissue-stage larvae in undercooked meat

Proglottids

T. solium only — human as intermediate host

Ingested

Larvae hatch in intestine

Larvae in bloodstream

Encysted larvae cause cysticercosis in liver, lung, brain and other tissues

Fig. 1 **Life-cycles of** *Taenia saginata* **and** *solium.*

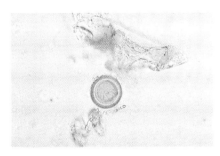

Fig. 2 *Taenia* **spp. egg from faeces.**

can exceed 6 metres in length, and the segments (proglottids) are similar to *T. saginata* except that the uterus has 7–12 lateral branches.

Life cycle and pathogenesis. The normal life cycle in which humans are the definitive host occurs exactly as with *T. saginata*. However, the larval stage, which would normally occur in the pig, can develop in humans if the eggs are ingested in faecally contaminated water or plants. Larvae from the egg go through the bowel wall to the blood stream, thence to brain, muscle, eye and other tissues. This results in serious disease

from the encysted larval tissue stage (**cysticercosis**) in brain (p. 91), liver or other tissues. Eventual degeneration of the cyst causes an inflammatory reaction with worsening of symptoms in the human dead-end host.

Clinical syndromes. Abdominal symp-toms are non-specific. CNS symptoms from encysted larvae include fits, strokes, and cranial nerve or visual impairment.

Confirmatory tests. Intestinal worms are diagnosed as for *T. saginata*. Cysti-cercosis is diagnosed by X-ray, CT or brain scans, MRI or by surgery.

Chemotherapy. Praziquantel is the first effective drug for cysticercosis. Niclosa-mide is effective for intestinal tapeworms.

Control is by education, sanitation, meat inspection and proper cooking of pork (or not eating pork).

ECHINOCOCCUS (HYDATID WORM)

The hydatid worm does not use humans as definitive hosts. Humans become (dead-end) intermediate hosts for the

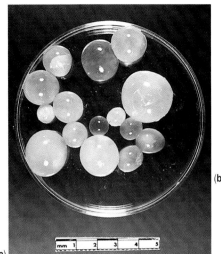

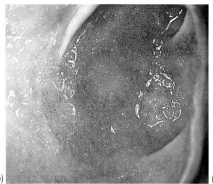

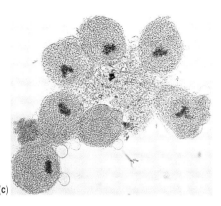

(a)

(b)

(c)

Fig. 3 *Echinococcus granulosus.* (a) Hydatid brood capsules removed from the cyst; (b) 'hydatid sand' within the cyst; (c) seven hydatid scolices within the laminated membrane (latter not visible).

encysted larvae after being infected by eggs in canine faeces. Canines are the definitive reservoir host and herbivores, such as sheep and cattle, are the usual intermediate hosts. Human disease is most common in sheep-rearing countries.

Two species infect humans: the more usual *E. granulosus* (Fig. 3), forming hydatid cysts, and the invasive *E. multilocularis*, spreading like a malignancy.

The egg is very similar to *Taenia* eggs. The adult worm is small, only 3–9 mm long, with only four proglottids.

Life cycle and pathogenesis. The life cycle is like that of *T. solium* but only the intermediate stage occurs in humans. Larvae hatch from eggs in the human gut and travel in the portal vein to the liver. Those not lodging there travel to the lungs, and, if still not filtered out, to the brain and other tissues. The larvae of *E. granulosus* develop to hydatid cysts with a host-derived laminated membrane and *inner* germinal layer, giving rise to brood capsules containing daughter cysts each containing an embryo larva as a scolex with hooklets (Fig. 3). In *E. multilocularis*, the germinal membrane is *outermost* so no cyst is formed, and the disease spreads like a malignancy.

Confirmatory tests. Serology particularly for arc 5 in a gel immunodiffusion test is highly specific; latex agglutination and other serology is used for screening or supplementary testing. X-rays, CT and liver and brain scans are used for localisation. Old liver cysts calcify.

Clinical syndromes. Symptoms are usually caused by a SOL in liver (p. 146), brain (p. 91), lung (p. 120) or other tissues.

Chemotherapy. Albendazole is more active than mebendazole but surgery is still often necessary, being very careful not to spill infective cyst contents.

Control. Treating adult worms in farm herbivores, preventing dogs from eating raw, infected animal (e.g. sheep) viscera, and hand washing after dog or soil contact are important means of control.

HYMENOLEPIS NANA (DWARF TAPEWORM)

Dwarf tapeworm (2–4 cm long) occurs worldwide especially in children. Ingested eggs develop to the tissue stage (cysticercoids) in the mucosa of the small bowel, releasing adults back into the lumen. Eggs released from proglottids can autoinfect the same host or be shed to reinfect via ingestion. Autoinfection leads to very high worm loads (hyperinfection).

Only local gut symptoms occur and stool microscopy identifies eggs (Fig. 4).

Chemotherapy. Praziquantel is the first choice in chemotherapy, with niclosamide as alternative.

Control depends on education, sanitation and handwashing after defecation.

DIPHYLLOBOTHRIUM LATUM (FISH TAPEWORM)

Fish tapeworm is unusual in having two intermediate hosts, and causing human disease by preventing absorption of vitamins, especially B_{12}. It occurs wherever infected raw (or smoked) fish are eaten.

The adult is huge, 3–10 metres long with 3000–4000 proglottids. The egg is operculated (with a cap over an opening), 40×70 µm. The eggs release ciliated larvae in water that are eaten by and develop in crustacea ('water fleas', *Cyclops* spp.) The larvae then encyst in the muscles of freshwater fish that eat the water fleas.

Pathogenesis in the human gut is mechanical by the worms' size and also by interfering with vitamin absorption.

Clinical syndromes. Symptoms include abdominal distress, passage of proglottids, and pernicious anaemia from vitamin B_{12} deficiency. Stool microscopy can show the characteristic egg or proglottids.

Chemotherapy. Niclosamide is the drug of choice in chemotherapy; praziquantel is the second choice.

Control depends on sanitation, cooking fish and treating infections.

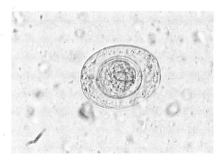

Fig. 4 *Hymenolepis nana* cyst.

Cestodes

- Cestodes infect humans either by ingestion of encysted larvae to produce intestinal worms, or by ingestion of eggs to produce large tissue masses from encysted larvae.
- Diagnosis is usually by stool microscopy and imaging techniques (X-ray, CT, etc.), sometimes by serology.
- Treatment is usually praziquantel or niclosamide, with surgery for tissue masses.
- Control is by sanitation, education and proper cooking.

TREMATODES

Trematodes (flukes) are flat, leaf-shaped worms. They have an oral muscular sucker and an incomplete digestive system. Schistosomes have separate sexes but other trematodes are hermaphrodites, i.e. have both male and female sexual organs in one body.

The life cycle of all trematodes occurs in part outside the human body (Fig. 1), and all have reservoir hosts and a first intermediate host in a snail (or other mollusc). Several have a second intermediate host stage in fishes (*Opisthorchis* spp.) or crabs or crayfish (*Paragonimus* sp.). Humans are infected by ingestion of larva, except schistosomes, which infect through the skin. The adults, as their common names indicate, live in bowel, bile duct, liver, lungs or blood vessels.

OPISTHORCHIS SINENSIS (CHINESE LIVER FLUKE)

O. (previously *Clonorchis*) *sinensis* and related species are found in China, Japan and nearby countries. Humans are infected by eating uncooked fish infected with the larvae (metacercariae). Eggs in human faeces are ingested by the first intermediate host (fresh-water snails) which are eaten by the second intermediate host, fish. Reservoir hosts include dogs, cats and fish-eating animals.

The egg is oval, operculated and small, $15 \times 25\ \mu m$ (Fig. 2). The larvae (cercariae) are $500 \times 100\ \mu m$, the adult is about 4×25 mm.

Associated clinical syndromes correlate with the life cycle (Fig. 1), and range from no symptoms to abdominal pain, diarrhoea and fever, to obstructive jaundice, cholecystitis, liver abscesses or bile duct carcinoma.

Laboratory identification is by microscopy of stools, which shows the distinctive eggs. Praziquantel is the drug of choice in treatment. Control of the parasite depends on education, sanitation and changed cooking or eating habits.

FASCIOLA HEPATICA (SHEEP LIVER FLUKE)

F. hepatica and related species are found where sheep and the snail hosts coexist: Japan, China, Russia, Egypt and South America. Humans are infected by eating water plants, like watercress, contaminated with the larvae (metacercariae). Reservoir hosts include sheep, cattle and humans.

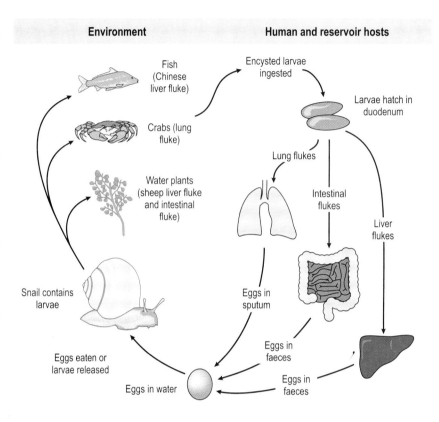

Fig. 1 **Life cycle of flukes.**

The egg is oval, operculated and large, 75×140 mm. The larvae (cercariae) are 0.7 mm, the adult is about 2×12 mm.

Associated clinical syndromes again correlate with the life cycle: initially there is tender hepatomegaly from larval migration; adults in the bile ducts cause cholangitis and obstruction, and finally adults in the liver cause necrosis (liver rot), bacterial infection and cirrhosis.

Stool microscopy shows the eggs, which are indistinguishable from *F. buski*.

Praziquantel is the drug of choice; bithionol is an alternative. Control depends on education, sanitation and avoiding uncooked water plants.

FASCIOLOPSIS BUSKI (GIANT INTESTINAL FLUKE)

F. buski and related species (e.g. *Heterophyes, Metagonimus* spp.) are found where the snail hosts exist: China, South East Asia and India. Humans are infected by peeling water plants like water chestnuts contaminated with the larvae (metacercariae). Reservoir hosts include pigs, dogs and humans.

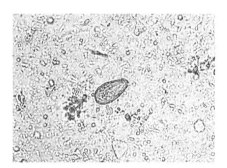

Fig. 2 *Opisthorchis sinensis* ovum.

The egg is oval, operculated and large, 75×140 mm. The larvae (cercariae) are 0.7 mm, the giant adult is about 3×12 cm.

The intestinal fluke produces abdominal pain, diarrhoea, malabsorption and even obstruction.

Stool microscopy shows the eggs, indistinguishable from *F. hepatica*. Eosinophilia is common, as in all fluke infections.

Praziquantel is the drug of choice in treatment; niclosamide is the alternative. Control depends on education, sanitation and avoiding uncooked water plants. The snails and definitive hosts are difficult to control.

PARAGONIMUS WESTERMANI (LUNG FLUKE)

P. westermani and related species are found where the snail hosts exist and the crab or crayfish hosts are eaten uncooked: Asia, India, Africa and Latin America. Humans are infected by eating uncooked crabs, crayfish or reservoir hosts infected with the larvae (metacercariae).

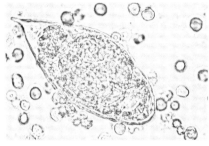

Fig. 3 *Schistosoma haematobium* ovum seen in faeces.

Reservoir hosts include pigs, boars and monkeys.

The egg is oval, operculated and quite large, 60 × 100 mm. The larvae (cercariae) are 70 × 200 μm; the adult is about 7 × 18 mm.

Migration of the larvae (adolescercariae) through the diaphragm, pleura and lung to the bronchioles forms cystic cavities where the adult fluke develops. This results in pulmonary cavitation and bronchiectasis, bronchitis, and pleural effusions (p. 121). Spread to the spinal cord, brain and other tissues can occur, causing fits or paralysis.

Sputum microscopy shows the eggs. Eosinophilia is common, as in all fluke infections, and chest X-rays are abnormal.

Chemotherapy involves praziquantel as the drug of choice; bithionol is the alternative. Control depends on education, sanitation and avoiding uncooked crabs and crayfish. The snails and definitive hosts are difficult to control.

SCHISTOSOMA SPP. (SCHISTOSOMES)

The three major pathogenic species of blood flukes are *S. haematobium*, causing vesicular (bladder) schistosomiasis with haematuria, and *S. mansoni* and *S. japonicum*, causing intestinal schistosomiasis with hepatosplenomegaly. They are found where humans and the snail hosts co-exist (Table 1). Humans are infected by larvae (cercariae) penetrating the skin. Reservoir hosts include primates, rodents, domestic animals and humans.

The diagnostic differences between the eggs (Fig. 3) of the three species are listed in Table 1. Unlike other trematodes, all have no operculum. The larvae (cercariae) are 0.8 mm, the adult is about 1 × 15 mm.

The life cycle has one intermediate host, freshwater snails (Fig. 4).

Symptoms again correlate with the life-cycle. Initially there is skin itch and allergy on skin penetration, then hepatitis, followed by either bladder symptoms including haematuria from *S. haematobium*, or abdominal symptoms including dysentery, portal hypertension and gross hepatosplenomegaly with the two other, mesenteric vein species (p. 131).

Eggs may spread to cause pulmonary, cerebral and spinal cord granulomas. Carcinoma of the bladder is common with chronic *S. haematobium* infection.

Stool or urine microscopy shows the distinctive eggs (Fig. 3).

Praziquantel is the drug of choice; oxamniquin is an alternative in *S. mansoni* infections. Control of infection is difficult and depends on education, sanitation, mass treatment or molluscicides against the snails in some circumstances. Vaccine development continues.

Table 1 Different characteristics of three major schistosomes

Characteristic	S. haematobium	S. mansoni	S. japonicum
Distribution	Africa, India, Middle East	Africa, Arabia, S. America, West Indies	Japan, China, SE Asia
Reservoir hosts	Monkeys	Primates/rodents	Many animals
Egg size (μm)	80–180 × 60	110–170 × 60	70–100 × 55
Egg spine	Terminal	Lateral	Tiny lateral
Veins infected	Urogenital	Inferior mesenteric	Both mesenteric
Egg excretion	Urinary	Faeces	Faeces
Chronic disease	Urinary (late cancer)	Hepatosplenic (spinal, lung)	Hepatosplenic (cerebral)

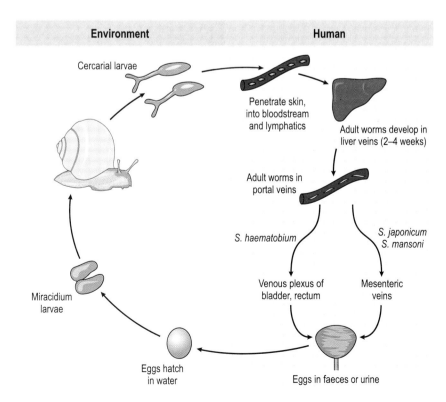

Fig. 4 Life cycle of schistosomes.

Trematodes

- Trematodes infect humans either by ingestion of encysted larvae to produce intestinal, bile duct or lung flukes (adult worms) or by skin penetration to produce intravascular schistosomiasis.
- Diagnosis is usually by stool, urine or sputum microscopy, sometimes assisted by imaging techniques (X-ray, CT, etc.), or serology.
- Treatment is by praziquantel or specialised alternatives.
- Control is by sanitation, education and proper food preparation or cooking, and minimising water and snail contact.

ACUTE MENINGITIS

Acute meningitis is acute inflammation, developing over a few hours or days, of the meninges (Fig. 1), particularly the arachnoid mater overlying the cerebrum, the adjacent area of which may also be inflamed (cerebritis). Meningitis is usually spontaneous in normal hosts but can occur after trauma, surgery, or delivery, or in the immunocompromised host. Organisms enter the CNS by three routes:

- via the blood stream after inhalation, through bites (Rickettsia) or through the placenta (congenital infections)
- via the olfactory bulb (amoebic infections)
- via direct inoculation (surgery or trauma).

CAUSATIVE ORGANISMS

The important causative agents are shown in Table 1; the first three account for three-quarters of all cases and have several virulence factors in common including a polysaccharide capsule.

Meningitis usually occurs in single episodes but can occur in epidemics in crowded conditions or where asymptomatic carriers are involved.

CLINICAL FEATURES

Headache, fever, clouding of consciousness and neck stiffness occur in all types of acute meningitis. Vomiting is common as intracranial pressure increases, which can cause VI nerve pareses (Fig. 2) as a false localising sign.

Certain **clinical clues** may suggest the causative organism:

- petechial rash (Fig. 3) and shock with the meningococcus, rarely with the pneumococcus or *S. aureus*
- otitis media with *S. pyogenes, S. pneumoniae* or *H. influenzae*
- rash in an endemic area of Rocky Mountain spotted fever (RMSF)
- myalgia and occupational exposure with leptospirosis
- skin rash of erythema chronicum migrans (ECM) in Lyme disease.

Special situations may also suggest causative organisms.

- CSF leak from ear or nose after closed cranial trauma: this is pneumococcal in about 75% of cases and is usually prevented by chemoprophylaxis
- CSF shunt for hydrocephalus: usually *S. epidermidis* (55%) or *S. aureus* (25%) implanted with the shunt
- immunocompromised patient: *Listeria, Pseudomonas* and other Gram-negative rods, fungi such as *Cryptococcus*, and parasites like *Toxoplasma*
- neonatal: *E. coli* and group B streptococci cause about 75%.
- postoperative: *Pseudomonas* and other Gram-negative rods can be more common than staphylococci
- recurrent meningitis:
  — underlying anatomic abnormality
  — parameningeal focus

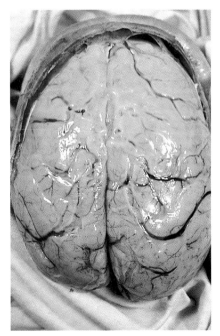

Fig. 1 **Pneumococcal meningitis.** Thick pus is seen over the vertex of this brain at autopsy.

Table 1 **Bacterial and 'aseptic' meningitis**

Causative organism (% of all cases)	Clinical clues	Chemotherapy	Control and prophylaxis
PYOGENIC MENINGITIS			
Most common			
Haemophilus influenzae	Children, severe	Ceftriaxone	Rifampicin, vaccine
Neisseria meningitidis	Rash and shock	Penicillin	Rifampicin
Streptococcus pneumoniae	Severe	Penicillin or ceftriaxone	Vaccine
Common bacteria			
Gp A streptococci	Otitis media	Penicillin	Penicillin
Gp B streptococci	Neonatal	Ceftriaxone	Obstetric care
Gram-negative rods	Neonatal, immunocompromised	Ceftriaxone +/– aminoglycoside	Obstetric care
Staphylococci	Trauma, surgery	Flucloxacillin	Operative
ASEPTIC MENINGITIS			
Uncommon bacteria			
Actinomyces and *Nocardia* spp.	Brain abscess	Penicillin or sulphonamide	Dental
Anaerobes	Abscess or focus	Metronidazole	Surgery
Brucella spp.	Encephalitis also	Tetracycline and streptomycin	Animal vaccination
Listeria monocytogenes	Immunocompromised	Ampicillin	–
M. tuberculosis	Rarely acute	Triple	BCG
Mycoplasma pneumoniae	Respiratory also	Tetracycline	–
Rickettsia (RMSF)	Rash and shock	Tetracycline	Avoid vector
Spirochaetes			
Borrelia burgdorferi	Very variable	Tetracycline	Avoid vector
Treponema pallidum	Rarely acute	Penicillin	Safer sex
Leptospires	Systemic, myalgic	Penicillin	Animal control
Viral	Usually mild	None	(Vaccines)
Fungal			
Cryptococcus neoformans	Immunocompromised	Amphotericin B and fluconazole	Fluconazole
Parasitic			
Toxoplasma gondii	Immunocompromised	Pyrimethamine and cotrimoxazole	Pyrimethamine and sulphonamide
Amoebic			
Naegleria fowleri	Swimming	Amphotericin B and fluconazole	Chlorination
Acanthamoeba spp.	Immunocompromised	Sulphonamide and flucytosine	–

— immunocompromised patient
— unrelated viral meningitis attacks (rare)
— Mollaret's meningitis: benign, unknown cause.

Important **differential diagnoses** which need different treatment include:

- cerebral abscess
- parameningeal focus: in middle ear, mastoid or sinuses, or subdural, extradural or paraspinal abscess, sometimes with adjacent osteomyelitis
- suppurative thrombophlebitis.

CONFIRMATORY TESTS

The pathogen involved is suggested by the clinical features and confirmed by urgent lumbar puncture for cell count, Gram stain and culture, bacterial antigens, and measurement of glucose and protein levels in CSF, even if empiric antibiotics have been given. Important findings include:

- Cell count above 1500 with >60% neutrophils favours bacterial meningitis, and one below 1000 with <10% neutrophils favours viral or partially treated bacterial meningitis. There is overlap, especially in early disease.
- Gram stain of centrifuged CSF shows organisms in about 80% of patients with untreated bacterial meningitis, and antigen detection (usually by latex agglutination, including for *Cryptococcus*) is specific.
- Culture, including chocolate agar for *Haemophilus*, is especially useful if Gram stain is unhelpful, and for sensitivity tests.
- Protein is highest and glucose lowest in bacterial meningitis.

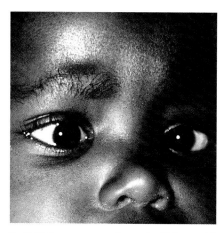

Fig. 2 **Sixth nerve pareses from raised intra-cranial pressure caused by *Haemophilus* meningitis.**

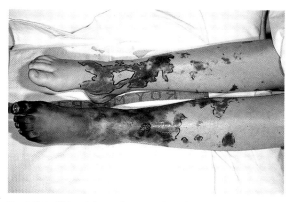

Fig. 3 **Petechial rash in meningococcal meningitis.**

Blood cultures and other cultures depending on special clinical situations (see above) should be done. Imaging by CT scan or MRI is necessary if focal neurologic signs are present, suggesting an abscess or other focal lesion (pp. 90–91). Special stains and serology are needed for specific pathogens.

A repeat LP is necessary in 6–12 hours if the first was non-specific, antibiotics were withheld and the patient worsens.

Aseptic meningitis

Aseptic meningitis is meningitis with increased cells but no growth in usual CSF cultures, and is caused by:

- unusual bacteria (Table 1)
- non-bacterial causes: viral, fungal, amoebic or parasitic
- partially treated bacterial meningitis
- chemical, carcinomatous or other non-infective causes.

TREATMENT

If the microbial diagnosis is known, the drugs of choice are listed in Table 1.

Microbial diagnosis unknown. Chemotherapy regimens are chosen by the clinically suspected cause (Table 1); usually a third-generation cephalosporin is initiated if the pathogen is not known except for:

- CSF shunt-related infection: vancomycin is used initially
- immunocompromised patients need specific rapid investigations (e.g. cryptococcal antigen), specialist advice and initial broad therapy, which may include amphotericin and fluconazole, or sulphas and pyrimethamine.
- neonatal treatment usually involves the use of gentamicin and/or ampicillin until the microbial diagnosis is known
- postoperative infection: flucloxacillin is added to ceftriaxone for better antistaphylococcal cover.

Other management. This often includes dexamethasone and monitoring of intracranial pressure, or removal of infected shunts. It is now rare to insert a reservoir to give intraventricular aminoglycoside, vancomycin or amphotericin B.

Control and prevention. Rifampicin should be given to the close contacts and the patient with *Haemophilus* (for 4 days) or meningococcal meningitis (2 days) to prevent secondary cases. Routine childhood vaccination is now available in developed countries against *H. influenzae* type b, and meningococcal vaccine is available for high-risk situations.

Acute meningitis

- The commonest causative organisms are *Neisseria meningitidis*, *Streptococcus pneumoniae* and *Haemophilus influenzae*, but other causes are other bacteria, fungi, parasites and viruses.
- The pathogen is suggested by certain clinical clues and special situations, and confirmed by lumbar puncture.
- Headache, fever, clouding of consciousness and neck stiffness occur in all types of acute meningitis.
- Clinical clues include rash, myalgia, shock, or a source such as otitis media or respiratory infection.
- Special situations include CSF leak, CSF shunt, immunosuppression, neonatally, or postoperatively.
- Urgent lumbar puncture is essential and guides treatment.
- Chemotherapy depends on the organism, and empiric chemotherapy in the absence of an organism depends on the clinical clues and special situation.

CHRONIC DIFFUSE CNS INFECTIONS

Chronic infections of the CNS are characterised by development over weeks or months. They may be classified into *diffuse* infections, meningitis and meningo-encephalitis, described here, and *focal* infections (following topic).

Diffuse infections are usually bacterial or viral in origin; three are three main non-viral causes:

- *Cryptococcus neoformans*: cryptococcosis
- *Treponema pallidum*: syphilis
- *Mycobacterium tuberculosis*: tuberculosis.

Other less common pathogens are *Borrelia burgdorferi*, *Trypanosoma* spp. (pp. 76–7, 131) and, rarely, *Angiostrongylus cantonensis* or *Brucella* spp.

CLINICAL FEATURES

In chronic meningitis, as with acute meningitis, headache, fever, clouding of consciousness and neck stiffness are usual. In meningo-encephalitis, cerebral disturbances occur. Intractable vomiting and visual changes signal raised intracranial pressure.

Differential diagnosis. This includes cerebral tumour, subdural or subarachnoid haemorrhage, multiple sclerosis, metabolic encephalopathies or autoimmune diseases.

Confirmatory tests. Lumbar puncture (LP) is the key to diagnosis (but should not be done if a focal infection is indicated by CT or MRI). The fluid is tested for cell counts, staining, culture, serology, plus CSF and serum glucose and protein. Mantoux and chest X-ray (tuberculosis) may be helpful.

SPECIFIC DISEASES

Cryptococcosis

The fungus *Cryptococcus neoformans* (p. 64), causes both meningitis and multiple abscesses in the CNS (Fig. 1). It also causes pulmonary, bone and skin lesions, and disseminated disease (cryptococcaemia) particularly in the immunocompromised, e.g. in AIDS, when it may be more acute.

Clinically, in CNS infection, fever and headache are not marked, while changes in mentation, behaviour and memory are prominent. Cranial nerve involvement can cause visual impairment or facial weakness; other focal lesions are rare. As with any cause of raised intracranial pressure, VI nerve paresis may be a false localising sign (Fig. 2).

Confirmation is by LP, which shows the characteristic capsulated fungi; antigen in CSF or blood is detected by latex agglutination.

Chemotherapy is usually by amphotericin B and fluconazole, and continuing secondary prophylaxis by lower-dose fluconazole is needed in the immunocompromised.

Syphilis

Syphilis is a systemic spirochaetal disease caused by *Treponema pallidum* (p. 52). In **primary syphilis**, asymptomatic invasion of the CNS may occur and current treatment aims to treat this. In **secondary syphilis**, acute 'aseptic' meningitis may occur. In **latent syphilis**, by definition there are no symptoms and CSF syphilis serology is negative, though serum is positive.

Tertiary neurosyphilis. There are five categories:

1. *Asymptomatic* with positive CSF serology.
2. *Meningo-vascular* with chronic meningitis or meningo-encephalitis from endarteritis obliterans of meninges, brain or spinal cord.
3. *General paresis of the insane* (GPI) with cerebral cortical destruction causing changes in *p*ersonality, *a*ffect, *r*eflexes

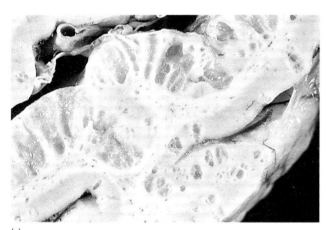

(a)

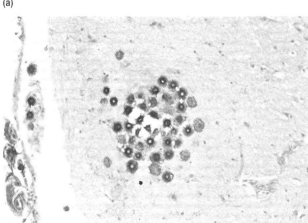

(b)

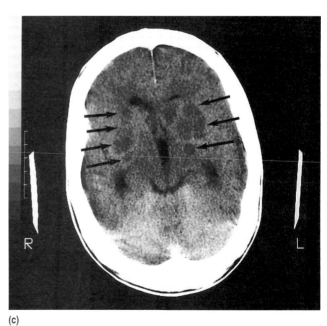

(c)

Fig. 1 *C. neoformans:* cerebral infection.
(a) Multiple gelatinous masses;
(b) Organisms stained blue in brain tissue;
(c) CT scan.

(hyperactive), *eye* (Argyll Robertson pupils, reacting to accommodation, not light), *sensorium* (delusions, hallucinations), *intellect* and *speech*.

4. *Tabes dorsalis* with spinal posterior column and dorsal root destruction causing bladder dysfunction, Romberg's sign, ataxia, impotence, and neuropathy (both cranial and peripheral) with loss of vibration and position sense.

5. *Gumma*, a localised necrosis, with symptoms and signs dependent on its position.

Diagnosis is clinical and serological, LP being essential. Treatment is by at least 3 weeks i.v. penicillin. Ceftriaxone or tetracycline are unproved alternatives in penicillin allergy. Prevention is by treatment of primary or secondary disease.

Tuberculosis

Tuberculosis is a chronic bacterial disease (Fig. 3) caused by *Mycobacterium tuberculosis* (p. 54). It causes a chronic meningitis with the usual symptoms or, rarely, localised masses called tuberculomata. Rarely, the meningitis is acute. Spread can cause cerebral, cerebellar or spinal symptoms and signs.

Diagnosis depends on LP; the usual findings are 100–1500 cells, mainly lymphocytes, elevated protein and low sugar; bacteria are only visible on AF stain about one third of the first LPs, which must be repeated if doubt remains.

Treatment is by intensive triple or quadruple therapy including isoniazid and rifampicin.

Eosinophilic meningitis

Eosinophilic meningitis is an unusual disease found in southeast Asia and the Pacific, and is caused by the parasite *Angiostrongylus cantonensis*, the rat lungworm, which has an unusual life cycle, from rats to slugs and snails to rats. Humans are accidentally infected by ingesting raw snails or contaminated raw vegetables. Human infection is often asymptomatic or the ingested larvae migrate to the CNS, causing severe headache, vomiting, neck stiffness and often cranial nerve lesions. Unusually for meningitis, fever is minimal or absent.

LP is unusual because the pleocytosis (500 or more) is eosinophilic (10–50%) and lymphocytic. Larvae or young adult worms are seldom found in the CSF, and there is no other specific test to distinguish it from other parasitic infections of the CNS (pp. 90, 91, 93). There is no specific therapy, and most patients recover in 3–8 weeks.

Lyme disease

Lyme disease is a systemic spirochaetal disease caused by *Borrelia burgdorferi* (p. 52). After infected tick bites, CNS involvement occurs, with initial acute meningismus (headache, neck pain and stiffness without demonstrable meningitis on LP). This is followed by fluctuating chronic meningitis or meningoencephalitis, often with cranial or peripheral nerve lesions, rarely myelitis (spinal cord).

LP shows mild (100–300) lymphocytosis. Diagnosis is clinical and serological. Treatment is by i.v. penicillin or ceftriaxone. Control is by avoidance of tick bite in endemic areas.

Brucellosis

Brucellosis is a chronic or relapsing systemic bacterial infection caused by one of the *Brucella* spp. (p. 50). It can cause a chronic meningitis, which gives the usual symptoms, or more commonly meningoencephalitis, where headache and lassitude, cranial nerve lesions and neuropsychiatric symptoms (particularly depression) are common. LP has an unusual picture, with mainly monocytes in the pleocytosis (elevated cell count) of some hundreds, and prolonged culture in CO_2 is positive in about 50%. Serology is helpful but not always diagnostic. Treatment is usually streptomycin (which penetrates poorly into CSF) and tetracycline, while rifampicin in addition may be useful. Control is aimed at infected animals and animal products including cheese.

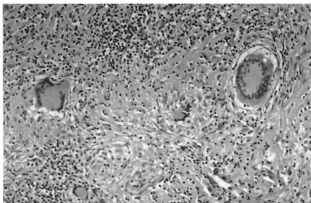

Fig. 3 *M. tuberculosis:* tubercles with giant cells.

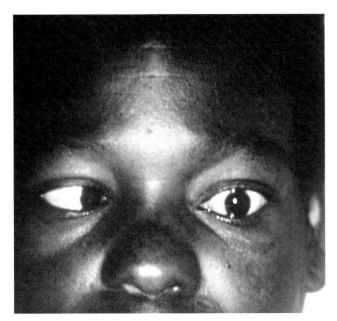

Fig. 2 *C. neoformans:* sixth nerve paresis.

Chronic diffuse CNS infections

- Chronic meningitis and meningoencephalitis are infections developing over weeks or months
- Main non-viral causes are cryptococcosis, syphilis and tuberculosis.
- Chronic meningitis causes headache and vomiting, changes in conscious state and neck stiffness, while meningoencephalitis causes changes in mentation.
- Diagnosis is principally by lumbar puncture showing increased cells (often not neutrophils), with special stains and cultures. Serology and imaging help with specific diagnoses.

CNS ABSCESSES AND OTHER FOCAL LESIONS

Focal infective lesions of the CNS are characterised by forming a mass lesion, also called a space-occupying lesion (SOL). The distinction from meningitis and diffuse infective cerebral lesions (preceding topics) is very important for two reasons, firstly because lumbar puncture is contraindicated with a mass lesion, and secondly because surgery is often needed.

Focal lesions may be classified into **abscesses** (with pus), usually bacterial or fungal and often needing surgery, and **cysts**, usually parasitic and seldom needing surgery.

Causative organisms are shown in Table 1.

Abscesses usually have one of three sources as predisposing factors:

- **parameningeal infection** (40–50%): paranasal sinuses, ear, mastoid or dental infections, with sub- or extradural abscess and/or osteomyelitis; less commonly, face or scalp infections (including head tongs or ICP or fetal monitors) or meningitis
- **distant infection with septicaemic spread** (25%): especially from lungs or with cardiac right–left shunts.
- **head surgery or trauma.**

Parasitic cysts occur from haematogenous spread.

CLINICAL FEATURES

Focal lesions present rather like meningitis or diffuse lesions with headache, fever and changes in conscious state, but with three important distinctions:

- neck stiffness is absent or slight
- focal signs are present
- evidence of raised intracranial pressure (severe vomiting, papilloedema) is much more prominent.

Differential diagnosis is from non-infective focal lesions, including tumour, infarcts or haemorrhage.

CONFIRMATORY TESTS

Lumbar puncture is contraindicated for three reasons: it is potentially dangerous because intracranial pressure (ICP) is usually raised even if papilloedema is not visible; it is seldom helpful, and it may be misleading.

Imaging by CT or MRI is the basis of diagnosis (Fig. 1). If not available, a brain scan (or skull X-ray if a calcified

Table 1 **Major causes of focal CNS infections**

Category and causative organism	Clinical features and incidence	Chemotherapy (usually with surgery)
Bacteria		
Actinomyces or *Nocardia*	Immunocompromised	Penicillin or sulphonamide
Anaerobes	Common, 50–75%	Penicillin or metronidazole
especially *Bacteroides*	Common, 20–30%	Metronidazole
Staphylococci	Postoperative	Flucloxacillin
Streptococci (especially micro-aerophilic)	Frequent 20%	Penicillin
Gram-negative rods	Ear or sinus	Ceftriaxone
Mixed	Common, 40–50%	Ceftriaxone + metronidazole
M. tuberculosis	Tuberculoma rare	Triple therapy
T. pallidum	Gumma very rare	Penicillin
Fungi		
Aspergillosis	Immunocompromised	Amphotericin + itraconazole
Candidiasis	Immunocompromised	Amphotericin + fluconazole
Cryptococcosis	Immunocompromised	Amphotericin + 5-flucytosine or fluconazole
Systemic mycoses	Geographic area	Amphotericin, ? + ketoconazole
Zygomycosis	Diabetes, leukaemia	Amphotericin + azole
Parasites		
Cysticercosis	Cyst, endemic area	Praziquantel
E. histolytica	Abscess, endemic area	Metronidazole
Hydatid disease	Cysts, endemic area	Albendazole
Toxoplasmosis	Immunocompromised	Pyrimethamine + sulphas
Viruses		
E.g. HSV are not discussed		

pineal gland is displaced) can confirm a mass lesion but give less detail.

The causative organism is obtained by aspiration or surgery or deduced from serology. Aspirate or surgically removed tissue is tested by staining, culture and biochemical tests for the suspected pathogen.

BRAIN ABSCESSES

Anaerobes

Anaerobes, such as peptostreptococci and *Bacteroides* spp. (pp. 48, 49), often in mixed infections with aerobes such as streptococci and Gram-negative enteric bacilli, are by far the commonest organisms in brain abscesses. Confirmatory laboratory tests depend on rapid transport of adequate anaerobic and aerobic samples followed by specific anaerobic and aerobic culture. Chemotherapy pending culture results is usually penicillin plus metronidazole; for abscesses where Gram-negative aerobes are common, e.g. secondary to ear or sinus infections, a third-generation cephalosporin, such as ceftriaxone, is added. Early surgical consultation concerning aspiration or incision and drainage is essential. The source, particularly parameningeal foci and septicaemic spread from distant foci, must be sought and treated.

Staphylococci

Staphylococci (p. 28) are particularly important in brain abscesses after trauma

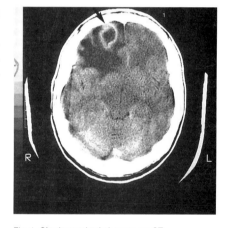

Fig. 1 **Single cerebral abscess on CT.**

or neurosurgery are diagnosed by routine techniques and are treated by flucloxacillin unless methicillin resistant, when vancomycin is used.

Actinomyces and *Nocardia* spp

Actinomyces and *Nocardia* (p. 56) are rare causes of brain abscess. Both are chronic bacterial infections, though more acute in the immunocompromised. Cervicofacial infection and sinus formation are characteristic of actinomycosis, while pulmonary disease resembling tuberculosis plus multiple, multiloculated abscesses are more typical of nocardiosis. Prolonged anaerobic culture (for *Actinomyces*) and aerobic culture (for *Nocardia*) are needed to grow the branching Grampositive, weakly acid-fast filaments. Chemotherapy

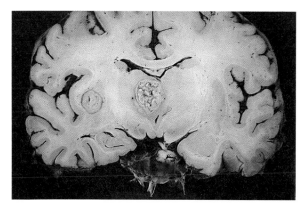

Fig. 2 **Cerebral tuberculomata.**

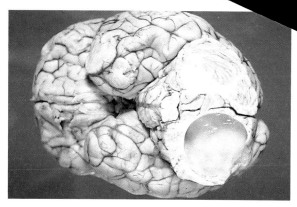

Fig. 4 **Hydatid cyst of brain.**

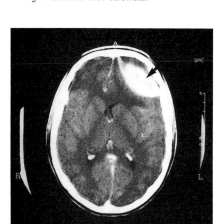

Fig. 3 **Extradural abscess.**

is long-term penicillin for actinomycosis, and sulphonamide for nocardiosis. Surgery is often necessary for the chronic thick-walled abscesses.

Syphilis
Syphilis (*Treponema pallidum*, p. 52) is a very rare cause of a focal lesion (a gumma) and is diagnosed by imaging and syphilis serology. Chemotherapy is high dose i.v. penicillin for 14 days.

Tuberculosis
Tuberculosis (pp. 54 and 118) also rarely produces one or more focal lesions (plural: tuberculomata, Fig. 2), which are diagnosed by imaging, a positive Mantoux, evidence of tuberculosis elsewhere, and AFB stain and culture if surgery is needed. Treatment is by triple therapy.

Fungal abscesses
Fungal abscesses (Table 1) are uncommon and, in aspergillosis, candidiasis or cryptococcosis, most often occur in immunocompromised patients with fungaemia. In aspergillosis, blood vessel invasion often causes infarction, and in cryptococcosis meningitis is common (pp. 64, 88). Of the systemic mycoses, blastomycosis, coccidioidomycosis and

rarely histoplasmosis (pp. 64–7) can cause brain abscesses. Treatment is specific chemotherapy, and surgery as indicated.

Rhinocerebral zygomycosis including mucormycosis (p. 70) is a rapidly progressive infection particularly in patients with diabetes or haematologic malignancy. Fungus spreads through the nose, sinuses, skull, meninges and brain, causing tissue destruction and necrosis rather than true abscess formation. Unless diagnosed early, the predisposing cause reversed if possible, and the fungus treated vigorously with amphotericin B plus an azole and surgery if possible, it is usually fatal.

SUB- AND EXTRADURAL ABSCESSES

Abscesses may form just inside the dura mater (subdural abscess: subdural empyaema) or outside it (extradural (Fig. 3): epidural abscess) or both at once. As with brain abscesses, there is usually one of the three predisposing factors (above).

Typically a patient with sinusitis or otitis worsens rapidly, with fever, severe headache, vomiting, altered mental state, and then focal signs with signs of raised intracranial pressure. The focal signs may

spread to involve a whole cerebral hemisphere. Imaging and surgery are urgent and essential. The usual causative organisms are aerobic and anaerobic streptococci, other anaerobes, and *S. aureus*. If the abscess has developed postoperatively, aerobic Gram-negative rods (GNRs) are common. Chemotherapy is high-dose i.v. flucloxacillin and metronidazole, with gentamicin or ceftriaxone for GNR.

CYSTS

When the haematogenous spread of parasites causes focal lesions in the CNS, these are usually cysts: of tachyzoites in toxoplasmosis (p. 72), of larvae of *Taenia solium* in cysticercosis (p. 82) and of larvae of *Echinococcus granulosus* within brood capsules within a hydatid cyst (Fig. 4) (p. 83). An exception is *Entamoeba histolytica* (p. 74), which does not form cysts but, as its name describes, 'lyses tissue' to form abscesses.

All present as mass lesions with focal signs, and little or no fever or other systemic upset.

Imaging, serology and epidemiology may lead to the diagnosis, or surgery may be needed for both diagnosis and treatment.

CNS abscesses and other focal lesions
- Focal infections of the CNS must be distinguished from meningitis and diffuse lesions because lumbar puncture is contraindicated and surgery is often essential.
- Focal lesions may be abscesses, usually bacterial or fungal, or cysts, usually parasitic.
- Factors predisposing are parameningeal infections, distal infection with haematogenous spread, or local trauma including surgery.
- Bacterial abscesses are usually anaerobic or mixed anaerobic–aerobic.
- Fungal infections are usually candidiasis, cryptococcosis or aspergillosis in immunocompromised patients.
- Parasitic infections include toxoplasmosis, hydatid cysts, cysticercosis and amoebiasis.
- Diagnosis is by imaging (usually CT), microscopy and special cultures of surgical specimens, or serology in special infections.
- Treatment is specific chemotherapy, and usually urgent surgery unless multiple small cysts are medically treatable, as in toxoplasmosis.

M: TROPICAL AND RARE INFECTIONS

...maged by bac-
...al and viral (not
...k) pathogens. Some
...issues: others produce
neu.....ich interfere with neurone
functio...

BOTULISM

Botulism is a rare but serious disease caused by the neurotoxin of *Clostridium botulinum* (p. 35); it is characterised by flaccid paralysis. *C. botulinum* is an anaerobic Gram-positive rod which is found almost everywhere in the environment; it provides heat-resistant spores which occur in soil and in food, particularly inadequately canned or bottled vegetables, honey and fish. Wound botulism, the rarest form, and infant botulism (from honey) are *infections*, with actual multiplication of the organism in the body before intoxication occurs. Adult foodborne disease, the commonest form, is an *intoxication* from ingestion of toxin.

The toxin enters neurones and prevents the release of acetylcholine.

Clinical features. These include blurred vision with fixed dilated pupils, dry mouth, constipation and abdominal pain, descending flaccid paralysis, respiratory arrest and death in 10–60% of adults, but only 1–5% in infants.

Differential diagnosis. This includes the Guillain–Barre syndrome of ascending polyneuritis, myasthenia gravis and polio.

Confirmatory tests. These are specialised, depending on anaerobic culture or on detection of botulinum toxin in mice.

Chemotherapy. Penicillin is relatively unimportant, treatment depending on antitoxin and respiratory support.

Control and prevention depends on sufficient time at the correct temperatures during food preparation, and not giving honey to infants less than 1 year old.

LEPROSY

Leprosy (Hansen's disease) is a chronic infection of the nerves and skin with *Mycobacterium leprae* (p. 54). Disease is classified by the cell-mediated immune (CMI) response (Fig. 1). Infection is by droplet spread from infected nasal secretions, direct skin contact and possibly from breast milk and insect bites.

Clinical features. These are extremely variable, depending on the classification and extent of disease. Extensive LL involves leonine facies, succulent ear lobes, nasal cartilage destruction, erythema

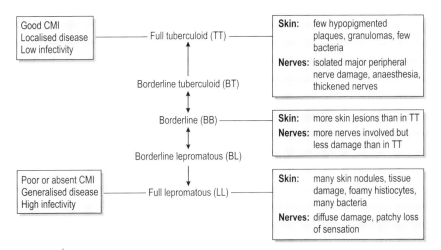

Fig. 1 **Classification of leprosy.**

nodosum leprosum (p. 165), polyarthritis and painful peripheral neuritis with, for example, foot drop, a neuropathic ulcer (Fig. 2) or ulnar nerve involvement causing a 'claw hand' (Fig. 3). Local loss of sensation makes the patient susceptible to secondary trauma and bacterial infections.

Confirmatory tests. These depend on seeing acid-fast bacilli in nasal scrapings or skin biopsies, and on histology, as culture is only possible in mouse footpads or armadillos (facilitated by the lower body temperature).

Chemotherapy. This is with daily dapsone with monthly rifampicin, plus clofazimine or ethionamide in LL disease. Duration varies from 6–9 months in TT to years or even lifelong in LL. Management of ulcers and disfigurement is important.

Control and prevention. This depends on treatment of infective (LL, BL) patients, and dapsone for their close contacts aged < 16 years.

MALARIA

The CNS is affected in two ways by malaria:

- cerebral malaria, which is severe and may be fatal (Fig. 4) is characterised by widespread changes in cerebral function from capillary plugging caused by *Plasmodium falciparum* (p. 72)
- mild cerebral symptoms from general tissue anoxia can occur in malaria from other species.

Clinical features. Cerebral malaria presents with severe headache and high

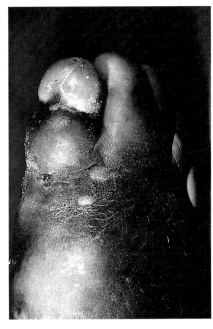

Fig. 2 **Neuropathy in leprosy.**

fever, rigors, disturbed consciousness, altered behaviour, impaired intellect, hallucinations and, sometimes, fits or focal lesions.

Confirmatory tests. These depend on urgent thick and thin stained blood films.

Chemotherapy. This is by i.v. quinine, begun immediately but given slowly. Monitoring of cardiac function and blood glucose is necessary. Follow-up therapy with oral antimalarials is needed, the drug depending on specialist advice.

Control and prevention involves measures against mosquitoes, and chemoprophylaxis with the appropriate antimalarial, which varies for different countries.

PARAGONIMIASIS

Paragonimiasis is caused by *Paragonimus westermani*, the oriental lung fluke (p. 85). Adult flukes develop in cystic cavities, particularly in the lungs, but also in the brain and other tissues after haematogenous spread. Cerebral paragonimiasis can produce severe disease with visual impairment, fits and motor weakness.

Confirmatory tests. Imaging by brain CT shows the cysts, and sputum microscopy may show the eggs. Eosinophilia is common, as in all fluke infections, and chest X-rays are abnormal.

Chemotherapy. Praziquantel is the drug of choice; bithionol the alternative.

Control depends on education, sanitation and avoiding uncooked crabs and crayfish. The snails and definitive hosts are difficult to control.

Fig. 3 **Leprosy: claw hand from ulnar nerve lesion.**

TETANUS

Clostridium tetani (p. 35) is a strictly anaerobic Gram-positive rod (with terminal spores) which is not invasive; it multiplies only in the infected wound. It produces a toxin, tetanospasmin, which spreads to the CNS where it causes severe tetanic (sustained) muscle spasm by blocking inhibitors of neurotransmission. The disease is classified into adult localised or generalised tetanus from soil-contaminated wounds, and neonatal tetanus from soil-contaminated umbilical wounds.

Clinical features. These are almost unmistakable, with progressive muscle stiffness, spasms, then spastic paralysis, giving trismus (lockjaw), sardonic smile (risus sardonicus, Fig. 5), and arching of the back (opisthotonos) in advanced cases. Autonomic involvement gives sweating, salivation, swinging blood pressure and supraventricular or other arrhythmias. Laryngeal or glottal spasm can cause death.

Confirmatory tests. The organism usually cannot be found in the often trivial wound causing this potentially fatal disease.

Management. This is by penicillin, plus anti-toxin and supportive measures.

Control and prevention. This is by active immunisation with tetanus toxoid, which gives excellent immunity.

TRYPANOSOMIASIS

African trypanosomiasis is a chronic encephalitis caused by blood and tissue flagellated protozoa (pp. 76, 77).

Causative organisms. There are two variants of *Trypanosoma brucei*, *T. brucei gambiense* and *T. brucei rhodesiense*, which live and multiply in the human blood stream and invade the CNS to cause 'sleeping sickness'. Both are spread by tsetse flies, but while the West African (Gambian) form has no known animal reservoir, the East African (Rhodesian) has cattle, sheep and wild animals as reservoir hosts.

Clinical features. These begin with a trypanoma (lump) or ulcer at the site of the tsetse fly bite, followed by acute systemic spread with fever, myalgia, arthralgia and lymphadenopathy. Finally progression to the third, chronic stage of encephalitis occurs with lethargy, tremors, mental impairment, convulsions, paralysis and incontinence and merciful death in 9–18 months.

Confirmatory tests. Microscopy and specific staining (Giemsa) for trypomastigotes in anticoagulated blood, blood films or lymph node aspirates.

Chemotherapy. Suramin is the drug of choice for early *T. brucei* infections, the alternative being pentamidine. Toxic organic arsenicals such as melarsoprol are needed for CNS infection.

Control and prevention. This depends on tsetse fly control, particularly around houses, protection from biting, and treatment of infected humans.

OTHER CAUSES

Numerous **congenital** infections of the fetus in utero can infect the CNS, including the 'Torch' group; *t*oxoplasmosis, *o*thers including HIV and syphilis, *r*ubella, *c*ytomegalovirus, and *h*erpes simplex (p. 186).

Postinfectious encephalomyelitis is not an infection but an immune response causing demelination subsequent to a viral or bacterial infection.

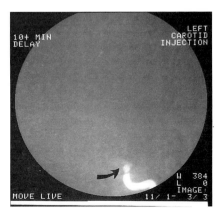

Fig. 4 **Cerebral malaria: angiogram showing no circulation, shortly before death.**

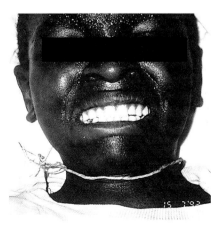

Fig. 5 **Tetanus: risus sardonicus.**

Tropical and rare infections

- Botulism is a flaccid paralysis from a neurotoxin.
- Leprosy causes hypopigmentation, anaesthesia, skin infiltration and nodules.
- Malaria causes fever, confusion, coma and death.
- Paragonimiasis causes multiple cysts, fits and focal signs.
- Tetanus causes spasms, spastic paralysis and death.
- Trypanosomiasis causes 'sleeping sickness' and death.

OTITIS, MASTOIDITIS AND SINUSITIS

OTITIS EXTERNA

Otitis externa (OE) is infection of the external auditory canal. It is classified into four types. The causative organisms vary; the commonest are given in parentheses:

- acute localised, often secondary to folliculitis elsewhere (*Staph. aureus*)
- acute diffuse, often secondary to swimming or spa baths (*P. aeruginosa*)
- chronic, usually secondary to chronic otitis media (as chronic OM plus *Candida* spp., rarely TB, syphilis)
- malignant otitis externa, usually secondary to diabetic microangiopathy (*P. aeruginosa*).

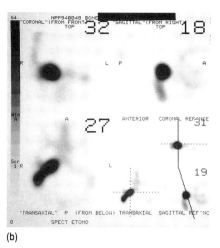

(b)

Fig. 1 **Malignant otitis externa.**
(a) Facial (7th) nerve paralysis;
(b) Bone scan showing 'hot spots' (black) in temporal bone.

(a)

Clinical features

- Acute localised OE is usually a painful pustule, often with local lymphadenopathy.
- Acute diffuse OE (swimmer's ear) is itchy, painful, oedematous and reddened with purulent discharge and at times perichondritis.
- Chronic OE is similar but has less redness and pus comes from a perforated ear drum. Often there is fungal or other chronic disease elsewhere, particularly in an immunocompromised host.
- Malignant OE is a serious disease. The name easily misleads, for it is not malignant meaning cancerous, but malignant in its progressive, often fatal course. Nor does the condition remain as OE but spreads to surrounding bone, blood vessels, facial nerve and meninges (Fig. 1). Consult with ID physician, ENT surgeon and/or neurosurgeon urgently.

Confirmatory tests

Tests are usually microscopy and culture of the discharge; malignant OE needs full investigation with X-ray, CT scans and MRI, as appropriate, to define the extent and to monitor progress (Fig. 1).

Management

Chemotherapy depends on the diagnosis and organism:

- acute localised: none or flucoxacillin
- acute diffuse: local neomycin or polymyxin
- chronic: local imidazole for fungi, treat OM
- malignant: i.v. tobramycin plus ticarcillin or ceftazidime for weeks or months.

Malignant OE may need surgery. Control and prevention depend on the predisposing factors: folliculitis, moisture, otitis media, chronic disease elsewhere and diabetes mellitus.

OTITIS MEDIA

Otitis media (OM) is infection of the middle (media) ear. It is classified into:

- acute suppurative, presenting with pus in the middle ear and often simply called 'otitis media'. It is extremely common in infants and children
- chronic, including both **recurrent OM** and the very common persisting middle-ear effusion after acute OM, called **secretory OM** (glue ear) because of the thick consistency of the middle ear fluid.

The two most important **causative organisms** are the usual respiratory pathogens: the pneumococcus *S. pneumoniae* (35%) and *H. influenzae* (20%). *Neisseria* spp., *Branhamella catarrhalis, S. aureus,* other streptococci, Gram-negative rods and viruses are less common.

Clinical features

Otitis media not uncommonly follows an upper respiratory infection, presumably spread up the Eustachian tube, and may be accompanied by sinusitis, also by direct spread.

Deep ear pain, headache and fever, then tinnitus and vertigo, are accompanied by a reddened, bulging and immobile ear drum (Fig. 2), hearing loss, then a perforated ear drum and ear discharge if treatment is delayed. Even with apparently adequate treatment, a middle ear effusion often (in 40%) persists for a month or more.

Confirmatory tests. These are unnecessary routinely, but aspiration through the ear drum (tympanocentesis) for microscopy and culture is done if response to treatment is slow.

Management

Chemotherapy is usually either amoxycillin or cotrimoxazole, with amoxycillin/clavulanate if beta-lactamase-producing *H. influenzae* are common. Tympanostomy tubes (grommets) or adenoidectomy may be needed for 'glue ear' persisting more than 3 months with adequate medical management.

Chemoprophylaxis may be used to prevent recurrent OM in winter and spring, with pneumococcal vaccination.

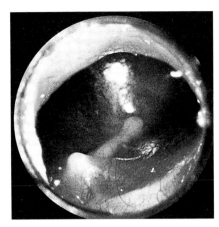

Fig. 2 **Otitis media: red eardrum.**

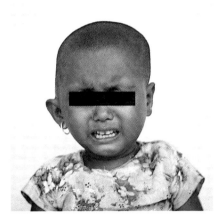

Fig. 3 **Mastoiditis: note displacement of ear.**

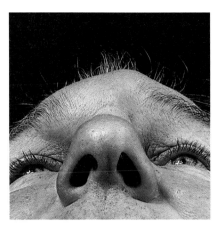

Fig. 4 **Pott's puffy tumour.**

MASTOIDITIS

Mastoiditis is infection in the air cells of the mastoid process of the temporal bone. These are connected to the middle ear, so they are extensions of the upper respiratory tract. Therefore mastoiditis is an extension, now rare, of OM. It may be acute or chronic. Causative organisms are the same as those causing OM (see above).

Clinical features are those of acute OM plus pain, tenderness, redness and then swelling behind the pinna of the ear which is displaced out and down (Fig. 3). Systemic symptoms become more marked. Involvement of nearby vital structures (facial nerve, meninges, venous sinuses) follows if treatment is delayed. Confirmatory laboratory tests are X-ray or CT, showing air-cell opacity and then bone destruction.

Chemotherapy uses the drugs for OM media, or ceftriaxone, given i.v. Drainage by mastoidectomy is needed for unresponsive abscesses. Control and prevention depends on early treatment of OM.

ACUTE SINUSITIS

Acute sinusitis is infection in one or more of the paranasal sinuses; maxillary (antral) sinusitis is the most prominent. This, like mastoiditis and OM, is an infection in a bony cavity extending from the upper respiratory tract. The predictable predisposing factors are:

- upper respiratory tract infection (URTI), either viral or bacterial
- dental infection, especially upper molar
- nasal deformities, obstruction or infection
- swimming, especially in 'summer sinusitis'
- chronic sinusitis.

The causative organisms are similar to those in OM.

Clinical features

Acute sinus pain and tenderness, purulent nasal discharge, nasal obstruction and headache are usual, with fever in 50%. Complete opacity of the sinus on trans-illumination is highly reliable unless chronic sinusitis preceded the acute attack. Acute frontal sinusitis may progress to osteomyelitis and even frontal lobe abscess. Subperiosteal pus causes forehead oedema and swelling called Pott's puffy tumour (Fig. 4) ('tumor' in Latin means swelling, not necessarily malignant).

Confirmatory tests

These include X-rays of the sinuses, with CT if extension to bone, meninges or brain is suspected. Culture of nasal pus may mislead because of normal flora, so antral puncture is needed to isolate the pathogen(s). This is only necessary in severe, unresponsive or recurrent disease.

Management

Chemotherapy is usually either amoxycillin or cotrimoxazole, with amoxycillin/clavulanate if beta-lactamase-producing *H. influenzae* are common. Nasal decongestants and analgesia alleviate symptoms but not the infection. Frequent recurrent attacks may be prevented by early treatment of URTIs.

CHRONIC SINUSITIS

Chronic sinus disease is mainly permanent mucosal damage and poor drainage, with infection a minor component. Causative organisms are similar to those of acute sinusitis except anaerobes are more prominent. Clinical features are minimal; mild pain and chronic nasal discharge are common, with acute sinusitis superimposed. The confirmatory test is an X-ray showing mucosal thickening, and partial or complete sinus opacification. Chemotherapy is given for acute attacks. Surgery may help the chronic state. Control and prevention depend on effective treatment of acute sinusitis.

Otitis, mastoiditis and sinusitis

- Otitis externa can be localised (folliculitis), diffuse (secondary to swimming or other moisture), chronic (fungal, or secondary to chronic otitis media), or so-called malignant (with extension to bone, venous sinuses, facial nerve or meninges). Malignant OE is a severe and often fatal disease requiring urgent treatment.
- Otitis media or acute sinusitis often follow an upper respiratory infection, and the pneumococcus and *H. influenzae* are the commonest causative organisms. Amoxycillin or cotrimoxazole are, therefore, the usual chemotherapy providing *H. influenzae* is sensitive.
- Persistent effusion (glue ear) may need tympanostomy tubes (grommets) inserted. Chronic sinusitis may need surgery.
- Mastoiditis is a complication of untreated otitis media and can progress to facial nerve paralysis, meningitis and intracranial thrombophlebitis if unrecognised. Treatment is i.v. antibiotics, and drainage if needed.

SUPERFICIAL OCULAR INFECTIONS

BLEPHARITIS

Blepharitis is inflammation of the eyelid and occurs in two forms:

- usually chronic, bilateral, marginal and associated with seborrhoea
- uncommonly acute, unilateral and preseptal, i.e. involving the whole lid anterior to the orbital septum and, hence, not extending into the orbit. It is usually post-traumatic in adults, or from bacteraemic spread in children.

Causative organisms. Infection is commonly with *Staph. aureus*, coagulase negative staphylococci, or *Strep. pyogenes* in adults, or *H. influenzae* in children under 5 years. Anaerobes infect after human or animal bites.

Clinical features. These include chronic itching and scaling with associated seborrhoea in adults, or acute purplish swelling in children after a respiratory infection. Full, painless ocular mobility distinguishes it from orbital cellulitis (see below).

Confirmatory tests. Infection is confirmed by local and blood cultures.

Chemotherapy. Tetracycline or chloramphenicol ointment can be administered locally, plus flucloxacillin orally if lid abscess develops in adults. Children need a third-generation cephalosporin such as ceftriaxone. Concomitant seborrhoea of the skin and scalp is treated with shampoos or ketoconazole cream.

Control and prevention. This depends on control of the seborrhoea.

STYE AND CHALAZION

A stye (hordeolum) is an acute infection of eye lash follicle glands. A chalazion is an infection of the Meibomian glands deeper in the lid stroma. Causative organisms are *Staph. aureus* and skin flora. Clinical features are pain and swelling, with visible pus in a stye. Confirmatory tests (Gram stain and culture) are seldom necessary. Chemotherapy is initially local, with incision and drainage of a chalazion as necessary.

CONJUNCTIVITIS

Conjunctivitis is inflammation of the conjunctiva and can be bacterial, viral, or chlamydial in origin, or non-infective.

Causative organisms. *S. aureus, S. pneumoniae, S. pyogenes, Haemophilus* spp., *C. trachomatis* (different serotypes causing inclusion conjunctivitis and trachoma, p. 99) and viruses are common infecting organisms in adults and children, plus *N. gonorrhoea* in neonates. Ophthalmia neonatorum (p. 187) was once only gonococcal but now is more often chlamydial.

Clinical features. These vary with the cause: discharge is usual, ranging from mild and thin in viral infections, through moderate in most bacterial infections to profuse in gonorrhoeal and chlamydial infections. Pain is not prominent. Follicles are most prominent in trachoma which progress to scarring, and the cornea is involved (p. 99).

Confirmatory tests. Infection is confirmed by microscopy (Gram stain and chlamydial IF), and bacterial culture. Viral and chlamydial cultures are seldom done.

Chemotherapy. This depends on the cause. Trachoma needs oral sulphonamide or tetracycline, gonorrhoea needs i.v. penicillin or ceftriaxone and local saline, while the other bacteria need local bacitracin, neomycin or gentamicin.

Control and prevention. This is only feasible in gonorrhoea and trachoma, depending on treatment of infected mothers, and hygiene.

KERATITIS AND CORNEAL ULCERS

Keratitis is inflammation of the cornea, usually caused by bacterial or viral infection, less commonly by fungal or parasitic infection (Fig. 1). Any of these can progress to corneal ulceration and, thence, to corneal perforation, loss of aqueous humour and potential permanent blindness.

Causative organisms. Over 60 bacteria, viruses, chlamydia, fungi and parasites have been described as causes of keratitis, but the most important are *S. aureus*, pneumococci and beta-haemolytic streptococci, *Bacillus* spp., and herpes simplex virus. *Pseudomonas* spp., other Gram-negative rods and *Acanthamoeba* spp. are important in infections associated with soft contact lenses. *Staph. aureus* can cause corneal abscesses, e.g. after corneal grafting.

Clinical features. Severe pain, marked conjunctival injection at the limbus, i.e. adjacent to the cornea, grey-white haze or ulceration and blurred vision from central ulcers are characteristic. Pus in the anterior chamber is called hypopyon (Fig. 2). It is important to involve an ophthalmologist.

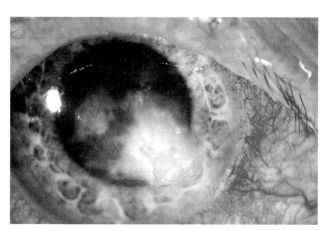
Fig. 1 **Fungal keratitis.**

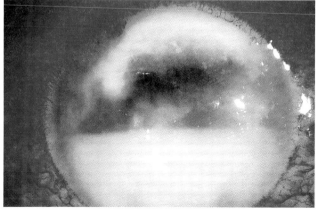
Fig. 2 **Corneal ulcer with hypopyon.**

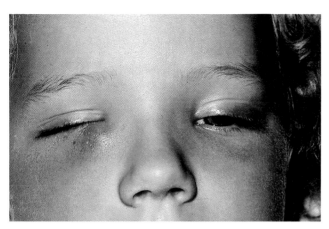

Fig. 3 **Dacryocystitis.**

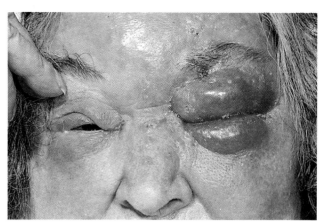

Fig. 4 **Orbital cellulitis with swelling and proptosis.**

Confirmatory tests. Infection is confirmed by microscopy, with special stains if necessary, and culture of corneal swabs or scrapings. Corneal staining with fluorescein can show the distinctive dendritic ulcer of herpes simplex. Corneal biopsy may be necessary for difficult pathogens.

Chemotherapy. This depends on the pathogen. Initially, broad-spectrum local antibiotics such as vancomycin and tobramycin are given for bacterial keratitis, acyclovir for viral, amphotericin B or flucytosine for fungal, and 'Brolene' plus oral ketoconazole for amoebic infections.

Control and prevention. Early attention to minor corneal trauma, and scrupulous sterile care with contact lens and their solutions, are important.

DACRYOCYSTITIS, CANALICULITIS AND DACRYOADENITIS

Dacryocystitis is inflammation of the lacrimal sac (cyst) just below the inner end of the lower eyelid (Fig. 3); canaliculitis is inflammation of the 'little canals' leading from the lacrimal puncta of the inner end of the lids to the lacrimal sac; and dacryoadenitis is inflammation of the main lacrimal gland, anteriorly in the upper–outer part of the orbit.

Causative organisms. Acute dacryocystitis and dacryo-adenitis are caused by pneumococci, other streptococci, staphy-lococci, and *P. aeruginosa*. Long or branching organisms, i.e. *Actinomyces, Aspergillus, Candida* or *Fusobacteria,* cause canaliculitis. Obstruction, whether congenital or from trauma, tumour or dacryoliths (tear stones) predisposes to infection.

Clinical features. Epiphora (overflow of tears) and pain, swelling and redness are characteristic in acute cases.

Confirmatory tests. Gram stain and culture of expressed pus should confirm infection.

Management. Treatment involves local antibiotic eye drops, systemic flucloxacillin in acute cases, local warmth, and consul-tation with an ophthalmologist concerning probing or incision.

ORBITAL INFECTIONS

These are infections in the orbit behind the orbital septum (contrast with preseptal blepharitis above). Analogous to brain abscesses, they have one of three sources as predisposing factors:

- adjacent infection, usually ethmoidal (in children) or frontal (in adults) sinusitis in 75%; less commonly from nearby infection such as facial cellulitis, otitis or dental infection
- distant infection with septicaemic spread (rare)
- eye surgery or local penetrating trauma.

Causative organisms. The two major respiratory pathogens are usually causative: *S. pneumoniae* or *H. influenzae. S. aureus* is common after trauma, anaerobes (with necrosis and foul smell) are common in chronic infections, and Gram-negative rods and fungi may occur.

Clinical features. These cumulate through the five stages.

1. *Orbital oedema* causes painless swelling of the lids.
2. *Orbital cellulitis* causes fever, pain, tenderness, redness and painful limited eye movement, with some proptosis (Fig. 4); call an ophthalmologist and infectious diseases physician.
3. *Subperiosteal abscess* with pus between the periosteum and bone causes displacement of the globe of the eye.
4. *Orbital abscess* causes ophthalmoplegia (paralysis of ocular movement) and some visual loss.
5. *Cavernous sinus thrombosis* is potentially fatal and causes severe headache and eye pain, further visual and retinal changes, paresis of 3rd, 4th and 6th cranial nerves, meningitis with neck stiffness, plus high fever, chills and systemic toxicity.

Confirmatory tests. Conjunctival, blood and sinus aspirate stains and cultures are essential. Orbital and sinus CT have replaced sinus X-rays. Lumbar puncture is needed if there are any meningeal signs.

Management. This depends on the source and the stage: initially high-dose i.v. flucloxacillin and ceftriaxone are usual, modified after culture results. Abscesses need drainage, and sinusitis may need further drainage.

Control and prevention. This depends on early treatment of possible sources.

Superficial ocular infections

- Blepharitis is usually a mild infection needing local treatment only.
- Styes usually need local ointment and lash removal only, while chalazions often need incision and drainage.
- Common bacterial conjunctivitis needs only local antibiotics, while trachoma or gonorrhoea need oral or i.v. treatment, respectively.
- Keratitis, corneal ulcer and corneal abscess are serious, have numerous causes and need specialist advice and treatment.
- Infections of the lacrimal apparatus need local antibiotics, systemic flucloxacillin, and specialist consultation concerning drainage.
- Orbital infections are usually secondary to sinusitis, are serious and spreading, and need specialist advice and systemic antibiotics.

TROPICAL OCULAR INFECTIONS

Infections in the tropics affect the eyes in one of four ways.

- **Severe prostrating infections**, often accompanied by malnutrition and inadequate medical care, result in exposure keratitis and conjunctivitis, corneal ulcers, iritis and even endophthalmitis, loss of eyesight or the eye. This sequence can occur with infections such as cholera, amoebic or bacillary dysentery, malaria or typhus plus viral diseases not included in this book, including measles and smallpox.
- **Wandering worms, larvae or parasites** invade the eye, e.g. *A. cantonensis* (p. 89), dracontiasis or filariasis (p. 177) and many rarer parasites.
- **Systemic diseases infect the eye**, e.g. brucellosis, the systemic mycoses, leishmaniasis (all three forms, p. 130, 177), leprosy, toxocariasis, toxoplasmosis, trypanosomiasis, tularaemia and yaws (see below).
- **Specific infection of the eye also occurs**, e.g. loiasis, onchocerciasis and trachoma (see below).

SYSTEMIC DISEASES

Leprosy
Leprosy (pp. 54, 92) affects the eye either indirectly through facial nerve paresis causing lagophthalmos (incomplete eyelid closure) or directly by invasion and infection, causing madarosis (loss of eyebrows and eyelashes), superficial punctate keratitis or deeper opaque stromal keratitis, episcleritis with limbal nodules or iritis with 1 mm 'pearls' (Fig. 1). Posterior chamber involvement is rare unless corneal perforation from keratitis leads to endophthalmitis. Confirmatory tests involve microscopy of skin and nose tissue. Treatment is local for the specific tissue(s) damaged, plus systemic anti-leprotics (dapsone plus rifampin). Surgery may be needed.

Toxocariasis
Visceral larva migrans is due to infection with *Toxocara canis* or *T. cati*, the dog and cat round worms (p. 81). Ingestion of ascarid eggs, usually by children, causes pneumonitis, hepatitis and retinitis, rarely irido-cyclitis. Clinical diagnosis is confirmed serologically, and treatment is by mebendazole or diethyl-carbamazine. Control depends on keeping children away from infected puppies and kittens, and on treating pets.

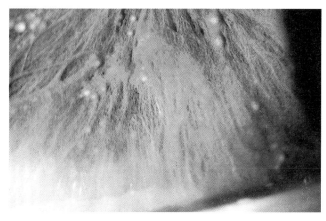

Fig. 1 **Leprosy: iritis pearls.**

Toxoplasmosis
Toxoplasma gondii infection (toxoplasmosis) (pp. 72, 185) can affect the eye, causing chorio-retinitis (Fig. 2) or, rarely, irido-cyclitis. Infection can be congenital or acquired (serious in immunosuppression) and is confirmed by serology. Treatment is with sulphadiazine and pyrimethamine. Control depends on good hygiene with cats and meat.

Trypanosomiasis
Unilateral oedema of the eyelids, Romana's sign, is diagnostic in South American trypanosomiasis (p. 124), caused by *T. cruzi*. Eyelid oedema, keratitis and irido-cyclitis occur with the African forms, caused by *T. brucei*, and cranial nerve pareses and papilloedema occur late from the encephalitis (p. 93). Infection is confirmed with serology and treated with nifurtimox (*T. cruzi*) or suramin/melarsoprol (*T. brucei*).

Tularaemia
Oculoglandular tularaemia (*Francisella tularensis*, p. 51) is one cause of conjunctivitis, with preauricular and parotid gland swelling (Parinaud's syndrome). Serology confirms the diagnosis, and treatment is with streptomycin/gentamicin.

Yaws
The secondary stage lesions (p. 175) resulting from infection with *Treponema pertenue* (p. 52) may involve the eyebrows and

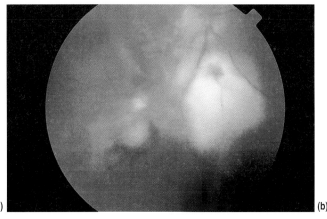

(a)

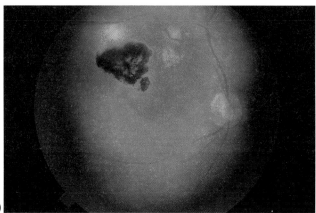

(b)

Fig. 2 **Toxoplasmosis. (a)** Active disease; **(b)** Healed scar.

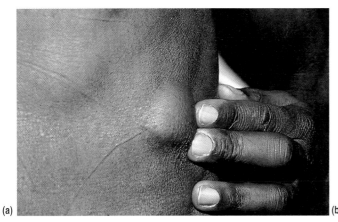

(a)

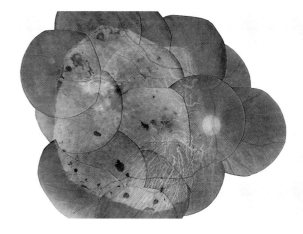

(b)

Fig. 3 **Onchocerciasis. (a)** Skin nodule; **(b)** Extensive chorioretinitis.

lids. Gangosa is a dreadful destructive rhino-pharyngitis which can spread to the eyelids and cause exposure keratitis. Treatment is with penicillin.

SPECIFIC INFECTIONS OF THE EYE

Loiasis

Loiasis is a chronic filarial disease in West and Central Africa that particularly affects the eyelids. *Loa loa*, a filarial nematode, is the causative organism (p. 80). Infection is spread between humans by the bite of *Chrysops* spp. flies.

Initially, palpable worms migrate through the subcutaneous tissues including the eyelids and conjunctivae. Subsequently, transient oedematous 'Calabar swellings' appear at various sites, including the eyelids, as a result of allergic reactions to filarial toxins. Confirmatory tests are microfilaria in the blood by day, eosinophilia and serology.

Chemotherapy is diethylcarbamazine. Surgical removal of the worms may be necessary. Control and prevention are by protection from flies, and diethylcarbamazine for 3 days monthly.

Onchocerciasis

Onchocerciasis is a chronic filarial disease in West and Central Africa and Central and South America which particularly affects the skin and anterior chamber of the eye. The causative organism is *Onchocerca volvulus*, a filarial nematode (p. 80). Infection is spread between humans by the bite of *Simulium* spp. black flies.

Clinical features include skin nodules with depigmentation and atrophy (Fig. 3a), lymphadenopathy and lymphoedema;

eventually, blindness from iritis and secondary glaucoma occur. Conjunctivitis, keratitis and chorioretinitis (Fig. 3b) also occur. Confirmatory laboratory tests are microfilaria in skin or conjunctival snips. Eosinophilia is common.

Chemotherapy is with ivermectin (less often with diethylcarbamazine or suramin). Control and prevention is by fly control, and possibly mass chemoprophylaxis.

Trachoma

Chlamydia have surface molecules or ligands that can bind to receptors on conjunctival cells; this enables them to avoid host defences and results in trachoma being the most important eye infection in the world. Transmission is eye to eye, directly by droplets or through flies. There are at least 15 different serotypes. Trachoma is a chronic kerato-conjunctivitis caused by serotypes A, B and C; D–K cause inclusion conjunctivitis in neonates and adults from genital infections (p. 187), and L1, L2 and L3 cause ocular lymphogranuloma venereum with keratitis, uveitis and optic neuritis.

Clinical features of trachoma are classified in four stages:

1. TF: trachomatous folliculitis (Fig. 4)
2. TS: trachoma scarring
3. CO: conjunctiva scarred, pannus (blood vessels) over cornea gives corneal opacity
4. TT: trichiasis with ingrowing eyelashes and corneal ulceration.

Confirmatory tests are direct antigen detection or culture.
Chemotherapy is with tetracycline, alternatively erythromycin or sulphas. Scarring may need surgery. Control and prevention occur through education and improved living and hygienic standards.

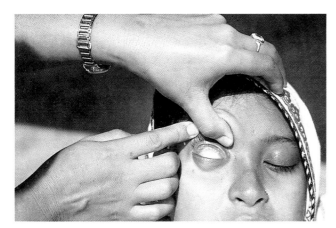

Fig. 4 **Trachoma: everting the upper eyelid to examine the tarsal conjunctiva.**

Tropical ocular infections

- Exposure keratitis can occur in any severe systemic infection, e.g. dysentery, typhoid, typhus.
- Worm infections can invade the eye accidentally in their migrations, e.g. dracontiasis and filariasis.
- Systemic infections, e.g. leprosy and toxocariasis, can cause severe eye disease.
- Specific infections include loiasis, onchocerciasis and trachoma.
- All need precise diagnosis and specialist treatment.

DEEP EYE INFECTIONS

Deep eye infections are all serious and are caused by a wide range of pathogens so specialist diagnosis and treatment are required to avoid blindness.

ANTERIOR UVEITIS (IRIDOCYCLITIS)

The iris and ciliary body (anterior segment of uveal tract) share the same blood supply and are commonly inflamed together: iridocyclitis. Many causes are not proven infections (e.g. rheumatoid arthritis).

Causative organisms. These are unusual organisms: *Brucella* spp., chlamydia and other causes of Reiter's syndrome (p. 154), mycobacteria (*Mycobacterium leprae*), rickettsia (Rocky Mountain spotted fever), spirochaetes (syphilis, Lyme disease and leptospirosis) and numerous viruses.

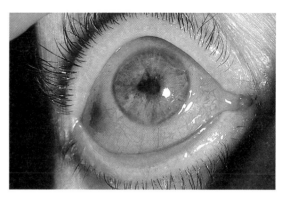

Fig. 1 **Iridocyclitis.**

Clinical features. These are predominantly photophobia, pain (deep, ocular), pupillary constriction, profuse tears and a red eye with intense limbal injection (Fig. 1). Call an ophthalmologist.

Confirmatory tests. Slit lamp shows typical keratic precipitates (KPs) and adhesions (synechiae). Serology, special culture and, at times, anterior chamber aspiration for microscopy and special (e.g. viral) cultures are used depending on the suspected pathogen.

Management. This is dictated by the cause of disease. Do not delay.

CHORIORETINITIS (POSTERIOR UVEITIS)

Inflammation of the choroid usually spreads to the retina, forming chorioretinitis. This is usually chronic and granulomatous and may be viral, e.g. CMV, or non-infective, e.g. sarcoid.

Causative organisms. As with granulomatous disease elsewhere, the usual causes are *Candida* spp., *Cryptococcus neoformans*, *Histoplasma capsulatum*, *M. tuberculosis* or *M. leprae*, *Toxoplasma gondii*, *Toxocara* spp. or *T. pallidum*, usually associated with systemic infection and/or immunoparesis.

Clinical features. Gradual visual loss with lack of pain are the main features.

The damaged retina is seen on fundoscopy (Fig. 2).

Confirmatory tests. These depend on the causative systemic disease and are usually serology, antigen detection for cryptococcosis, special cultures and chest X-ray.

Management. This is dictated by the cause of disease. Do not delay: specialist care is essential.

ENDOPHTHALMITIS

Endophthalmitis is a feared infection of the vitreous body which causes loss of vision or loss of the eye. Unlike the other two deep infections above, it is commonly bacterial, leading to abscess formation after surgery, trauma, corneal ulceration or bacteraemic spread (compare with the sources of cerebral and parameningeal abscesses, p. 90).

Causative organisms. Over 60 causative organisms have been described. Bacteria include staphylococci, streptococci, *Bacillus cereus*, Gram-negative rods and anaerobes. Fungi are involved particularly in the immunocompromised; genera

include *Candida*, *Aspergillus*, *Cryptococcus* and *Zygomycetes*. The parasites *O. volvulus*, *T. gondii*, *T. solium* and *Toxocara* spp. can also infect the vitreous body.

Clinical features. Ocular pain and visual loss, with photophobia, headache, fever and systemic illness are characteristic. Signs are swelling of the lids and conjunctivae (chemosis), limited eye movement, pus in the anterior chamber (hypopyon) and iritis from local spread and vitreal opacities (Fig. 3).

Confirmatory tests. These are essential because of the wide range of virulent pathogens. Microscopy and culture of anterior and posterior chamber fluid and any wound is urgent, as is specialist care.

Chemotherapy. Treatment is initially empiric, both intravitreal and systemic, and often subconjunctival or topical also. Initial anti-bacterial therapy usually includes vancomycin and gentamicin, while anti-fungal therapy is usually amphotericin B and flucytosine. Vitrectomy is usually needed, and steroids diminish the destructive immune response when infection is controlled.

Control and prevention. These depend on operative technique and rapid effective treatment of penetrating wounds and of bacteraemia and fungaemia.

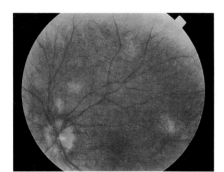

Fig. 2 **Chorioretinitis.**

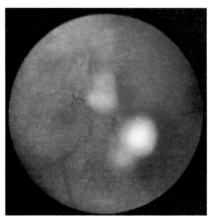

Fig. 3 **Endophthalmitis.**

Deep eye infections

- Deep eye infections are serious, requiring specialist care to save sight.
- Laboratory identification is important because of the wide range of potential pathogens.
- Areas infected are the iris and ciliary body, the choroid and retina, and the vitreous body.

STOMATITIS

The oral cavity (mouth) contains a normal flora which is controlled by the flushing action of saliva. Infection of the oral cavity is usually acute and is called stomatitis.

- Infections of the teeth are described on page 102.

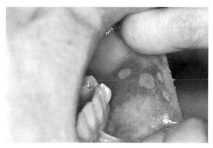

(a)

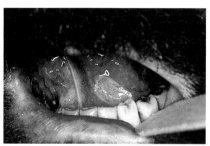

(b)

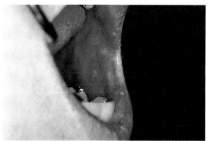

(c)

Fig. 1 **Aphthous ulcers. (a)** Minor; **(b)** Major; **(c)** Herpetiform ulcers.

- Viral infections are not described in this book.
- Oral manifestations of systemic disease are described in the relevant sections.

APHTHOUS STOMATITIS

Aphthous stomatitis is characterised by painful recurrent ulcers of the oral mucosa.

Causative organisms. These are not known. Immune mechanisms and local trauma may be involved, as may viruses.

Clinical features. The ulcers are classified in three types (Fig. 1):

- Minor ulcers: small in size (3–5 mm), very painful and often anterior and multiple. The ulcers have a yellow-grey base and a bright red edge; they heal in 4–14 days.
- Major ulcers: larger, occur throughout the mouth and are more chronic, taking 4–8 weeks to heal, then relapsing months later.
- Herpetiform ('shaped like a serpent' and *not* caused by herpes virus) ulcers: pinhead in size and multiple, appearing in crops.

Confirmatory tests. These are not available, but the clinical appearance is diagnostic (Fig. 1).

Chemotherapy. This is of no use. Local measures and dental attention are indicated. Local steroids are a last resort.

STOMATITIS IN THE IMMUNOCOMPROMISED

Immunocompromised patients are prone to stomatitis if their oral flora is disturbed (e.g. by broad-spectrum antibiotics), their oral mucosa damaged (e.g. by anti-cancer drugs causing 'mucositis'), or their other lines of defence impaired (e.g. by neutropenia). The stomatitis is classified as fungal, viral or bacterial, though in 'neutropenic stomatitis' only normal flora may be cultured.

Causative organisms. The commonest fungi are *Candida* spp., followed by *Aspergillus* spp. and the feared *Zygomycetes*; the commonest bacteria are anaerobes and mixed Gram-negative rods, while herpes simplex is the commonest virus.

Clinical features. These are variable and may suggest the causative organism: candidiasis ('thrush') shows white patches like a thrush's breast (Fig. 2); aspergillosis or zygomycosis cause progressive ulceration and necrosis, while anaerobic infection causes foul odour and tissue destruction.

Confirmatory tests. Microscopy and culture of mouth swabs may confirm infection, but oral flora often make culture difficult to interpret; biopsy is therefore necessary in progressive infection.

Chemotherapy. This depends on the pathogen and the duration of the predisposing factor(s), e.g. local nystatin or clotrimazole is sufficient for candidiasis if broad-spectrum antibiotics are stopped, while oral ketoconazole is needed if neutropenia persists.

Control and prevention. This depends on avoidance or removal of the predisposing factor(s).

GANGRENOUS STOMATITIS (NOMA, CANCRUM ORIS)

Gangrenous stomatitis is a rare and rapidly destructive infection caused by mixed anaerobes and spirochaetes, and occurs particularly in debilitated children. It begins on the gums and spreads outwards, destroying lips and cheeks, and exposing bone and teeth. Immediate i.v. penicillin is essential. Reparative surgery is usually needed.

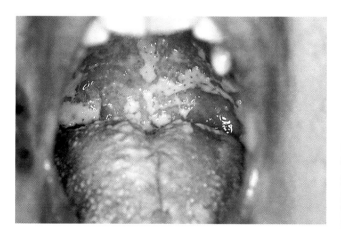

Fig. 2 **Candidial stomatitis.**

Stomatitis

- Aphthous ulcers are painful, recurrent and of uncertain cause.
- Fungal, bacterial and viral infections occur when oral flora is altered, oral mucosa damaged or general defence mechanisms are impaired.
- Gangrenous stomatitis is a rare, rapidly destructive disease needing urgent penicillin therapy and specialist care.

DENTAL AND PERIODONTAL INFECTIONS

The structure of teeth is shown in Fig. 1. The periodontal structures – ligament and gingiva – support the teeth and maintain the attachment to the bones of the jaw.

Dental, periodontal and related infections (Fig. 1) are classified into:

- **dento-alveolar infections** of the teeth and adjacent alveolar bone
- **periodontal disease**
- **deep fascial space infections**
- **other local spread**
- **metastatic spread by bacteraemia.**

Densely packed bacteria called **bacterial plaque** cause the three major infections of the teeth and periodontal tissue: supragingival plaque causes dental caries, and subgingival plaque causes gingivitis and periodontitis. Hyper-responsiveness of some immune mechanisms, and deficiency of others, also contribute to periodontal infections.

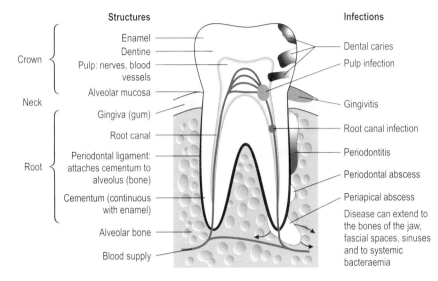

Fig. 1 **Dental and periodontal structures and infections.**

DENTO-ALVEOLAR INFECTIONS

Infections of the teeth and adjacent alveolar bone progress from dental caries ('tooth decay'), through pulp infection to periapical abscess and acute alveolar abscess.

Causative organisms

If teeth are not brushed for 1–3 days, oral streptococci, particularly *S. mutans*, and many other oral bacteria form dense plaque attached to the thin glycoprotein layer of pellicle on the tooth enamel, especially on the crown and at the gingival margin. Such plaque is removed by brushing, but completely mineralised plaque called dental calculus is not. In unmineralised dental plaque, bacterial metabolism of sucrose, glucose and fructose produces glucans and fructans, which are anaerobically converted to organic acids. The consequent low pH leads to enamel demineralisation and hence to dental caries.

Fig. 2 **Dental caries: extensive decay.**

Clinical features

Caries (Fig. 2) converts hard enamel to soft leathery tissue. Extension to the pulp causes toothache and temperature sensitivity, while further extension to a periapical or bone abscess (Fig. 3) produces severe throbbing pain, fever and eventually swelling and a discharging sinus (Fig. 4). X-rays show the extent of tooth and bone destruction.

Management

Chemotherapy is unhelpful for uncomplicated dental caries, which needs mechanical removal and repair by a dentist; endodontic infection (in the root canal) needs an endodontist. Periapical and bone infection need drainage and chemotherapy such as penicillin.

Control and prevention depend on oral or local fluoride, and regular brushing.

PERIODONTAL INFECTIONS

Infection of the periodontal structures is classified into four types:

Gingivitis. This is a reversible infection of the soft tissues of the gingiva, with no bony infection. All gingivitis is primarily caused by plaque (Fig. 5) but in addition, secondary factors including anaerobic and spiral bacteria (in 'trench mouth', Vincent's acute necrotising ulcerative gingivitis (ANUG)), hormones, drugs such as phenytoin or cyclosporin, and malnutrition worsen plaque effect.

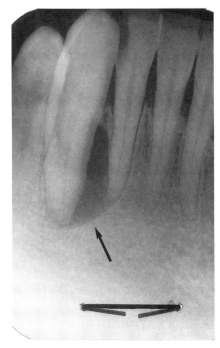

Fig. 3 **Periapical abscess.**

Periodontitis. This is irreversible infection of hard tissues, i.e. periodontal ligament, cementum and alveolar bone. It is classified into adult (the most common), rapidly progressive, juvenile and prepubertal types.

Periodontal abscess. This abscess complicates periodontitis when infection is trapped between tooth and gum.

Pericoronitis. This is an acute local infection beneath gum flaps covering the crown ('corona') of a partially erupted molar.

Causative organisms

In gingivitis, as bacterial plaque extends into the pocket between tooth and gingiva, the predominant streptococci and *Ancinomyces* are partly replaced by *Bacteroides intermedius* and other anaerobic Gram-negative rods. In ANUG there is infection with mixed fusiform anaerobes and spirochaetes (Fig. 6). In adult periodontitis, with deeper extension, anaerobes particularly *B. gingivalis* predominate. In juvenile periodontitis, *B. gingivalis* is replaced by *Actinobacillus actinomycetemcomitans* and *Capnocytophaga* spp. In periapical and alveolar abscesses, there is usually a mixture of anaerobic oral flora (*Actinomyces*, *Bacteroides*, *Fusobacteria* and *Peptostreptococci* spp.) and aerobic streptococci. Enteric Gram-negative rods and staphylococci are uncommon. In pericoronitis, there is (again) mixed oral flora.

Clinical features

Gingivitis shows tender, swollen, bleeding, usually painless, gums. In ANUG, there is also foul mouth odour, acute pain and a purulent exudate. There is no bony resorption. Periodontitis shows progressive gum recession with exposed teeth roots and bony resorption. Abscesses as usual cause throbbing pain, swelling and fever.

X-rays show the extent of bone destruction.

Management

Antibiotics such as penicillin and metronidazole are only useful chemotherapy in gingivitis (particularly ANUG) and abscesses. Local debridement with chlorhexidine gluconate mouthwash, and removal of plaque, calculus and secondary factors, such as drugs and malnutrition, are essential.

Dental hygiene by regular brushing and flossing minimises plaque formation.

OTHER INFECTIONS

- **Deep fascial space infections.** Direct spread of disease can occur between the deep fascia (p. 105).
- **Other local spread.** Although rare, this can occur to give:
  – sinusitis (p. 95)
  – osteomyelitis (p. 178)
  – cavernous sinus thrombosis (p. 97)
  – suppurative jugular thrombophlebitis (p. 105)
  – carotid artery erosion (p. 105)
  – mediastinitis (p. 105).
- **Metastatic spread by bacteraemia.** Metastatic spread is rare but can cause distant abscesses, infective endocarditis (p. 126) or infected joint prostheses (p. 180).

Fig. 4 **Jaw sinus from dental abscess.**

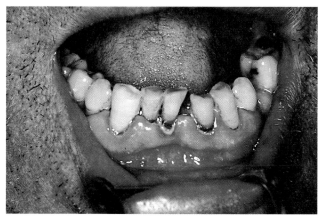

Fig. 5 **Gingivitis with calculus from poor hygiene.**

Fig. 6 **Vincent's infection (ANUG) Gram stain of mouth swab.**

Dental and periodontal infections

- Bacterial plaque builds up on teeth that are not cleaned; bacterial action produces acid, which attacks tooth minerals. Calculus is mineralized plaque.
- Dental caries is caused by supragingival bacterial plaque, while subgingival plaque causes gingivitis and periodontitis.
- All may be complicated by bacterial abscesses and spread to adjacent fascial spaces, to adjacent tissues and, rarely, to distant tissues.
- Chemotherapy is only useful for specific infections like ANUG, for abscesses and for bacterial infections in adjacent fascial spaces and tissues.
- Control depends on prevention by oral hygiene, including regular brushing and flossing.

THROAT INFECTIONS

PHARYNGITIS AND TONSILLITIS

Pharyngitis is inflammation of the pharynx, usually caused by infection, and may occur without tonsillitis. Tonsillitis is infection of the tonsils and is often accompanied by pharyngitis, i.e. pharyngotonsillitis. Classification is by the causative organisms.

Causative organisms

The two most common causes are viruses (about 40%) and streptococci (about 25%), of which a third to half is the important Group A *S. pyogenes*. Less common (only 1–2%) but therapeutically important are *Corynebacterium diphtheriae* (diphtheria), *Neisseria gonorrhoeae* (gonorrhoea), and *Fusobacteria* spp. and spirochaetes (Vincent's angina). *Mycoplasma* spp. or *Chlamydia pneumoniae* may well account for many of the 30% of patients with pharyngitis in whom no cause is found by conventional bacteriological techniques.

All are spread person-to-person by direct contact or aerosol, except Vincent's angina which is endogenous.

Clinical features

Streptococcal infection. This varies from mild to severe pharyngitis, characterised by fever, sore throat, enlarged regional lymph glands and a bright red pharynx (Fig. 1) and uvula, with purulent exudate on the pharyngeal wall, and tonsillitis with pus in the follicles (Fig. 2).

Diphtheria. There is a characteristic adherent grey-white membrane and, at times, a musty odour. Untreated, it progresses to systemic toxaemia and respiratory obstruction, especially with laryngeal diphtheria (p. 107).

Pharyngeal gonorrhoea. There may be no symptoms (hence the importance of culture) or mild pharyngitis with throat discomfort.

Vincent's angina is an infection with mixed anaerobes and spirochaetes, causing foul mouth odour and a purulent exudate. It is often associated with Vincent's infection of the gums, called acute necrotising ulcerative gingivitis (ANUG) or 'trench mouth' (p. 102).

Viral pharyngitis is usually mild, with variable associated symptoms and signs depending on the causative virus; e.g. in glandular fever. It is not reliably distinguished from streptococcal infection.

Confirmatory tests

Culture and microscopy of throat swabs will identify most bacterial pathogens. Streptococcal and viral causes can be

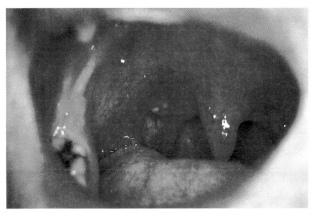

Fig. 1 **Streptococcal pharyngitis.**

distinguished serologically using rapid specific latex agglutination for the former and immunofluorescence for the latter.

Complications

Complications are uncommon if treatment is prompt. They are:

- local spread to give peritonsillitis or peritonsillar abscess (quinsy) and fascial space infections (see below)
- distant spread to give sinusitis, otitis media, mastoiditis (pp. 94–5); postanginal septicaemia (Lemierre's disease); or extension along the carotid sheath to give mediastinal infection (p. 105)
- toxin production: scarlet fever (pp. 31, 165)
- immune mechanisms: rheumatic fever (p. 125) and acute glomerulonephritis (p. 152).

Chemotherapy and control

There is debate whether penicillin should be given before diagnosis of a 'strep throat' is proved by the laboratory. On balance, it is better than progressing to rheumatic fever. Penicillin for 10 days is the drug of choice for streptococcal infections, diphtheria and Vincent's angina; ceftriaxone is more effective for gonorrhoeal pharyngitis. Ampicillin should not be given for any pharyngitis or tonsillitis, as it is less effective and provokes a rash in mononucleosis. Antitoxin is essential for diphtheria, with appropriate measures to maintain oxygenation and circulation. Viral infections are treated symptomatically.

Control and prevention is only really effective for diphtheria, by immunisation, and for Vincent's infection, by good oral and dental hygiene. All others depend on avoidance of sources of infection.

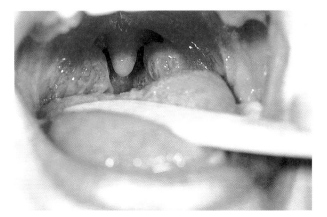

Fig. 2 **Tonsillitis.**

Fig. 3 **Quinsy: puncture mark over swelling from diagnostic aspiration.**

Structures

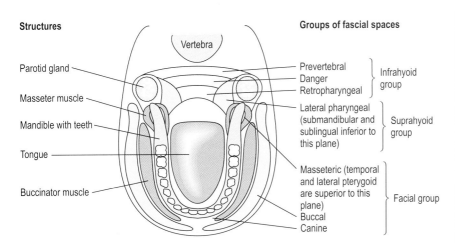

Groups of fascial spaces

Fig. 4 **Fascial spaces.** Horizontal section showing three groups of spaces.

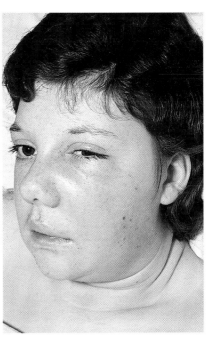

Fig. 5 **Buccal space infection showing facial swelling.**

PERITONSILLITIS AND PERITONSILLAR ABSCESS (QUINSY)

Untreated tonsillitis can lead to cellulitis then abscess formation in the peritonsillar tissues by local spread. Rarely there is direct extension along the carotid sheath to the mediastinum. Mixed anaerobes and streptococci are the usual causative organisms.

Clinical features. In addition to tonsillitis, there is more severe pharyngeal pain and fever, with dysphagia (difficult, painful swallowing) and even some respiratory obstruction. There is peritonsillar redness and swelling (Fig. 3), displacing the tonsil medially.

ENT specialist opinion is necessary, for peritonsillitis needs penicillin and metronidazole, while an abscess must be aspirated or drained surgically.

FASCIAL SPACE INFECTIONS

The primary *source* of infections in the fascial spaces is either a dental or periodontal infection (odontogenic infection), or a non-odontogenic infection, such as stomatitis, pharyngotonsillitis, parotitis or sinusitis.

Many of the fascial spaces (Fig. 4) between fascia, muscles, bones and other structures communicate with each other, but the spaces and their infections can be *classified* into three groups (Fig. 4).

Causative organisms
These are usually a mixture of anaerobic oral flora (*Actinomyces, Bacteroides, Fusobacteria* and *Peptostreptococcus* spp.) and aerobic streptococci. Enteric Gram-negative rods and staphylococci are uncommon.

Clinical features
These obviously differ somewhat with the site, but all show fever and pain, and

symptoms from the source (see above). Trismus naturally occurs with masticator space infections. Swelling and brawny induration is early with upper and superficial spaces (temporal, parotid, buccal (Fig. 5), canine, submandibular and sublingual), but late and internal with the other, deeper spaces. These deep and infrahyoid space infections, and Ludwig's angina (p. 107) in the submandibular and sublingual spaces, thus endanger the airway, cause dysphagia and can extend down into the mediastinum. Specialist consultation is essential.

Confirmatory tests
Microscopy and culture of aspirated or operative pus can confirm the causative organisms. Imaging by CT may be necessary.

Chemotherapy
This is important, particularly in the early cellulitic stage, but urgent drainage is necessary if swelling and abscess formation threatens the airway, plus dental treatment for odontogenic infections. Penicillin i.v. is usually sufficient in normal hosts, though metronidazole may be added. Clindamycin can be used in those with an allergy to penicillin. Cefoxitin, timentin or even imipenem may be needed in immunocompromised patients, e.g. with leukaemia.

Control and prevention
This depends on proper oral hygiene and dental treatment.

POSTANGINAL SEPTICAEMIA (LEMIERRE'S DISEASE)

Lemierre's disease is a rare but serious complication of Vincent's angina (anaerobic pharyngitis) which occurs from local spread to cause septic jugular vein thrombophlebitis, then bacteraemic spread

to cause lung and other metastatic abscesses. Sometimes there is direct extension along the carotid sheath to the mediastinum. *Fusobacterium necrophorum* is the usual causative organism.

Clinical features
In addition to tonsillitis and/or peritonsillar abscess, neck pain, stiffness, tenderness and dysphagia occur with fever, septicaemia, metastatic infection and death if untreated.

Management
This is usually with penicillin i.v. and metronidazole; appropriate circulatory and respiratory support is needed.

Throat infections

- Pharyngitis and tonsillitis are commonly caused by viruses or streptococci, rarely by gonorrhoea, diphtheria or Vincent's angina.
- The clinical features of streptococcal and viral infections are similar, and a swab for culture and antigen detection is advised.
- Complications of throat infections arise from local spread, distant spread, toxin production and immune mechanisms.
- Antibiotic treatment depends on the pathogen; penicillin is usually given on suspicion of streptococci. Anti-toxin is given for diphtheria. Fascial space and other abscesses need drainage plus chemotherapy.

EPIGLOTTITIS AND DIPHTHERIA

EPIGLOTTITIS

Epiglottitis is an infection of the epiglottis and the surrounding tissues, particularly around the larynx and subglottic area. It can be very acute in children and can cause respiratory arrest in 4 hours or less. It also affects adults.

Causative organisms

The causative organism is nearly always *Haemophilus influenzae* type b (p. 38), rarely other *Haemophilus* spp., pneumococci, streptococci or staphylococci. Epiglottitis develops when virulent *H. influenzae* infects those without specific antibodies, especially children aged 2–4 years who have lost maternal antibodies but not yet developed their own.

Clinical features

The distinctive features are dysphagia, drooling, dyspnoea, distress (respiratory, circulatory and mental), developing obstruction with stridor, and death if untreated. Sore throat is usual, and the child leans forward and drools because it cannot swallow its secretions. Auscultation early shows inspiratory stridor and tachycardia, then decreased breath sounds and bradycardia as death approaches. The throat should *not* be examined for the classical 'cherry red' epiglottis (Fig. 1), for throat examination may provoke respiratory arrest.

The differential diagnosis includes:

- **croup**, where the child usually has had a preceding URTI, lies supine and has a barking cough
- **diphtheria**, where the adherent grey-white membrane is typical
- **angioneurotic laryngeal oedema** or **foreign body**, where the history should be helpful
- **pharyngeal infections**, including severe tonsillitis, peritonsillar abscess, or retropharyngeal abscess, revealed if the throat can be examined
- **Ludwig's angina**, which shows submandibular swelling (see below).

Confirmatory tests

There may be no time to do any investigations before securing the airway. If feasible, a lateral X-ray shows the swollen epiglottis, 'like an adult thumb' (Fig. 2). A high white cell count and positive blood culture may later confirm diagnosis.

Management

Antibiotics are less urgent than securing the airway, usually by an expert in intubation. Tracheostomy may be necessary if intubation is impossible or unsafe. Ceftriaxone is now the antibiotic of choice, or chloramphenicol if it is unavailable.

Control and prevention. Rifampicin should be given at once to all household contacts and to the patient on recovery to eradicate carriage and prevent secondary cases. Vaccination should then be given.

DIPHTHERIA

Diphtheria, caused by toxin-producing strains of *Corynebacterium diphtheriae*, is a rare but important infection of the upper airways that causes respiratory obstruction and distant effects, particularly on myocardium and nervous tissue. It is classified by the major site of infection, i.e. nasal, pharyngeal and tonsillar, nasopharyngeal, laryngeal, bronchial or cutaneous diphtheria. 'Pseudo-diphtheria' means similar but milder infections with *C. haemolyticum* or *C. ulcerans*. The bacteria multiply locally without spreading. The exotoxin destroys epithelial cells and polymorphs causing a local ulcer. The toxin enters the blood stream causing fever, myocarditis (within the first 2 weeks) and polyneuritis (many weeks later).

Causative organism

C. diphtheriae (p. 32), is a Gram-positive aerobic rod. It has three colonial variants, small smooth 'mitis' (mild), through 'intermedius', to large rough 'gravis' (grave, severe), that in spite of their names do not correlate with virulence. It is spread person-to-person by respiratory droplets from patients or asymptomatic carriers.

Clinical features

Nasal diphtheria is usually mild, with purulent nasal discharge and few systemic symptoms. **Pharyngeal diphtheria** is the initial and commonest clinical presentation, with sudden onset of fever, malaise and pharyngitis, followed by a prominent

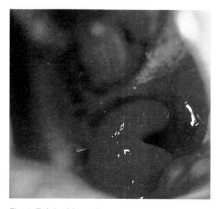

Fig. 1 **Epiglottitis — direct view.** NB: this is potentially very dangerous.

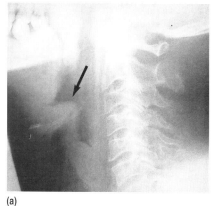

(a)

Fig. 2 **Epiglottitis (lateral X-ray).**
(a) Before treatment; (b) after treatment.

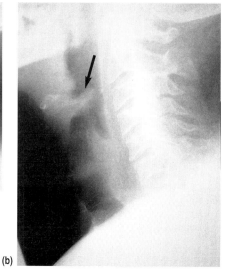

(b)

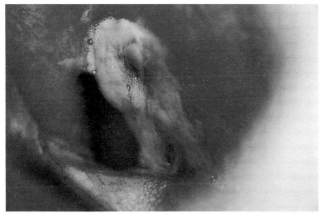

Fig. 3 **Diphtheria: membrane over tonsils.**

Fig. 5 **Respiratory paralysis in diphtheria treated in tank respirator.**

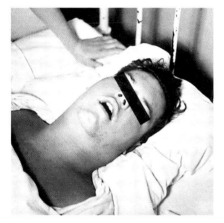

Fig. 4 **Bull neck diphtheria.**

pseudomembrane of bacteria and necrotic tissue cells on the tonsil and posterior pharynx (Fig. 3). This can spread upwards, with marked systemic symptoms, as **naso-pharyngeal diphtheria**, and/or downwards as **laryngeal** and **bronchial diphtheria**. Respiratory obstruction (Fig. 4) and death may follow rapidly, or arrhythmias and myocarditis or neurological complications, including peripheral neuritis, follow over 1–6 weeks.

Cutaneous diphtheria (Veldt sore, Barcoo rot) is now very rare. It forms a chronic indolent ulcer, either spontaneously or in a wound. Systemic symptoms rarely follow.

Pseudo-diphtheria. Rarely other corynebacteria cause 'pseudo-diphtheria', i.e. pharyngitis with scarlatiniform rash (*C. haemolyticum*) (p. 32) or severe diphtheria-like pharyngitis (*C. ulcerans*).

The differential diagnosis of diphtheria includes severe pharyngitis from streptococci, Vincent's angina (anaerobic pharyngitis), Ludwig's angina and infectious mononucleosis (pp. 104–5).

Confirmatory tests

Treatment cannot wait on laboratory tests, but throat and nasopharyngeal swabs should be taken for later confirmation. Gram stain is difficult to interpret without great experience, but culture on non-selective and special (tellurite, Loeffler's) media grow the typical colonies for biochemical confirmation. Elek's specific immunodiffusion test shows precipitation where toxin from the patient's organism meets anti-toxin from filter paper.

Management

Diphtheria is a life-threatening illness. As soon as the diagnosis is suspected the patient is isolated to reduce spread and treatment with anti-toxin is started. The airway must be secured. Penicillin i.v. is the drug of choice, and erythromycin is second choice. Cardio-respiratory support may be necessary (Fig. 5).

Control and prevention. Immunisation with killed vaccine, usually as DPT triple antigen with pertussis and tetanus, is highly effective for 10 years. Patient contacts need a booster if immunisation was more than 10 years ago. Erythromycin usually eradicates asymptomatic carriage.

LUDWIG'S ANGINA

Ludwig's angina is a severe inflammation of both sides of the floor of the mouth in the submandibular and sublingual spaces (p. 105). It results in massive swelling of the neck and, if untreated, can lead to airway obstruction and death.

Causative organisms. Oral flora are thought to be involved, both anaerobic and aerobic.

Clinical features. Submandibular swelling can spread to the neck. Fever is common.

Confirmatory tests. Culture of blood and operative specimens will indicate the particular pathogen(s).

Management. Drainage of the area may be necessary, and tracheostomy if the airway is blocked. Penicillin and metronidazole are antibiotics of choice. Prevention is through oral and dental hygiene.

Epiglottitis

- Epiglottitis usually affects children; it is usually caused by *Haemophilus influenzae* and is a serious infection with respiratory distress quickly leading to respiratory obstruction.
- The distinctive features are dysphagia, drooling, dyspnoea, distress (respiratory, circulatory and mental), developing obstruction with stridor, and death if untreated.
- The throat should not be examined before the airway is secured. Ceftriaxone is now usual treatment.

Diphtheria

- Diphtheria is a serious infection of the upper (and sometimes the lower) airways causing respiratory obstruction and, at times, arrhythmias, myocarditis, peripheral neuritis and death.
- It is caused by *Corynebacterium diphtheriae*, with a potent exotoxin. It is diagnosed clinically and confirmed later by culture, Gram stain, biochemical tests and immunodiffusion for toxin production.
- Treatment is anti-toxin and penicillin, with respiratory and cardiac support. Prevention is by immunisation.

TROPICAL AND RARE ORO-FACIAL INFECTIONS

ACTINOMYCOSIS

Actinomyces spp. are found in normal oral flora (p. 56), and cervico-facial actinomycosis develops when oral or dental infection, trauma or surgery gives a portal of entry into the tissues.

Clinical features. Typically, there is a painful lump in the face or neck, with purple-red overlying skin, which necroses to form a sinus. Mild fever and systemic symptoms occur, and the infection may spread to mandible (Fig. 1), sinuses or orbit. It may mimic tuberculosis.

Confirmatory tests. These are microscopic examination and prolonged anaerobic culture of pus, or histopathology on excised tissue.

Chemotherapy. Prolonged penicillin therapy is recommended, initially intra-

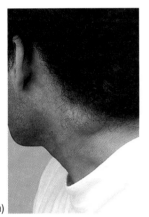

(a)

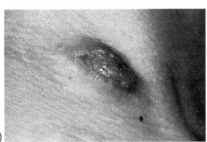

(b)

Fig. 2 **Tuberculous cervical gland (a) enlargement and (b) ulceration.**

venous, followed by oral penicillin for 6–12 months. **Control** is by dental care and appropriate chemoprophylaxis.

CERVICAL LYMPHADENITIS

Enlarged cervical lymph nodes are seen in local infections and as a consequence of generalised lymphadenopathy.

Mycobacterial infection

This is an important cause of enlargement, often without other signs of disease. Clinically, the nodes are painless and fluctuant under purplish thin skin (Fig. 2a) and ulcerate if untreated (Fig. 2b).

Confirmatory tests. Staining for acid-fast bacilli, culture of the pus and chest X-ray are confirmatory.

Management. Treatment is specific (usually triple) therapy including isoniazid and rifampicin for tuberculosis itself. Atypical mycobacterial infections are more antibiotic resistant and usually need excision. Fluctuant tuberculous nodes should never be incised, or a chronic sinus and, later, scarring results.

Control and prevention are by pasteurisation of milk, immunisation by BCG and prevention of spread from open infections.

OTHER INFECTIONS

Parotitis

Inflammation of the parotid gland can be bacterial or viral (mumps) in origin.

Acute suppurative parotitis is now uncommon; it follows oral neglect and dehydration from any cause. Local pain, swelling and redness (usually unilateral, Fig. 3) are accompanied by fever; spread is serious, causing osteomyelitis, blockage of airways and bacteraemia.

Chronic bacterial parotitis is recurrent acute exacerbations of chronic low-grade infection, causing gradual parotid destruction.

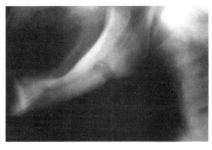

Fig. 1 **Actinomycosis of mandible.**

Confirmatory tests are microscopy and culture of aspirated or operative pus.

Management involves flucloxacillin and metronidazole for both forms of parotitis. Acute infection needs surgical drainage but advanced chronic destruction or loculated abscesses (Fig. 3b) necessitate parotidectomy.

Paracoccidioidomycosis (South American blastomycosis)

This systemic fungal infection is caused by *P. brasiliensis* (p. 67). Chronic 'mulberry' oral or nasal mucous membrane lesions or warty skin ulcers develop years after the primary pulmonary infection. Treatment is with ketoconazole.

Rhinoscleroma

This rare infection is caused by *Klebsiella rhinoscleromatis* (p. 43). Clinically, there is progressive painless nasal deformity and distortion or destruction of respiratory passages. It is treated with streptomycin, oral tetracycline or cotrimoxazole.

Rhinosporidiosis

This is a rare subcutaneous mycosis (p. 69) that usually involves nasal mucosa. The causative fungus is *Rhinosporidium seeberi*. Clinically, there are large vascular, friable warts or polyps which are excised.

Gangrenous stomatitis

This rare infection, also called noma or cancrum oris ('cancer of the mouth'), destroys lips and cheeks (p. 101).

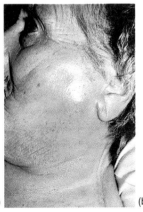

(a)

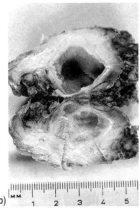

(b)

Fig. 3 **(a) Acute parotitis with swelling; (b) parotid abscess.**

Tropical and rare oro-facial infections

- Cervico-facial actinomycosis is a chronic bacterial infection from oral or dental disease or trauma.
- Mycobacterial cervical lymphadenitis is either tuberculous or 'atypical' and causes relatively painless fluctuant matted nodes that ulcerate and discharge caseous pus to leave a chronic sinus.
- Parotitis is acute after oral neglect and dehydration, chronic with gradual gland destruction, or viral (mumps).
- Paracoccidioidomycosis is a systemic mycosis with oral or nasal warty ulcers.

LARYNGITIS, TRACHEITIS AND PERTUSSIS

LARYNGITIS

Laryngitis is inflammation of the larynx and may occur alone, or with proximal pharyngitis, or distal tracheitis and bronchitis. It is classified by the causative organisms.

Causative organisms. Laryngitis alone is usually caused by viruses, but bacterial laryngitis occurs in about 10% of patients with streptococcal pharyngitis (p. 104). Laryngitis without pharyngitis can occur in diphtheria, tuberculosis and *Branhamella catarrhalis* infections.

Laryngo-tracheo-bronchitis (croup), while common and important, is almost always caused by viruses.

Clinical features. Hoarseness is the major symptom, which may progress to aphonia. Fever suggests a bacterial cause. The distinctive membrane of diphtheria (p. 106) is not easily visible in the larynx until almost too late. Infection of the larynx by sputum was seen in 30% of patients with advanced open pulmonary TB and is now sometimes caused by metastatic spread. *B. catarrhalis* produces acute hoarseness and mild fever.

Confirmatory tests. Throat swabs and sputum need special microscopy and culture if diphtheria, TB or even rarer causes like syphilis are suspected.

Chemotherapy. Diphtheria needs specific anti-toxin and penicillin urgently, and tuberculosis needs specific, usually triple, therapy. *B. catarrhalis* is not uncommonly beta-lactamase producing, and so needs amoxycillin plus clavulanate, or ceftriaxone.

Control and prevention. Diphtheria is preventable by immunisation, and BCG gives some protection from tuberculosis.

TRACHEITIS

Tracheitis is inflammation of the trachea and may occur alone or with proximal laryngitis or distal bronchitis.

Causative organisms. Tracheitis alone is usually caused by viruses, but sometimes acute bacterial tracheitis occurs in adults and children, and must be distinguished from epiglottitis or croup. It is usually caused by *S. pyogenes*, *H. influenzae* type b or *S. aureus*.

Clinical features. Substernal pain and cough are the major symptoms. Bacterial tracheitis usually follows tracheal injury or intubation, with acute onset of pain, hoarse cough and fever. Mucosal swelling and copious sputum can cause airway obstruction.

Fig. 1 **Tenacious sputum after cyanotic spasm in pertussis.**

Fig. 3 **Subconjunctival haemorrhages in pertussis.**

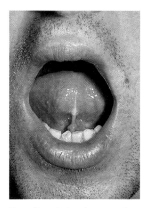
Fig. 2 **Frenal ulcer in pertussis.**

Confirmatory tests. Throat swabs and sputum are used for microscopy and culture.

Chemotherapy. Ceftriaxone is suitable initial treatment for bacterial tracheitis until the specific pathogen is identified.

Control and prevention. This depends on avoiding viral infections and on tracheal care in hospital.

PERTUSSIS

Pertussis is a severe preventable bacterial infection of the ciliated epithelium of large airways, characterised by spasmodic attacks of coughing.

Causative organism. Most attacks of pertussis are caused by *Bordetella pertussis* but a minority are caused by *B. parapertussis* (p. 39).

Clinical syndromes. The clinical features occur in three stages:

1. Catarrhal stage: a very infectious stage lasting 1–2 weeks with non-specific symptoms similar to the common cold.
2. Paroxysmal stage: lasts 2–4 weeks with the classic whooping spasms of uncontrollable repetitive cough until breathless or even cyanotic (Fig. 1), then a gasping inspiratory 'whoop' through the narrowed glottis for air, repeated 30–50 times a day. This is often followed by vomiting and expectoration of tenacious sputum (Fig. 1). Airway obstruction can occur. The forceful cough may

produce fernal ulcers (Fig. 2) or subconjunctival (Fig. 3) or other haemorrhages.
3. Convalescent stage: the paroxysms gradually disappear over 3–6 weeks, but pneumonia, fits or encephalopathy may appear.

Confirmatory tests. Confirmatory tests are scarcely necessary in a classic case; the diagnosis may be proved by rapid direct immunofluorescence, or by a pernasal swab. Blood films show lymphocytosis.

Chemotherapy. Erythromycin decreases infectivity though it does not shorten the clinical course. It should also be given to susceptible, unvaccinated household contacts.

Control and prevention. The vaccine, usually with tetanus and diphtheria in triple antigen, is moderately effective, though unwarranted scare campaigns about its side-effects have diminished acceptability.

Laryngitis, tracheitis and pertussis

- Laryngitis is usually viral, but *B. catarrhalis* infection, diphtheria and tuberculosis can be causative.
- Tracheitis is usually viral, but a bacterial cause should be excluded in acute tracheitis after tracheal trauma.
- Pertussis is a severe, preventable bacterial infection characterised by recurrent paroxysms of coughing with 'whooping' inspirations, tenacious sputum and uncommon but serious complications. Erythromycin decreases infectivity, and vaccination is moderately protective.

BRONCHIAL INFECTIONS

Bronchial infection may be a primary disease or secondary to underlying broncho-pulmonary disease.

ACUTE BRONCHITIS

Bronchitis is inflammation of the bronchi and may occur alone, or with proximal tracheitis or distal pulmonary infection. It is classified by the causative organisms.

Causative organisms

Bronchitis alone is often caused by viruses, but about 40% is bacterial, usually one of the five common respiratory pathogens: *S. pneumoniae, H. influenzae* type b, *Branhamella (Moraxella) catarrhalis, Mycoplasma pneumoniae* or *Chlamydia pneumoniae* ('TWAR'). Rarely, other bacteria are the cause.

Clinical features

Cough and purulent sputum are the major symptoms. Fever and noisy respiration are less common. Dyspnoea indicates pre-existing or current lung disease.

Confirmatory tests

Bacterial infections need sputum microscopy and culture. Mycoplasma or chlamydial infections need special cultures or specific serology.

Chemotherapy and control

Erythromycin is suitable initial treatment for bacterial bronchitis until laboratory results show the specific bacteria. Amoxycillin is inactive against mycoplasma, and *C. pneumoniae* needs a tetracycline.

    Control and prevention depends on avoiding contact with similar infections. Influenza vaccine is particularly useful in epidemic years.

ACUTE INFECTIONS IN CHRONIC OBSTRUCTIVE PULMONARY DISEASE

Chronic obstructive pulmonary disease (COPD) includes emphysema and chronic bronchitis: the latter is defined as the production of sputum on most days for at least 3 months for more than 2 years (if wheeze and bronchospasm accompany the sputum production, it is called asthmatic bronchitis). The role of infection in *chronic bronchitis* is complex: although it is seldom a causative factor, it may be a perpetuating factor and often is an exacerbating factor. As infection is only one factor,

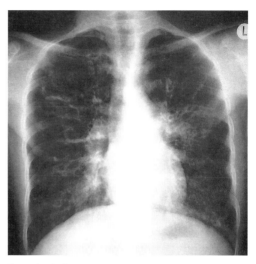

Fig. 1 **Cystic fibrosis: chest radiograph showing bronchiectasis and peripheral air trapping.**

along with smoking and occupational dust or fumes, so mucus gland hyperplasia and intraluminal mucus and pus are often more prominent than bronchial wall infection, though the usual bacterial respiratory pathogens, particularly pneumococci or *H. influenzae*, are common in intraluminal or expectorated sputum. The chronic, multifactorial inflammatory response in the bronchial wall leads to scarring, obstruction and bronchiectasis ('dilated bronchi', see below).

Causative organisms

Acute exacerbations are caused by viruses (including influenza) in about 50%, and by pneumococci or *H. influenzae* in about 40%.

Clinical features

Cough and purulent sputum are the major symptoms. Dyspnoea and wheeze are common because of the pre-existing lung disease. Fever is less common.

Confirmatory tests

Bacterial infections need sputum microscopy and culture. Chest X-ray helps to distinguish bronchial from pneumonic infection.

Chemotherapy and control

Oral amoxycillin, amoxycillin–clavulanate, cephalosporin or doxycycline are reasonable empiric treatment for acute

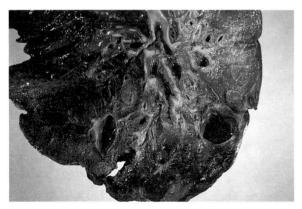

Fig. 2 **Cystic fibrosis lung at autopsy showing cystic spaces and lung destruction.**

Fig. 3 **Bronchiectasis secondary to tuberculosis at autopsy showing grossly dilated bronchi.**

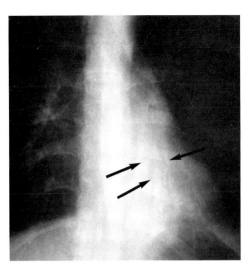

Fig. 4 **Bronchiectasis: chest radiograph showing dilated bronchi behind the cardiac shadow.**

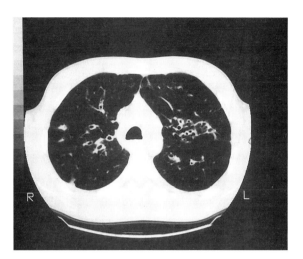

Fig. 5 **Bronchiectasis: CT scan of same patient reveals more extensive disease.**

bacterial exacerbations. Long-term suppressive antibiotics are not often of use.

Influenza vaccine is particularly useful in epidemic years.

ACUTE INFECTIONS IN CYSTIC FIBROSIS

Cystic fibrosis (mucoviscidosis) is a congenital disease of the pancreas and lungs where abnormally viscid mucus causes bronchial and bronchiolar plugging; this then produces bronchial and pulmonary infection, which causes more mucus plugging, hence more infection, with eventual bronchiectasis (Fig. 1), lung destruction and death (Fig. 2).

Causative organisms
In childhood *H. influenzae* and *S. aureus* are the prominent pathogens, replaced between age 5 and 18 by *Pseudomonas aeruginosa*, which may be accompanied in late disease by *Burkholderia* (previously *Pseudomonas) cepacia*.

Clinical features
Cough and large amounts of thick purulent sputum are the major symptoms, followed by fever and wheeze. Dyspnoea and cyanosis occur with progression of the lung disease. Systemic symptoms include tiredness, weakness, anorexia and weight loss with bulky malodorous stools.

Confirmatory tests
Acute exacerbations need sputum microscopy, culture and antibiotic sensitivities, for the organisms change with age and antibiotic treatment. Chest X-ray helps to distinguish bronchial from pneumonic infection and shows the development of bronchiectasis and air-trapping from bronchial obstruction (Fig. 1).

Chemotherapy and control
H. influenzae is usually treated with amoxycillin, amoxycillin–clavulanate or cephalosporins, *S. aureus* with amoxycillin–clavulanate or flucloxacillin, and *P. aeruginosa* with gentamicin or tobramycin with ticarcillin or ceftazidime; other, reserve antibiotics such as aztreonam and imipenem are needed as resistance develops.

The use of chronic suppressive antibiotics has considerable disadvantages as well as advantages, so active physiotherapy and postural drainage are very important. Lung transplantation is now done for late disease.

BRONCHIECTASIS

Bronchiectasis means dilated bronchi and is not a disease of itself but the result of bronchial obstruction by secretions or scarring, e.g. after measles, chronic bronchitis, cystic fibrosis or tuberculosis (Fig. 3).

Causative organisms
These are primary from the causative disease, or secondary following the obstruction and dilatation, including particularly the common respiratory pathogens *S. pneumoniae* and *H. influenzae*.

Clinical features
Cough and considerable purulent sputum are the major symptoms. Fever and haemoptysis are less common. Dyspnoea results from the underlying lung disease.

Confirmatory tests
Bacterial infections need sputum microscopy, culture and sensitivity tests. Chest X-ray may show the dilated bronchi (Fig. 4), but CT is more sensitive and specific (Fig. 5).

Chemotherapy and control
Oral amoxycillin, amoxycillin–clavulanate, cephalosporin or doxycycline are reasonable empiric treatment for acute exacerbations.

Control and prevention depend on diagnosis and treatment of the predisposing causes.

Bronchial infections

- Bronchial infection may be primary (acute bronchitis), or secondary to underlying broncho-pulmonary disease, such as chronic bronchitis, cystic fibrosis or bronchiectasis.
- Causative organisms include viruses and the five common respiratory pathogens: *S. pneumoniae, H. influenzae* type b, *Branhamella (Moraxella) catarrhalis, Mycoplasma pneumoniae* or *Chlamydia pneumoniae* ('TWAR'). In cystic fibrosis, *S. aureus* and then *P. aeruginosa* are most important.
- Clinical features are cough and sputum, with variable wheeze, fever, haemoptysis and dyspnoea.
- Treatment should be empirical antibiotics initially, with changes when the pathogen has been identified by sputum microscopy and culture and sensitivities. Chest radiographs and CT scan indicate the extent of disease.

PNEUMONIA IN THE NORMAL HOST

Pneumonia means an infection of the lung tissue. There are over 60 different types and causes, and it can be classified in many ways. For example:

- Anatomically: left, right, upper lobe, lower lobe.
- Bacteriologically (actually, microbiologically), by the causative organism: pneumococcal, chlamydial, tuberculous, etc.
- Clinically, by the patient's situation, presentation and associations:
  —neonatal
  —previously healthy child or adult
  —complicating chronic pulmonary disease
  —acquired by aspiration instead of inhalation
  —hospital acquired (nosocomial)
  —in an immunocompromised patient.

Pneumonia is commonly described as:

- lobar: involves distinct region of lung
- broncho-pneumonia: diffuse patchy, spreading throughout lung
- interstitial: invasion of interstitium
- necrotising, cavitation and destruction of parenchyma, forming abscesses.

Here the primary classification used is clinical.

NEONATAL PNEUMONIA

Neonatal pneumonia can be of three types.
Congenital pneumonia. Congenital infection is transplacental and can result from syphilis, toxoplasmosis or viruses (including cytomegalovirus, herpes simplex or rubella). Babies are small with specific signs of the causative organism. If possible, treatment is for the pathogen implicated.

Intrapartum pneumonia. This occurs as the result of aspiration of infected amniotic fluid or bacteria from the maternal birth canal. The pathogens are usually group B streptococci (*S. agalactiae*), Gram-negative rods or *Chlamydia trachomatis*. Intrapartum pneumonia presents with rapid respirations, asphyxia or generalised sepsis; fever is often absent. High-dose empirical antibiotics are essential immediately: usually an aminoglycoside with either ampicillin or a third-generation cephalosporin. Oxygen and supportive therapy are important.

Postpartum pneumonia. Infection in the period after birth is nosocomial from staff or equipment and is usually caused by staphylococci, Gram-negative rods or viruses. The clinical syndrome is one of respiratory distress, including tachypnoea or apnoea, rib retraction and grunting respiration. Treatment is the same as for intrapartum pneumonia.

Confirmatory tests
These include culture of respiratory secretions and blood, chest X-ray, blood gases and biochemistry. Co-existent meningitis should be considered, with lumbar puncture as necessary. Specific serology is done in congenital pneumonia.

Control and prevention
Congenital disease can be prevented or treated more effectively by screening of mothers for syphilis and rubella antibodies. Treatment of maternal infections, and caesarean sections to avoid birth canal pathogens reduce intra- and postpartum risks of infection. Neonatal respiratory suction and prophylactic antibiotics are used in high-risk infants.

PNEUMONIA IN THE PREVIOUSLY HEALTHY

This is conventionally classified into:

- classical or typical pneumococcal pneumonia
- atypical pneumonia.

However, the clinical distinction in real life is not always easy. Classical pneumococcal pneumonia is 'typical' in being of abrupt onset, severe, lobar and often with pleurisy. It has one microbial

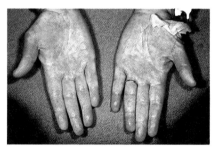

Fig. 2 **Erythema multiforme rash in mycoplasma pneumonia.**

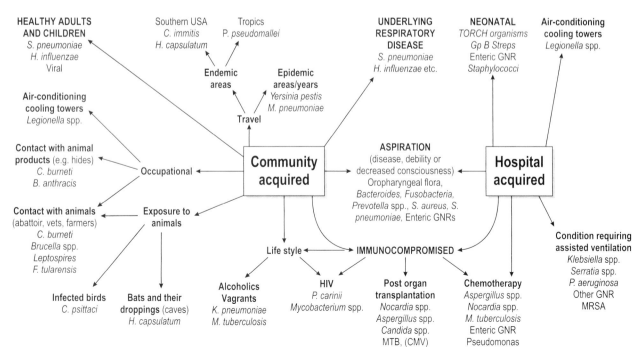

Fig. 1 **Some causes of pneumonia** (see also Fig. 1, p. 114).

cause and has been penicillin responsive. Atypical pneumonia, in contrast, is usually of more gradual onset, less severe, affects one or more segments, seldom gives pleuritic pain, has many microbial causes and is seldom penicillin responsive.

Causative organisms

Classical lobar pneumonia is caused by the pneumococcus (*S. pneumoniae*), while 'atypical' pneumonia has many causes for which the setting in which disease occurs — the epidemiology — may indicate a likely pathogen (Fig. 1).

Clinical syndromes

Classical lobar pneumococcal pneumonia characteristically begins abruptly with a rigor and fever, then dry cough followed by sharp chest pain on inspiration (pleuritic pain, which is shoulder-tip if diaphragmatic pleura is infected). Sputum is initially absent or minimal, streaked with blood ('rusty'). Systemic signs are severe, and the pneumonia is often fatal if untreated.

While 'atypical' pneumonia has many causes and hence variable clinical features, in most temperate areas there are five or six important causes (*Chlamydial, Legionella, Mycoplasma*, staphylococcal, TB, viral), which may be severe but are in general of more gradual onset than pneumococcal pneumonia, less severe, affecting one or more segments rather than a whole lobe, seldom giving pleuritic pain and are not penicillin responsive.

Similarly, each developing country or tropical area has its particular spectrum of common pneumonias, as do particular occupations or hobbies (Fig. 1).

There are few distinguishing signs, especially in temperate areas, though mycoplasma pneumonia may cause Raynaud's phenomenon, due to cold agglutinins, or erythema multiforme (Fig. 2).

Confirmatory tests

Difficulties arise in deciding what pathogen is involved and therefore, how best to treat. The choice of test will depend on the epidemiological clues and geographic area but in general will include sputum microscopy by Gram stain (which can guide initial treatment: Table 1) and special stains, sputum and blood cultures. Chest X-ray (Figs 3 and 4), specific serology, and sometimes histopathology of tissue are contributory to diagnosis.

Chemotherapy

It is often difficult to choose the initial empiric therapy. Penicillin by injection is the drug of choice for pneumococcal pneumonia unless high-level resistance is known or suspected. Most atypical pneumonias do not respond to penicillin, so alternative treatments are used in particular cases. These are:

- adults: erythromycin plus a third-generation cephalosporin
- children: *H. influenzae* and staphylococci are more likely, so flucloxacillin is substituted for erythromycin
- endemic areas or with epidemiological clues: specific therapy should begin.

In all cases, initial therapy should be reviewed after laboratory results and clinical progress (or deterioration) are assessed.

Control and prevention

Vaccination is available against pneumococci, *H. influenzae*, influenza and TB for the general population, and against plague, Q fever and scrub typhus in specific situations. The animal host can be vaccinated or killed in anthrax, brucellosis, plague and TB.

Table 1 **The use of Gram stain to guide initial therapy in pneumonia**

Major organism in purulent sputum	Probable pathogen	Empiric treatment
Gram-positive cocci	Diplococci, probably *S. pneumoniae*	Penicillin or ceftriaxone
	Cocci in clusters, probably *S. aureus*	Flucloxacillin
Gram-negative cocco-bacilli	May have capsules, probably *H. influenzae*	Ceftriaxone or cefotaxime (Amoxycillin now unreliable)
Gram-negative rods	Resemble 'enterics'	Third-generation cephalosporin +/– gentamicin
	Resemble pseudomonads	Ticarcillin + tobramycin

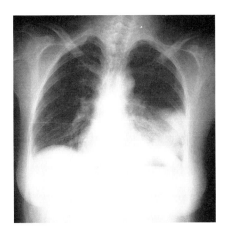

Fig. 3 **Pneumococcal lobar pneumonia.**

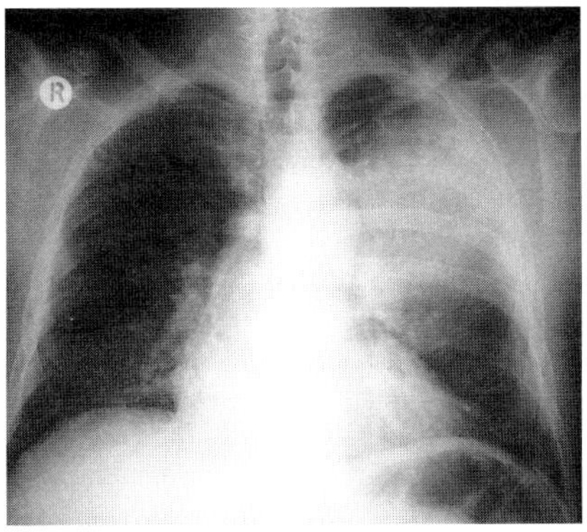

Fig. 4 **Legionella pneumonia in left mid-zone on chest radiograph.**

Pneumonia in the normal host

- Neonatal pneumonia needs urgent investigation and treatment. It can be:
  - congenital, by transplacental infection
  - intrapartum by aspiration of infected amniotic fluid or maternal birth canal organisms.
  - postpartum by nosocomial infection from staff or equipment.
- Previously healthy adults and children can be considered to have either typical or 'atypical' pneumonia.
- Classical pneumococcal pneumonia is 'typical' in being of abrupt onset, severe, lobar, often with pleurisy, has one microbial cause and is often penicillin responsive.
- Atypical pneumonia is usually of more gradual onset, less severe, affects one or more segments, seldom gives pleuritic pain, has many microbial causes and is seldom penicillin responsive.
- Atypical pneumonia in most temperate areas has five or six important causes: *Chlamydia, Legionella, Mycoplasma*, staphylococcal, TB and viral pneumonia.
- The cause of atypical pneumonia can be suggested by epidemiological clues including occupation, travel or residence, animal exposure and hobbies.

PNEUMONIA IN THE ABNORMAL HOST

Pneumonia is often a complication of a previously existing condition (either acute or chronic) which predisposes to the disease (Fig. 1). Because a wide range of pathogens can be causative, confirmatory tests are urgent and initial treatment is empirical.

Confirmatory tests

Sputum microscopy with Gram stain (plus special stains) and sputum culture are essential as is blood culture if bacteraemia is likely. Chest X-ray indicates areas of consolidation; some particular pathogens show typical features (see below). Blood gases and biochemistry are needed in severe disease, and sometimes serology may be useful. The different groups of predisposing diseases tend to show specific identifying features (see below).

PNEUMONIA COMPLICATING UNDERLYING PULMONARY DISEASE

Three main groups of pulmonary disease can be complicated by pneumonia:

- chronic obstructive pulmonary disease (COPD) (p. 110)
- cystic fibrosis (p. 111)
- viral respiratory infections.

The occurrence of COPD in many elderly people makes them vulnerable to pneumonia (the most common cause of death in the UK and USA). In addition to *S. pneumoniae* and *H. influenzae* (Fig. 1), enteric Gram-negative rods and *S. aureus* infect the elderly, especially in nursing homes.

Clinical syndrome

Abrupt onset and severe disease are usual with pneumococcal pneumonia (p. 112), while other pathogens generally cause more gradual onset of increased cough and sputum, fever and dyspnoea. Haemoptysis and pleuritic pain are less common. Severe pneumonia causes hypoxia and confusion, hypotension and circulatory failure.

Confirmatory tests

Chest X-ray in staphylococcal pneumonia characteristically shows multiple small abscesses which may leave cysts called pneumatocoeles (Fig. 2). Blood gases and biochemistry are needed in severe pneumonia.

Chemotherapy

Initial empiric therapy may be guided by the Gram stain on purulent sputum and is

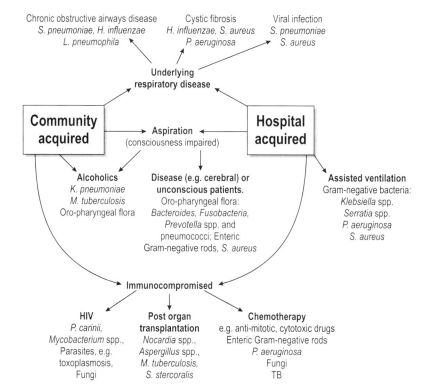

Fig. 1 **Pneumonia as a complication of pre-existing conditions.**

aimed at the likely causes. In COPD, intravenous penicillin or a third-generation cephalosporin is used, in cystic fibrosis ticarcillin and tobramycin, and in post-viral pneumonia penicillin or flucloxacillin. Subsequent specific therapy is guided by the sputum culture results in conjunction with the clinical response, i.e. penicillin or ceftriaxone are the drugs of choice in pneumococcal pneumonia, flucloxacillin in staphylococcal pneumonia, erythromycin in legionellosis, etc.

Control and prevention

Pneumococcal and influenza vaccine should be given to patients with COPD, and haemophilus vaccine to children.

ASPIRATION PNEUMONIA

Aspiration of oro-pharyngeal secretions occurs in any disease state where consciousness and hence normal gag and swallowing reflexes are impaired. It may be community acquired, e.g. in alcoholics, or nosocomial, e.g. in cerebral disease. Infection is often with oro-pharyngeal flora including anaerobes (Fig. 1).

Clinical features

Cough, sputum, fever and tachypnoea develop, with dyspnoea, hypoxia, cyanosis and respiratory and circulatory failure in severe cases.

Confirmatory tests

Sputum examination is less helpful than in other pneumonias, as the principal causative organisms are normal oropharyngeal flora that will be present in expectorated sputum from any pneumonia. Chest X-ray may show unilateral peripheral opacities in all lobes from aspiration while lying on one side (Fig. 3). If aspiration occurs while lying on the back (supine), most secretions enter the right upper lobe bronchus, causing upper lobe pneumonia.

Chemotherapy and control

As sputum examination is unhelpful, chemotherapy is empiric with metronidazole, often with penicillin or timentin, particularly if aspiration occurred in hospital.

Control and prevention depends on adequate care of patients with impaired gag and swallowing reflexes, e.g. the unconscious patient or those with some nervous system diseases such as pseudobulbar palsy.

HOSPITAL-ACQUIRED PNEUMONIA

Pneumonia may occur in any hospitalised patient but is particularly seen in those requiring assisted ventilation. Enteric bacteria often colonise such patients without causing disease, and their presence in respiratory cultures without

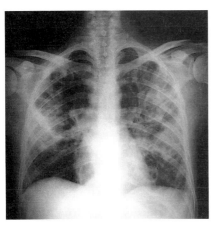

Fig. 2 **Staphylococcal pneumonia.** Chest radiograph showing bilateral abscesses and pneumatocoeles.

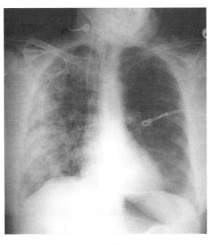

Fig. 3 **Aspiration pneumonia.** Chest radiograph showing right-sided peripheral opacities in all lobes from aspiration while unconscious, lying on the right side.

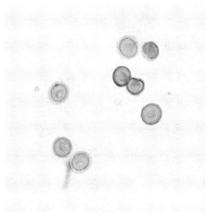

Fig. 4 *Pneumocystis carinii* from 'induced sputum'.

chest radiograph changes, neutrophilia or worsening pulmonary function does not mean infection.

Clinical features

Cough, sputum, dyspnoea and often fever occur in non-intubated patients, while those on ventilators show increased respiratory secretions, increased ventilatory requirements and fever.

Confirmatory tests

Neutrophilia is less common in old, ill patients. Further tests may be needed to exclude pulmonary embolism or other non-infectious lung pathology.

Chemotherapy and control

A beta-lactam such as a third-generation cephalosporin or timentin is often combined with an aminoglycoside like gentamicin for initial empiric treatment; they are replaced by specific antibiotics when cultures reveal the probable causative

organism. Respiratory and cardiac support are often necessary.

Physiotherapy with aseptic airway and ventilator care help in prevention.

PNEUMONIA IN THE IMMUNOCOMPROMISED PATIENT

This is usually a particular subgroup of hospital-acquired pneumonia, though sometimes it is community acquired. It is a special hazard for leukaemic and cancer patients on anti-mitotic chemotherapy, and for organ transplant patients, as well as for patients on high-dose steroids or cytotoxic drugs. Infection can occur with a wide range of organisms including enteric Gram-negative rods, fungi, protozoa, parasites and viruses.

Clinical features

The course of the pneumonia is often rapid, with dyspnoea, cough, sputum and haemoptysis, followed by respiratory failure.

Confirmatory tests

Alveolar sputum induced with nebulised saline is necessary to diagnose *P. carinii* pneumonia ('PCP') (Fig. 4). PCP has typically a uniform 'ground glass' appearance in a chest radiograph, while *C. neoformans* may cause a localised 'cryptococcoma' (Fig. 5). Tuberculosis, nocardiosis and staphylococcal pneumonia produce cavities, while aspergillosis often infects pre-existing cavities.

Management

Urgent empiric broad-spectrum cover usually includes a beta-lactam such as a third-generation cephalosporin or timentin combined with an aminoglycoside like gentamicin. Specific drugs are used when the pathogen is identified or probable. Respiratory and cardiac support are often necessary.

Chemoprophylaxis is used against TB in Mantoux-positive immunocompromised patients, and against PCP and atypical mycobacteria, while pure air and airway care can help in prevention.

Fig. 5 **Cryptococcoma in lung at post-mortem.**

Pneumonia in the abnormal host

- Pneumonia complicates several pre-existing pulmonary diseases:
  - chronic obstructive pulmonary disease: often pneumococcal
  - cystic fibrosis: staphylococci, *H. influenzae* and *P. aeruginosa*
  - viral infection: pneumococcal or staphylococcal.
- Aspiration pneumonia is caused by aspirated oro-pharyngeal anaerobes and streptococci in unconscious or incoordinate patients. The position during aspiration determines the lobe(s) infected.
- Hospital-acquired pneumonia particularly occurs in ventilated patients, and is usually caused by enteric bacteria, *P. aeruginosa* or *S. aureus.* Chest radiograph changes, neutrophilia or worsening pulmonary function indicate infection rather than simple colonisation.
- Pneumonia in the immunocompromised patient is a special hazard for leukaemic, cancer and transplant patients as well as patients on high-dose steroids or cytotoxic drugs. There is a wide range of causative organisms.

LUNG ABSCESS AND EMPYEMA

LUNG ABSCESS

Lung abscesses are pus-containing cavities within the lung; they may be single or multiple, uni- or bilateral. They arise by five routes (Fig. 1). The most common route is from above through the bronchi by inhalation (from sinusitis, dental or oral sepsis, bronchitis or bronchiectasis) or by aspiration (lost gag/swallowing reflexes, coma or anaesthesia, oesophageal reflux, foreign body or tumour).

Causative organisms

The likely pathogen is indicated by the route of infection:

1. from above: usually oro-pharyngeal flora of mixed anaerobes and mixed streptococci, plus Gram-negative rods in hospitalised patients
2. from below: mixed enteric flora including anaerobes and aerobic Gram-negative rods and cocci; *E. histolytica*; *P. westermani*, causing paragonimiasis, is described on page 121
3. from within, following pneumonia: *K. pneumoniae*, other Gram-negative rods and *S. aureus* are common; *Legionella* spp., *Actinomyces* and *Nocardia* spp. are rare. *Pseudomonas pseudomallei*, causing melioidosis, is described on page 120
4. from without: *S. aureus*, skin flora, soil organisms
5. from remote sources: *S. aureus*, *E. coli*, and anaerobes, especially *Fusobacterium necrophorum* and *B. fragilis*.

Clinical features

Copious foul sputum (indicative of anaerobes) and persistent fever are prominent, with anorexia, malaise and weight loss. Chest pain and dyspnoea are uncommon. Differential diagnosis is from a necrotic cavitating tumour or infarct, infected lung bullae, or rare infections including actinomycosis, nocardiosis, hydatid cyst (Fig. 2) or aspergillosis (p. 110).

Confirmatory tests

Chest X-ray confirms the diagnosis (Fig. 3). Sputum culture is unhelpful for abscesses from above (aspiration) but may help in other causes. Cultures from blood, wounds or remote sources can be helpful. Direct aspiration under CT control is now safe and specific.

Chemotherapy

Metronidazole (because of the usual involvement of anaerobes) plus penicillin is the usual empiric treatment. Clindamycin is active against most anaerobes and staphylococci. Metronidazole plus a third-generation cephalosporin is appropriate if aspiration occurred in hospital. Specific therapy is given if the abscess follows specific pneumonia or septic emboli. Posturing may help drainage, and surgery often hastens cure.

Control and prevention

This depends on preventing infection by the five routes: airway control, treatment of intra-abdominal abscesses, of pneumonia, of chest wounds, and of septicaemia. If treatment is delayed, infection may spread to the pleural cavity, resulting in empyema.

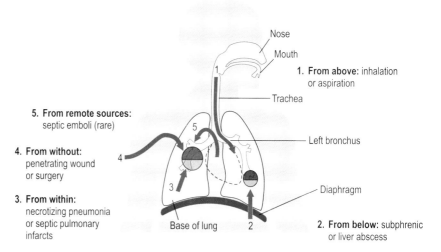

5. **From remote sources:** septic emboli (rare)

4. **From without:** penetrating wound or surgery

3. **From within:** necrotizing pneumonia or septic pulmonary infarcts

Nose

Mouth

1. **From above:** inhalation or aspiration

Trachea

Left bronchus

Diaphragm

Base of lung

2. **From below:** subphrenic or liver abscess

Fig. 1 **Pathogenesis of lung abscess: routes of infection.**

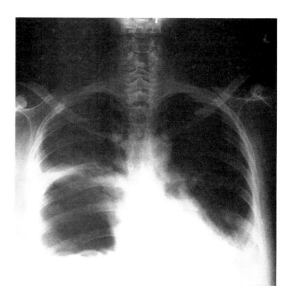

Fig. 2 **Hydatid cyst — note daughter cysts at base.**

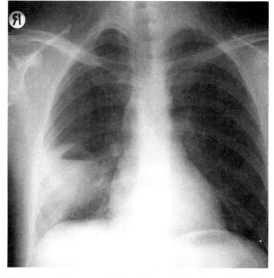

Fig. 3 **Lung abscess on chest radiograph, showing cavity with fluid level in right lower zone.**

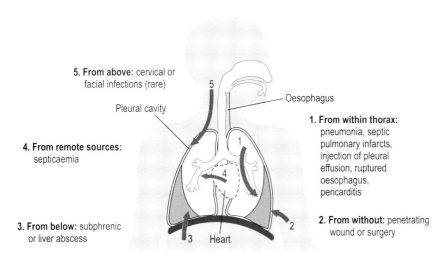

5. **From above:** cervical or facial infections (rare)

Pleural cavity

4. **From remote sources:** septicaemia

3. **From below:** subphrenic or liver abscess

Oesophagus

1. **From within thorax:** pneumonia, septic pulmonary infarcts, injection of pleural effusion, ruptured oesophagus, pericarditis

2. **From without:** penetrating wound or surgery

Heart

Fig. 4 **Pathogenesis of empyema: routes of infection.**

Causative organisms

Again, the common causative organisms depend on the route of infection:

1. from within: anaerobes, particularly *Bacteroides* spp., *F. nucleatum* and peptostreptococci; aerobic streptococci and, after nosocomial pneumonia, enteric Gram-negative rods
2. from without: *S. aureus* or enteric Gram-negative rods, soil or skin flora
3. from below: anaerobes and enteric GNR
4. from remote sources: *S. aureus*
5. from above: anaerobes and streptococci.

EMPYEMA

An empyema (= pyothorax) is a collection of pus within the pleural cavity, outside the lung. Like a lung abscess, it can arise by five routes, in different order of importance (Fig. 4).

Clinical features

Dyspnoea, chest pain and a dry cough are usual, with a productive cough and sputum only when the underlying lung is infected. There are sometimes few systemic symptoms and only a low-grade fever; usually major systemic symptoms with hectic swinging fevers occur.

Confirmatory tests

Chest X-ray shows the empyema (Fig. 5), though CT may be needed to differentiate it from a lung abscess. Microscopy of aspirated pus with Gram and special stains, plus aerobic and anaerobic culture reveal the pathogens.

Chemotherapy

Empiric initial treatment is often with metronidazole and a third-generation cephalosporin (or flucloxacillin if staphylococci are probable), then specific antibiotics when the pathogens are known. Adequate medical or surgical drainage is essential, with re-expansion of the lung.

Control and prevention

This depends on preventing infection by the five routes: treatment of pneumonia, of chest wounds, of cervical infections, of intra-abdominal abscesses and of septicaemia.

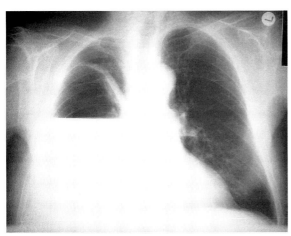

(a)

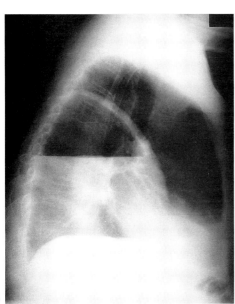

(b)

Fig. 5 **(a) A huge empyema on chest X-ray; (b) the lateral view shows it is behind the right lung.**

Lung abscess and empyema

- Lung abscesses arise by one of five routes:
  — most commonly from above, by inhalation or aspiration of oro-pharyngeal flora
  — from below, from intra-abdominal abscesses
  — from within the lung following pneumonia
  — from without the chest, by wounds or surgery
  — from remote sources (rare) through septic emboli.
- An empyema arises by one of the same five routes as for lung abscess though infection from above is from cervical infections, not by aspiration into the bronchial tree. Infection from within the thorax is the most common cause.
- Chest X-ray and CT define the abscess(es) or empyema. The pathogens are identified in aspirate for both diseases.
- Chemotherapy is empirical initially, guided by route of infection.
- Adequate medical or surgical drainage is essential, with re-expansion of the lung.

TUBERCULOSIS AND ATYPICAL MYCOBACTERIAL INFECTIONS

TUBERCULOSIS (TB)

Tuberculosis is a chronic infection of humans and animals. It affects healthy as well as immunocompromised people. It is primarily a disease of the lung but it may spread to other sites or form a generalised (miliary tuberculosis) infection. It has three stages:

- primary
- latent
- post-primary (secondary).

Primary tuberculosis

Primary tuberculosis usually occurs in childhood by person-to-person aerosol spread of *Mycobacterium tuberculosis*, rarely by *M. bovis* (p. 54). It is characterised by the **primary complex**, which has two parts, a primary focus of infected tissue plus the associated regional lymphadenopathy. Macrophages, multiplication, dissemination and damage are the essential features. The mycobacteria are engulfed by macrophages, in which they can survive and multiply. They are then disseminated via the lymph to the lymph nodes where a damaging cell-mediated immune response is initiated. The body reacts to contain the pathogen in infected tissue in **tubercles**, which are small granulomas of epithelioid cells and giant cells: multiplying *M. tuberculosis* make macrophages appear as multinucleated giant (Langhans) cells. Central cheesy necrotic damage in tubercles is called caseation and any progression to macroscopic damage causes **cavitation** (p. 54). This is an example of a disease where the pathology of the disease is a consequence of the cell-mediated immune response. This response is essential to prevent spread of the disease; the bacteria causes little direct damage itself and does not produce toxin.

The commonest primary focus is in the best aerated (middle and lower) zones of the lungs, resulting from inhalation. If infection is by ingestion of unpasteurised milk infected with *M. bovis*, the primary complex is in the small bowel wall and mesenteric lymph nodes. Very rarely, infection is by implantation of the skin.

Primary tuberculosis can evolve in three ways.

- Healing may occur, often with calcification (Fig. 1), without spread elsewhere. The development of cell-mediated immunity is shown by a positive Mantoux test, when purified protein derivative (PPD) of *M. tuberculosis* produces a palpable lump > 10 mm 48–72 hours after intradermal injection. When a pulmonary primary focus heals and calcifies it is named a Ghon focus.
- Progressive primary tuberculosis may develop by local spread through the lung as progressive primary tuberculous pneumonia, or to the pleura as a primary tuberculous effusion (Fig. 2).
- Miliary spread (like millet seeds on chest X-ray) occurs through the blood stream, particularly to lymph nodes, the upper zones of the lungs, to kidneys, bones, central nervous system and genital tract. This may cause severe clinical disease called *miliary tuberculosis* (Fig. 3) or may heal without symptoms, passing to the next, latent, stage.

Latent tuberculosis

Latent tuberculosis means there are no symptoms and signs of infection, though the Mantoux test usually remains positive for many years. Latency may last 10–80 years.

Secondary tuberculosis

Post-primary, 'adult' (secondary) tuberculosis develops by reactivation of dormant infection seeded by miliary spread of primary tuberculosis in the upper zones of the lungs (Fig. 4), lymph nodes (p. 108), kidneys (p. 153), bones and joints (pp. 181, 183), central nervous system (p. 89) or genital tract (p. 160).

Clinical syndromes

- The primary complex is usually asymptomatic, but mild fever and malaise may occur.
- Progressive primary pulmonary tuberculosis causes fever, cough, sputum, haemoptysis and, if extensive, dyspnoea.
- Primary tuberculous effusion causes dyspnoea and fever, with little cough. Sputum and haemoptysis reflect the degree of underlying lung infection.
- Miliary spread in the primary stage may be asymptomatic, but miliary disease is usually serious, often with pulmonary and meningeal disease together, at times with bone or renal disease as well.
- Latent disease by definition is asymptomatic with a positive Mantoux test. (Note broad similarities but differences of

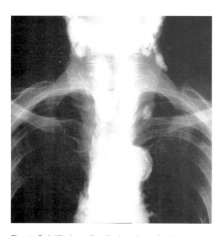

Fig. 1 **Calcified mediastinal and cervical tuberculous lymph nodes.**

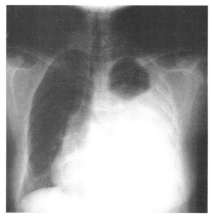

Fig. 2 **Pleural effusion from pulmonary TB, on chest radiograph.**

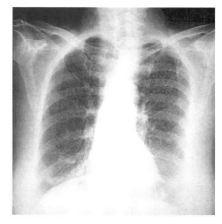

Fig. 3 **Miliary TB on chest radiograph, showing multiple fine opacities 'like millet seeds' in all zones.**

detail with the stages of the other classic chronic infectious disease, syphilis.)

- Post-primary pulmonary TB causes chronic cough, sputum with haemoptysis and fever, and anorexia and progressive wasting, hence the old name, 'consumption'. Dyspnoea and chest pain usually occur late.
- Secondary extrapulmonary TB occurs mainly in lymph nodes (25%) and pleura (20%), genitourinary tract (15%), bone and joint (10%), disseminated (10%) and meninges (5%).

Confirmatory tests

A positive Mantoux test indicates previous exposure to the tubercle bacillus, not necessarily active infection. Conversely, in overwhelming TB, the Mantoux may revert to negative.

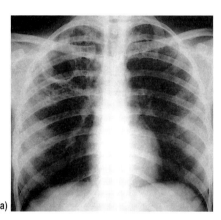

(a)

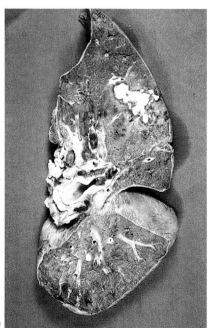

(b)

Fig. 4 **Pulmonary fibro-caseous TB. (a)** Apical fibrosis and cavitation on chest radiograph; **(b)** autopsy specimen showing fibrosis and cavities.

Acid-fast bacilli in clinical specimens are presumptive evidence of TB, but as other mycobacteria and *Nocardia* spp. are acid fast, full identification needs culture, which is also essential for sensitivity testing.

Chest X-ray (Fig. 4) shows the extent and monitors progress or the development of complications like pleural effusion or bronchiectasis.

Chemotherapy

This is specialised: triple therapy usually involves rifampicin and isoniazid (INAH) for 9 months, with a third drug, often pyrazinamide, for the first 2 months. Quadruple therapy with ethambutol, for example, added to the above three can shorten treatment to 6 months in uncomplicated disease caused by sensitive organisms. Incomplete or inadequate treatment leads to drug resistance in the index patient and those subsequently infected. Adjunctive surgery is still sometimes necessary

Control and prevention

BCG vaccination is used where TB is prevalent, so vaccinees react quickly to limit infection when it occurs. Chemoprophylaxis with isoniazid for 1 year is given to close contacts of new patients or those recently converting from Mantoux negative to positive, i.e. recent asymptomatic infections.

ATYPICAL MYCOBACTERIAL INFECTIONS

Many species of mycobacteria are associated with occasional disease: infection is well known in immunosuppressed patients and the association with AIDS has acquired prominence in the post-1980s. Mycobacteria other than *M. tuberculosis* and *M. bovis* cause a variety of diseases (Table 1). The species are now differentiated precisely by biochemical tests rather than using the Runyon classification based on pigmentation and rate of growth (p. 55).

Management

Chemotherapy is dependent upon species identification and sensitivity tests, although the latter are often unreliable. Atypical mycobacteria are often resistant to many antibiotics and usually combination therapy is required. *M. scrofulaceum* varies in susceptibility, and triple therapy or surgery is usually necessary. *M. chelonei* and *M. fortuitum* are relatively resistant to drugs; amikacin, doxycycline, cefoxitin and rifampicin may be useful. Surgery to remove all infected tissue is important. Infection from surgery or prostheses is minimised by aseptic techniques.

Table 1 **Atypical mycobacterial infections**

Species	Disease
M. leprae	Leprosy (p. 54)
M. marinum	Skin infections and deeper infections, 'fish-tank granuloma' (p. 170)
M. kansasii	Lung infections (rare) resembling TB
M. avium-intracellulare	Disseminated infections in AIDS patients
M. scrofulaceum	Cervical lymphadenitis (rare)
M. ulcerans	Skin infections: Bairnsdale or Buruli ulcers (p. 174)
M. fortuitum and M. chelonae	Infections associated with implantation by trauma or surgery

Tuberculosis and atypical mycobacterial infections

- Tuberculosis is a chronic disease caused by *M. tuberculosis* (rarely *M. bovis*). Bacteria multiply in macrophages both at the site of infection, usually in the lung by inhalation, and in the lymph nodes.
- The primary stage involving the primary complex may heal, may spread locally or may spread systemically (miliary).
- After a latent period, post-primary (secondary) disease develops, again most often pulmonary, but at times renal, glandular, skeletal, cerebral or genital.
- Because mycobacteria are often drug resistant, chemotherapy is specialised quadruple or triple therapy for 6–12 months longer.
- Control is by BCG vaccination, and chemoprophylaxis.
- The atypical mycobacteria can be contaminants or colonising organisms or cause infections: pulmonary (*M. kansasii*), skin and soft tissue ulcers (*M. ulcerans, M. marinum*), systemic (*M. avium-intracellulare* in AIDS), or opportunist (*M. chelonae, M. fortuitum*).

TROPICAL OR RARE RESPIRATORY INFECTIONS

These rare infections are caused by a range of organisms which may cause disease of other organ systems too. The features of lung disease are shown in Table 1.

ACTINOMYCOSIS

Actinomycosis is a rare, chronic infection by filamentous Gram-positive anaerobic slowly growing *Actinomyces*, usually *A. israelii* (p. 56). Thoracic actinomycosis occurs by inhalation of *Actinomyces* from normal oral flora. The pathological features are slow progression, sinus formation (late), sclerosis, scarring and sulphur yellow granules in sinus or cavitary pus.

The clinical features are cough, sputum, haemoptysis and chest pain. Fever and systemic symptoms are mild, yet the infection may spread to pleura, mediastinum or chest wall. It may mimic tuberculosis.

Confirmatory tests are Gram's stain of any granules, prolonged anaerobic culture of washed sputum or pus, or histopathology on excised tissue.

Chemotherapy and control. Chemotherapy is prolonged intravenous penicillin then oral penicillin for 6–12 months. Control is by dental care and appropriate chemoprophylaxis.

ASPERGILLOSIS

Infection of the lung by *Aspergillus* spp. (p. 62) occurs by inhalation; the species most often involved are *A. fumigatus*, *A. niger* and *A. flavus*. Infection is becoming more common because of the increase in immunocompromised patients. Aspergillosis of the lung is classified into three clinical types.

- **Allergic broncho-pulmonary aspergillosis.** This occurs in asthmatics and causes expectorated bronchial plugs (Fig. 1) and some increase in dyspnoea. Episodes may recur for years.
- **Aspergilloma.** This is a fungus ball in a pre-existing cavity or cyst, typically causing haemoptysis, plus the features of the underlying disease.
- **Fulminant pneumonia.** This usually occurs in the immunocompromised host and causes high fever, cough, progressive dyspnoea, respiratory failure and death in 1–3 weeks.

Confirmatory tests. Microscopy will identify hyphae and conidiophores (p. 7). Culture is needed for identification.

Table 1 **Features of rare lung infections**

Disease	Course in lungs	Characteristics	Cavitation	Complications
Actinomycosis	Chronic	Lower lobes often	++	Pleura/empyema
Aspergillosis				
Allergic	Recurrent	Fleeting	–	Bronchiectasis
Aspergilloma	Progressive	Fungus ball	Always	Bleeding, spread
Pneumonia	Fulminant	Expanding, >1 lobe	–	Spread, death
Hydatid disease	Silent	'Cannon ball'	–	Secondary infection
Melioidosis	Fulminant	Like TB, any lobe	+	Sepsis, death
Nocardiosis	Chronic	Like TB, any lobe	+	Pleural spread
Paragonimiasis	Progressive	Mixed, cysts, calcification	+	Fibrosis
Systemic mycoses				
Blastomycosis	Nil or progressive	Dense, segmental	+/–	Rarely spread
Coccidioidomycosis	Acute or progressive	Variable infiltrate	++	Rarely spread
Cryptococcosis	Nil or cryptococcoma	Globular mass	–	Spread in AIDS
Histoplasmosis	Acute or progressive	Like TB, any lobe	++	Rarely spread, calcification
Paracoccidioidomycosis	Nil or progressive	Fluffy infiltrates	–	Oro-nasal disease

TB, tuberculosis.

Chest X-ray and CT scan define the extent of disease.

Chemotherapy. Amphotericin B and itraconazole are helpful but seldom curative. Aspergillomas are removed surgically.

Control and prevention are not feasible except by filtered air in transplant units.

HYDATID DISEASE

The causative organism in pulmonary hydatid disease is usually the cestode *Echinococcus granulosus*, rarely *E. multilocularis* (p. 82). The clinical features are sometimes cough or dyspnoea, but often the cyst is asymptomatic, found incidentally on X-ray examination (Fig. 2).

Confirmatory tests are serological, particularly for arc 5 in a gel immunodiffusion test, which is highly specific. X-rays, CT, and liver and brain scans are used for localisation.

Chemotherapy. Chemotherapy with albendazole is more effective than mebendazole, but surgery is still often necessary, being very careful not to spill the infective cyst contents (Fig. 2).

Control and prevention is by treating adult worms in farm herbivores, preventing dogs from eating raw infected animal (e.g. sheep) viscera, and hand washing after dog or soil contact.

MELIOIDOSIS

Melioidosis is caused by *Pseudomonas pseudomallei* (p. 46). It causes cutaneous, pulmonary and systemic disease. It is found especially in Southeast Asia, Northern Australia and tropical Africa.

The clinical features are acute or chronic suppurative infection or acute septicaemia. Pulmonary disease occurs most often; this may be inapparent, or an acute fulminant infection with cough, sputum, fever, chest pain, tachypnoea, upper lobe consolidation and cavitation.

Confirmatory tests are microscopy for bipolar staining rods, like 'safety pins', and careful culture.

Chemotherapy. Chemotherapy has been with tetracycline, chloramphenicol or sulphonamide with an aminoglycoside; it is now more often with a third-generation cephalosporin with an aminoglycoside for 30 days, followed by cotrimoxazole for 2–6 months.

Control and prevention are by early energetic wound care.

NOCARDIOSIS

Characteristics of nocardiosis are chronic abscess formation with suppuration and spread, especially in skin and subcutaneous tissues, producing a mycetoma (p. 174) or cavitating pulmonary disease mimicking tuberculosis or actinomycosis. It occurs particularly in immunocompromised patients, e.g. after organ transplantation. Many other organs have been involved by systemic spread.

Causative organisms are usually *Nocardia asteroides*, sometimes other species (p. 56).

Clinical features of lung disease are chronic cough, thick sputum, chest pain, dyspnoea and fever with moderate systemic symptoms.

Confirmatory tests are Gram stain to show characteristic thin, beaded, branching Gram-positive filaments, a modified

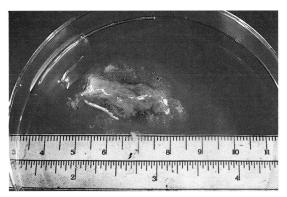

Fig. 1 **Bronchial plug in allergic aspergillosis.**

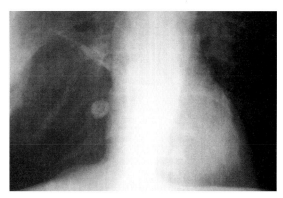

Fig. 2 **Small calcified hydatid cyst in right lung near cardiac border.**

acid-fast stain to show weakly acid fast filaments, and aerobic culture for at least 1 week.

Chemotherapy. Chemotherapy with sulphonamides has been prolonged but the newer beta-lactams may replace them.

Control and prevention involves care of soil-contaminated wounds and of immunocompromised patients.

PARAGONIMIASIS

Paragonimiasis is caused by the lung fluke *Paragonimus westermani* (p. 85). Humans are infected by eating encysted larvae in crabs or crayfish. These excyst in the stomach and migrate to cause human pulmonary cavitating infection (Fig. 3). Expectorated sputum infects snails, the other intermediate host.

Clinical features are fever, cough, sputum, haemoptysis, dyspnoea and severe chest pain. Fibrosis, bronchiectasis and pleural effusions follow. Worms spread to other organs, especially the central nervous system, causing fits, paralyses and blindness.

Confirmatory tests are microscopy of sputum or faeces for the characteristic large operculated eggs, chest X-ray, and CT or other scans of other infected organs.

Chemotherapy. Chemotherapy is with praziquantel, or bithionol as an alternative.

Control and prevention depends on education, sanitation and avoiding uncooked crabs and crayfish. The snails and definitive hosts are difficult to control.

SYSTEMIC MYCOSES

A number of fungi cause systemic mycoses (pp. 64–7). Respiratory disease is acute at the time of infection in coccidioidomycosis and histoplasmosis; progressive pulmonary disease is rare.

Blastomycosis

Blastomyces dermatitidis is a dimorphic soil fungus endemic in parts of N. America and Africa. It infects by inhalation. Asymptomatic pulmonary infection is common, but symptomatic disease is rare. Chronic bone and skin disease is usual, with broad-based budding yeasts and granuloma formation. Ketoconazole is used for mild infections, amphotericin for systemic infections.

Coccidioidomycosis

Coccidioides immitis is a dimorphic soil fungus endemic in areas of the Americas and infects by inhalation. Asymptomatic pulmonary infection is most common, followed by symptomatic pulmonary disease; progressive or systemic disease is rare. Arthroconidia and tissue spherules filled with spores

are characteristic; granulomata resembling tuberculosis occur. Amphotericin is needed for systemic or severe disease.

Cryptococcosis

Cryptococcus neoformans is a monomorphic yeast with a characteristic capsule, found particularly in pigeon droppings. It infects humans by inhalation. It can infect humans with a normal immune system, though 50% of those infected have a detectable immune deficiency. Infection causes pulmonary masses, (often asymptomatic), chronic meningitis or cerebral masses. Classical treatment is amphotericin plus flucytosine, though fluconazole is now used both acutely and for long-term suppression in the immunocompromised patient.

Histoplasmosis

Histoplasma capsulatum is a dimorphic fungus; the filamentous mould form is found in soil with bird or bat droppings, and certain endemic areas are notorious for histoplasmosis. Spores are inhaled and germinate to the yeast, which is ingested by macrophages. The clinical disease depends on the dose inhaled, the immune state and local lung disease. Amphotericin B is usual treatment, but ketoconazole has an increasing role.

Paracoccidioidomycosis

Paracoccidioides brasiliensis is a dimorphic fungus endemic in S. America. It is probably a soil organism and may infect by inhalation or possibly by implantation. Chronic progressive skin and mucous membrane ulcers are usual; pulmonary and systemic disease is rare. Multiple budding gives a 'pilot's wheel' appearance microscopically, with granulomata. Ketoconazole is replacing amphotericin and sulphonamides in treatment.

Tropical or rare respiratory infections

- Actinomycosis is a rare chronic cavitating pneumonia.
- Aspergillosis can involve:
  — allergic broncho-pulmonary aspergillosis in asthmatics
  — aspergilloma (fungus ball) in a pre-existing cavity or cyst
  — fulminant pneumonia, usually in the immunocompromised host.
- Hydatid disease has rounded 'cannon ball' lung cysts.
- Melioidosis is a fulminant fatal cavitating pneumonia.
- Nocardiosis is a chronic cavitating pneumonia similar to tuberculosis.
- Paragonimiasis involves progressive cavitating lung damage.
- Of the systemic mycoses, only coccidioidomycosis and histoplasmosis are characterised by acute pulmonary disease at the time of infection; progressive pulmonary disease is rare in all mycoses.

SUPPURATIVE THROMBOPHLEBITIS / LYMPHANGITIS / LYMPHADENITIS

SUPPURATIVE THROMBOPHLEBITIS

Suppurative or septic thrombophlebitis is venous infection with thrombosis ('thrombophlebitis' without an adjective is venous thrombosis with inflammation rather than infection). It is classified into:

- superficial
- pelvic
- portal (called 'pylephlebitis')
- intra-cranial suppurative thrombophlebitis.

It arises from intravenous catheters and other intra-vascular devices, from intravenous drug use (IVDU), or from local skin or tissue infections. It gives rise to local spread, or distant spread by bacteraemia or septic emboli (Table 1).

Causative organisms

In superficial infection, *S. aureus* and coagulase negative staphylococci are now less common than enteric Gram negative rods (GNRs), especially *Klebsiella Enterobacter* spp. *Candida albicans* occurs particularly with parenteral nutrition or IVDU. In pelvic infection, streptococci and anaerobes predominate. In pylephlebitis, enteric GNRs, enterococci and anaerobes are dominant. In intra-cranial infection, *S. aureus* and/or respiratory flora are usual.

Clinical features

Superficial septic thrombophlebitis when acute causes local redness, swelling, tenderness and pain (Fig. 1) with pus from any puncture or sinus. Subacute infection, especially in burns patients, may have mainly systemic manifestations with fever and shock, and few local signs. Both lead to bacteraemia, septic emboli and distant abscesses, especially pulmonary.

Pelvic thrombophlebitis follows childbirth, abortion or pelvic surgery, and again causes mainly systemic signs with high fever, chills, and vomiting, few abdominal signs, then septic pulmonary emboli and lung abscesses.

Portal thrombophlebitis follows appendicitis or other intra-abdominal infection, causes high fever and signs of sepsis, and leads to septic emboli and liver abscesses.

Intra-cranial thrombophlebitis can be in the cortical veins or the cavernous, lateral, sagittal or petrosal venous sinuses. It follows local facial, oral, ear or paranasal sinus infection, and causes (depending on the venous sinus involved), peri-orbital swelling (Fig. 2), various cranial nerve lesions, limb weakness, sensory impairment, fits and coma. Urgent neurological and surgical consultation is essential. Uncontrolled infection is fatal.

Confirmatory tests

Microscopy and culture of pus if present, and several blood cultures are essential. Chest X-ray may show lung abscesses and pleural fluid. In pelvic, portal or intra-cranial thrombosis, CT scanning or other special imaging is usually needed.

Chemotherapy

- Superficial suppurative thrombophlebitis needs surgical excision as well as appropriate antibiotics for cure.
- Pelvic thrombophlebitis responds slowly to antibiotics such as penicillin and metronidazole, while abscesses need drainage, and the

infected veins may need ligation.
- Pylephlebitis needs appropriate antibiotics against gut flora, such as ampicillin, gentamicin and metronidazole, plus removal of the cause and drainage of abscesses.
- Intra-cranial suppurative thrombophlebitis needs flucloxacillin plus metronidazole, surgery for the source and any abscesses, and control of intra-cranial pressure and of fits.

Control and prevention

This depends on early recognition and treatment of the cause.

LYMPHANGITIS

This is characterised by inflammation of the lymphatics, usually subcutaneous. It is classified into acute, which is usually bacterial, or chronic, usually fungal, mycobacterial or filarial.

Causative organisms

Acute lymphangitis (Fig. 3) is usually due to *S. pyogenes* or rarely to *S. aureus* after injury, *Pasteurella multocida* after animal bites, or *Wuchereria bancrofti* after mosquito bites. Chronic lymphangitis is most often due to *Sporothrix schenckii*, rarely to *M. marinum*, *Nocardia* spp. (Fig. 4), or *W. bancrofti* (p. 82).

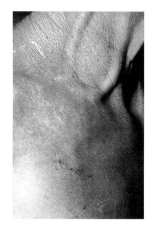

Fig. 1 **Jugular vein thrombophlebitis with thrombosis.**

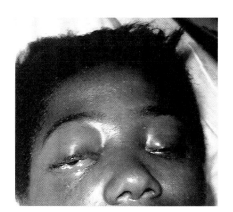

Fig. 2 **Cavernous sinus thrombosis with peri-orbital swelling obscuring 3rd, 4th and 6th nerve paralyses.**

Table 1 **Suppurative thrombophlebitis**

Classification	Causative organism	Clinical features	Confirmatory tests	Chemotherapy	Control
Superficial	*S. aureus*, GNRs, anaerobes	Pus, local signs	M&C of pus, blood cultures	Flucloxacillin and excision	Asepsis
Pelvic	Streptococci, anaerobes	Systemic signs, lung abscess	Blood cultures, CT, imaging	Broad spectrum, drain, ligate	Prevent cause
Portal ('pylephlebitis')	Enteric GNRs, anaerobes	Systemic signs, lung abscess	Blood cultures, CT, imaging	Broad spectrum, drain, excise	Prevent cause
Intra-cranial	*S. aureus*, respiratory flora	Orbital oedema, cranial nerve and cerebral lesions Fits and coma	Blood cultures, CT, imaging	Flucloxacillin and metronidazole Surgery for cause, drain	Prevent cause

Table 2 Infective lymphadenitis

Classification	Causative organism	Clinical features	Confirmatory laboratory tests	Chemotherapy	Control
Bacterial					
Pyogenic	S. pyogenes	Cervical	Throat culture	Penicillin	Avoidance
Vincent's angina	Anaerobes	Cervical	M&C, anaerobic	Penicillin	Avoidance
Diphtheria	C. diphtheriae	Cervical	Specific culture	Penicillin	Immunise
TB	M. tuberculosis	Cervical	AFB M&C of pus	Triple	BCG
Chancroid	H. ducreyi	Inguinal	Special culture	Ceftriaxone	Safer sex
Plague	Y. pestis	Inguinal	Gram stain	Streptomycin and tetracycline	Rats + fleas
Syphilis 1°	T. pallidum	Inguinal	Dark field micro	Penicillin	Safer sex
LGV	C. trachomatis	Inguinal	Serology	Tetracycline	Safer sex
Anthrax	B. anthracis	Regional	M&C	Penicillin	Of zoonosis
Listeriosis	L. monocytogenes	Regional	M&C, motility	Penicillin	Food hygiene
Rat-bite fever	S. moniliformis	Regional	Culture	Penicillin	Avoid rats
	S. minor	Regional	Dark field micro	Penicillin	Avoid rats
Tularaemia	F. tularensis	Regional	Serology	Streptomycin	Rodents/ticks
Brucella	Brucella spp.	General	Serology, culture	Streptomycin and tetracycline	Of zoonosis
Leptospirosis	Leptospira spp.	General	Serology	Penicillin	Rat control
Melioidosis	P. pseudomallei	General	Culture	C'taz+Cotrimoxazole	Wound care
Miliary TB	M. tuberculosis	General	CXR, Mantoux, AFB	Triple	BCG
Syphilis 2°	T. pallidum	General	Serology	Penicillin	Safer sex
Rickettsia					
Rickettsialpox	R. akari	Regional	Serology	Tetracycline	Mice control
Scrub typhus	R. tsutsugamushi	General	Serology (OX-K)	Tetra, chloro.	Mite control
Fungi					
Histoplasmosis	H. capsulatum	General	Serology, CXR	AmB, Ketoconazole	No bird faeces
Paracoccidioidomycosis	P. brasiliensis	Regional	M&C, histo, ID	Sulphas, AmB	None known
Parasites					
Filariasis	W. bancrofti	General/local	Blood film	Hetrazan	Mosquito control
Kala-azar	L. donovani	General	Bone marrow	Antimony	Sandfly control
Toxoplasmosis	T. gondii	General	Serology (histo)	Su + pyrimethamine	Avoid hosts
Trypanosomiasis	Trypanosoma spp.	Local/general	Micro, serology	Suramin, MelB	TseTse control

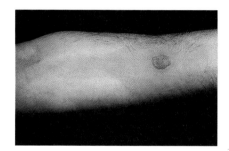

Fig. 3 **Streptococcal infection of a Mantoux test, with lymphangitis.**

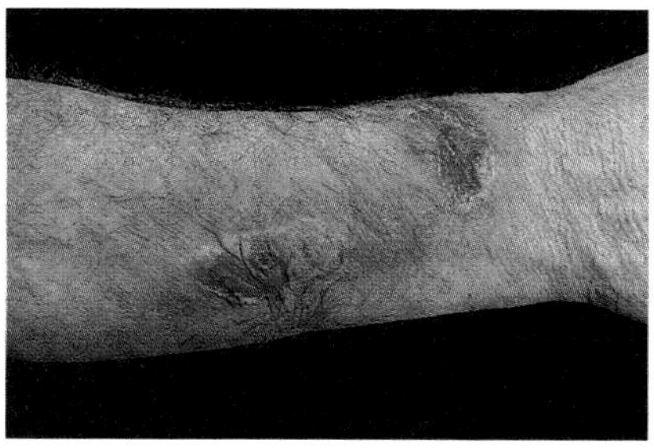

Fig. 4 **Nocardial lymphangitis and nodules in a gardener.**

Clinical features

Acute lymphangitis is characterised by a red streak (Fig. 3) from the primary site of infection towards the enlarged, tender regional lymph nodes. The onset is usually very abrupt with a rigor and high fever before the local infection appears. Recurrent episodes occur if scarring results, and are common in filariasis with chronic oedema, due either to streptococci or the parasite itself.

    Chronic lymphangitis follows a granuloma or ulcer at the site of implantation by trauma from plants, wood or soil in sporotrichosis or nocardiosis, from swimming pools or aquaria in *M. marinum*. Reddish nodules appear and may ulcerate along the proximal lymphatics, palpable as a thickened cord.

Confirmatory tests

In acute infection, Gram stain and culture of the local lesion, and blood cultures are needed. Nocturnal blood films are more helpful than serology in filariasis. Biopsy is rarely necessary.

    In chronic infection, specific fungal and mycobacterial stains and cultures are done.

Chemotherapy

In acute infection intravenous penicillin G is needed, or flucloxacillin if staphylococci are likely. In filariasis, di-ethyl carbamazine is given in addition. In sporotrichosis, potassium iodide is used. In *M. marinum* infections, rifampicin plus ethambutol, or minocycline or cotrimoxazole have all been used with some success, as has surgical excision.

Control and prevention

This depends on avoidance of penetrating injury, or mosquitoes in filarial areas.

LYMPHADENITIS

Lymphadenitis is infection of lymphatic glands, while lymphadenopathy is enlarged lymph glands from any disease, infectious or non-infectious. Lymphadenitis is not a disease of itself, but a manifestation of many infections: the causative organisms, clinical features, confirmatory laboratory tests, chemotherapy, and control are summarised in Table 2, and detailed in the relevant spreads.

Suppurative thrombophlebitis / lymphangitis / lymphadenitis

- *Suppurative thrombophlebitis* is bacterial infection of veins with thrombosis; the 4 major sites (superficial, pelvic, portal and intracranial) determine its features, treatment and control.
- *Lymphangitis* is inflammation of the lymphatics. Acute bacterial lymphangitis is usually streptococcal, while chronic is usually fungal, parasitic or mycobacterial.
- *Lymphadenitis* (infected lymphatic glands) is cervical, inguinal, regional or general, from a wide variety of bacterial, rickettsial, fungal and parasitic infections, of which streptococcal and tuberculous infections are common and important.

MYOCARDITIS, PERICARDITIS AND RHEUMATIC FEVER

Infections of the heart are classified by the anatomical areas affected (Fig. 1).

MYOCARDITIS

Myocarditis is inflammation, usually caused by infection, of cardiac muscle. Apart from non-infectious causes, it is classified by the causative organism into bacterial (rare), viral (commonest) and parasitic. The major infections are:

- Chagas' disease
- diphtheria
- Lyme disease
- viral myocarditis.

In addition, rheumatic fever produces a pancarditis, including myocarditis.

Clinical features. Myocarditis of any type produces chest discomfort, with dyspnoea and oedema from cardiac failure. Cardiomegaly can give functional mitral or tricuspid incompetence. The conduction system is also often involved, producing arrhythmias. Whatever the cause of myocarditis, any arrhythmias and cardiac failure must be treated. In fatal cases the heart is dilated with an abnormal myocardium (Fig. 2).

Chagas' disease

Chagas' disease is caused by *Trypanosoma cruzi* (p. 79), a flagellated protozoan that is transmitted by reduviid bugs. Clinically, Chagas' disease has two phases:

- acute infection, often in childhood, causes mild fever, malaise, an indurated red 'chagoma' at the site of infection with regional lymphadenopathy, or periorbital oedema (Romana's sign) from conjunctival infection. Myocarditis is rare but may be fatal.
- chronic Chagas' disease, years or decades later, causes cardiac arrhythmias, thromboembolism or cardiac failure, often right-sided. It may also cause mega-oesophagus or megacolon.

Confirmatory tests in acute disease are microscopic examination of fresh blood or blood films, while chronic disease is diagnosed serologically.

Chemotherapy and control. Nifurtimox is most useful in the acute stage. Benznidazole is used for Brazilian strains. Control and prevention are by insecticides and better housing.

Diphtheria

Diphtheria is caused by *Corynebacterium diphtheriae*. Systemic spread of the toxin

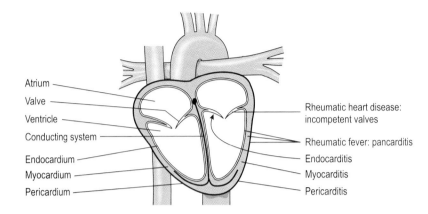

Fig. 1 **The infections of the heart.**

results in damage to the heart 1–2 weeks after infection. Diphtheria (pp. 106–107) is treated with penicillin and antitoxin. Vaccination has played a major role in reducing the incidence of disease.

Lyme disease

Lyme disease is caused by *Borrelia burgdorferi* (pp. 52–3), a spirochaete transmitted by ticks. An initial rash is seen at the site of infection; disseminated disease (pp. 130–1) can involve meningitis and arthritis as well as myocarditis. The diagnosis is clinical and serological. Oral doxycycline or amoxycillin is given for 10–21 days for mild myocarditis, with i.v. penicillin or ceftriaxone for 10–21 days for severe myocarditis.

Viral myocarditis

Myocarditis is most commonly caused by viruses, including Coxsackie group B (less often A), other enteroviruses, mumps, influenza and congenital rubella; it can lead to heart failure.

PERICARDITIS

Inflammation of the pericardium is classified into purulent, tuberculous, viral, other infections and non-infectious causes. Pericarditis of any type produces chest discomfort, usually with fever. Dyspnoea can occur from co-existent myocarditis, from cardiac tamponade (when cardiac function is impaired by pericardial fluid), or later from constriction by fibrosed or calcified pericardium. Unlike myocarditis, oedema and arrhythmias are usually absent.

Purulent pericarditis

Purulent pericarditis was classically linked to spread of the pneumococcus after pneumonia; it is now more commonly caused

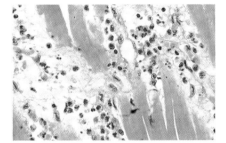

Fig. 2 **Myocarditis.**

by staphylococci, enteric flora or other organisms after surgery or septicaemia.

Clinical features include high fever, severe chest pain and signs of sepsis. Urgent chest X-ray, echocardiograph or CT scan define the anatomical changes. Gram stain and culture of the pericardial aspirate define the microbial cause.

Chemotherapy is guided by the Gram stain, then sensitivity tests; initial empiric therapy may be flucloxacillin plus gentamicin. Prevention is by asepsis and treatment of septicaemia.

Tuberculous pericarditis

M. tuberculosis, the cause of tuberculosis (pp. 54, 118), can infect the heart by local spread, usually from lung or glands. Clinically, the pericarditis may be acute with little effusion, but it is usually chronic, often with tamponade. Chronic constrictive pericarditis is very characteristic (Fig. 3). The extent of disease and the pathogen are defined as for purulent pericarditis but using AFB stain and culture. Anti-tuberculous triple therapy is required for at least 9 months. Surgery is essential for constriction.

Viral pericarditis

Viral pericarditis (the most common form) is caused by Coxsackie group B

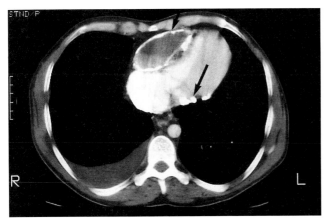

Fig. 3 **Constrictive pericarditis: CT scan.**

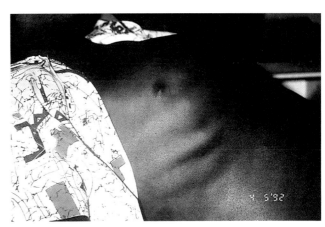

Fig. 5 **Rheumatic heart disease: chest deformity.**

Fig. 4 **Rheumatic fever: erythema marginata.**

Table 1 **Modified Jones criteria for diagnosis of rheumatic fever: two major criteria or one major plus two minor criteria and evidence of group A streptococcal infection are required**

Criteria	Characteristics
Major criteria	
Carditis	Pancarditis
Polyarthritis	Migratory around large joints
Erythema marginata	Faint fleeting rare rash (Fig. 4) with red irregular margins which move in minutes or hours like 'smoke rings'
Subcutaneous nodules	Rare, felt over bony prominences or tendons and last 1–2 weeks
Sydenham's chorea (St Vitus dance)	Rare, rapid involuntary jerking of limbs, face or body
Minor criteria	
Previous ARF or RHD	
Arthralgia	
Fever	
Acute phase reactants	Erythrocyte sedimentation rate, C-reactive protein, white cell count
Evidence of infection	Anti-streptolysin O title raised, DNAase B raised, positive culture, recent scarlet fever

ARF, acute rheumatic fever; RHD rheumatic heart disease

(less often A), other enteroviruses, mumps, influenza and varicella. Clinically, there is mild fever and precordial discomfort, sometimes preceded by a viral respiratory or gastrointestinal illness. Infection usually resolves in 2–6 weeks.

RHEUMATIC FEVER

Rheumatic fever is a post-infectious complication of streptococcal (*S. pyogenes*) infection that results from the immune response to infection. Antigens in the streptococcal cell wall stimulate antibodies that also react with the sarcolemma of the heart and with other tissues. The Jones criteria are diagnostic indicators for rheumatic fever (Table 1). Acute rheumatic fever (ARF) occurs 2–4 weeks after a streptococcal throat infection. Rheumatic heart disease (RHD) results from repeated attacks of *S. pyogenes* with different M antigen types; it is still common in developing countries.

Clinical features
The diagnostic criteria have distinct features (Table 1 and Fig. 4). Complications of rheumatic fever include heart block and cardiac failure in the acute attack, or later rheumatic heart disease, with stenosis or incompetence of aortic, mitral, tricuspid or pulmonary valves. Untreated, cardiac failure, cardiomegaly and even chest deformity follow (Fig. 5),

but timely prosthetic valve replacement prevents these problems.

Confirmatory tests. Because of the complications, children with fever and a sore throat should have a throat swab tested for streptococci (p. 30). A rising antibody titre to streptococci is also indicative. ECG often shows prolongation of the P–R interval or, sometimes, higher degrees of block.

Management. Penicillin is given for 10 days with anti-inflammatory drugs (especially aspirin and corticosteroids), bed rest and treatment of cardiac failure and arrhythmias. Penicillin treatment of streptococcal pharyngo-tonsillitis is preventative. Long-term penicillin prophylaxis is given after rheumatic fever; the duration is debated, to age 25, or 30, or lifelong?

Myocarditis, pericarditis and rheumatic fever
- Myocarditis is usually viral, but may be caused by Chagas' disease, diphtheria, Lyme disease or acute rheumatic fever. The infection and any resulting arrythmias and cardiac failure must be treated.
- Pericarditis is also usually viral but may be purulent, tuberculous or rheumatic. The cause is identified by culture and serology, plus imaging and aspiration or surgery.
- Acute rheumatic fever comprises pancarditis, migrating polyarthropathy, chorea and other specific, rare signs. It occurs after group A streptococcal infection, which it is important to confirm by laboratory tests. Chemotherapy with penicillin must be accompanied by anti-inflammatory drugs, and cardiac failure therapy as necessary. Prophylactic penicillin is usually needed to protect against repeat attacks, which lead to rheumatic heart disease.
- The pathogens causing cardiac disease have distinct geographical locations, leading to varying prevalences.

INFECTIVE ENDOCARDITIS

Endocarditis is infection of the cardiac valves, or sometimes of the mural endothelium. A similar clinical picture is seen with rare infections of a coarctation of the aorta or a patent ductus arteriosus.

Endocarditis has been classified by three major factors.

1. Severity and rapidity of infection (old, imprecise but still useful in early management):
    — acute, developing in days, (often 3–6)
    — subacute, in weeks, (often 3–6)
    — chronic, over months (often 3 or more)
2. Valve type:
    — native valve of the patient
    — prosthetic valve
3. Cause of bacteraemia, if known:
    — procedures, such as dental, medical or surgical
    — IVDU (intravenous drug use)
    — disease elsewhere, e.g. abscesses
    — devices, e.g. intravascular catheters.

Endocarditis is further described by:

- the infecting microorganism, usually bacterial
- the infected valve(s), usually mitral or aortic, except for right-sided, tricuspid valve, as expected from IVDU.

The commonest type has been streptococcal subacute bacterial endocarditis (SBE) on native mitral or aortic valves (Fig. 1).

Causative organisms
There are two major causative factors. The first is an abnormal heart valve or endocardium, from congenital, rheumatic or other acquired heart disease, or a prosthesis. These can all lead to platelet–fibrin deposition, called non-bacterial thrombotic endocarditis (NBTE). The second is bacteraemia (or fungaemia) from procedures, devices, disease or IVDU.

The five major groups of pathogens are:

- oral streptococci and enterococci
- *S. aureus*
- coagulase-negative staphylococci
- enteric Gram-negative rods (GNRs)
- fungi, mainly *Candida* spp.

However almost any microbe can cause endocarditis, including rare GNRs (pp. 46–7), anaerobes (p. 48), *C. psittaci*

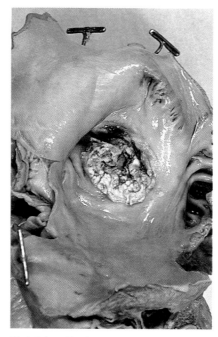
Fig. 1 **Infected 'native' valve.**

(p. 59), *Coxiella burnetii* (p. 61) and various fungi (p. 62).

Pathogenesis
Clinical features are very varied but are basically the result of four mechanisms.

Valve infection produces proliferative 'vegetations', destruction and incompetence, while intracardiac spread can give septal abscesses, heart block and, rarely, septal rupture.

Bacteraemia produces fever and systemic symptoms, while metastatic spread can give distal abscesses in, for example, the brain or kidney.

Embolisation produces distal infarction (especially seen in streptococcal and fungal infection) in the periphery (splinter haemorrhages under finger or toe nails (Fig. 2), or petechiae seen in the skin and conjunctiva) or in vascular organs, including the spleen (Fig. 3), kidney or brain. Embolisation also causes mycotic aneurysms in arteries (Fig. 4) or distal abscesses including micro abscesses ('Janeway lesions').

Immune complex deposition causes glomerulonephritis commonly, and Osler's nodes rarely; the latter are small painful red lumps in fingertips, palms or soles.

Clinical features
Patients fall into one of five groups.

Acute bacterial endocarditis is commonly caused by *S. aureus* (80%), enteric

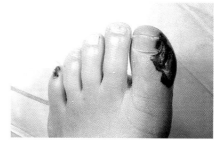
Fig. 2 **'Splinters' and infarcts produced by embolisation.**

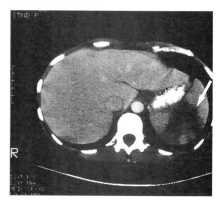
Fig. 3 **Splenic infarct (dark wedge) on CT scan.**

Gram-negative rods or *P. aeruginosa*. It occurs with high fever, sepsis and rapid valve destruction.

Subacute bacterial endocarditis is commonly caused by viridans streptococci, enterococci, staphylococci or unusual Gram-negative rods. It occurs with weeks of fever, emboli, immunologic phenomena and a cardiac murmur.

Chronic endocarditis (rare) is caused by fungi or coagulase-negative staphylococci. Typically it involves months of fever, with splenomegaly and a murmur.

Prosthetic endocarditis occurs either *early* after operation and is usually the result of infection with staphylococci or enteric Gram-negative rods or it occurs *late*, 2 months or more after operation, usually due to oral streptococci, enterococci or coagulase-negative staphylococci. Fever, bacteraemia and a new murmur occur earlier than emboli. Perivalvular leaks are serious; an occluded valve is rapidly fatal.

IVDU endocarditis is commonly caused by *S. aureus*, streptococci, enteric Gram-negative rods *P. aeruginosa* or *Candida* spp. Fever occurs for 1–2 weeks and there is tricuspid valve infection and pulmonary emboli.

Criteria for diagnosis

As the diagnosis is unproven until surgery or autopsy, various clinical diagnostic criteria have been used. Lord Horder proposed four: a changing or new murmur, emboli, pyrexia and positive blood cultures. Von Reyn from Boston in 1981 proposed further criteria and categories. Echocardiograms, especially transoesophageal, have so improved diagnosis that the Duke Endocarditis Service proposed a new set of criteria (Table 1) which define three diagnostic categories (definite, possible or rejected) that correlate well with surgical or autopsy confirmation.

Blood cultures are essential to confirm bacteraemia; one to three within 2 hours are sufficient in acute disease, while three or four in 12 hours are usual in sub-acute disease. Media should be rich and may need supplementing for fastidious organisms. Sensitivity testing with both MIC and MBC (p. 210) is required because body defences in cardiac valves are few, and bactericidal drugs must be used. The organism should be stored for 3 months in case of relapse.

Echocardiography is now essential to show vegetations or valve dysfunction. Transoesophageal echocardiography (TOE) is needed if transthoracic examination is negative or inconclusive. Anaemia, neutrophilia or thrombocytopenia are common and acute phase reactants are high. Serum biochemistry detects renal and/or hepatic impairment.

Management

In principle, appropriate bactericidal antibiotics are given intravenously in high dose for 4–6 weeks, though uncomplicated patients with very sensitive organisms now have shorter courses, with the latter part of their treatment orally and/or at home. The usual antibiotics for major pathogens are shown in Table 2.

Response to therapy is monitored by resolution of fever, fall in acute phase reactants, and improved well-being. Serum levels of gentamicin and vancomycin should be monitored.

Surgery is usually needed for prosthetic valve endocarditis (Fig. 5), chronic infection (including fungal endocarditis), culture-negative endocarditis and uncontrolled embolisation. It is essential for uncontrolled infection, including septal abscess, or for uncontrolled cardiac failure.

Control and prevention. Prophylaxis before dental and other procedures likely to cause bacteraemia has not been proven as a useful measure, but logical regimens before risky procedures are widely advocated in patients at risk. High-risk patients include those with a prosthetic valve or previous endocarditis, while high-risk procedures are those highly likely to cause bacteraemia, including dental extractions and periodontal, gastrointestinal and genitourinary procedures.

Oral amoxycillin is usually advocated for low-risk patients and procedures, replaced by clindamycin if the patient is hypersensitive to, or on, long-term penicillin. Intravenous ampi/amoxycillin plus gentamicin is advocated in some countries for high-risk patients and/or procedures; vancomycin replaces ampi/amoxycillin if the patient is hypersensitive to, or on long-term, penicillin.

Table 1 Duke Endocarditis Service diagnostic criteria

Major criteria	Typical blood culture
	Positive echocardiogram
Minor criteria	Predisposition
	Fever
	Vascular phenomena
	Immune phenomena
	Suggestive echocardiogram
	Suggestive microbiology

Table 2 Antibiotic treatment

Microbe	Primary antibiotic[a]	Duration	Low dose gentamicin
Sensitive streptococci	Benzylpenicillin	4 weeks	2 weeks
Relatively resistant streptococci	Benzylpenicillin	6 weeks	6 weeks
Enterococci	Ampi- or amoxycillin	6 weeks	6 weeks
Staphylococci	Flucloxacillin	6 weeks	1–2 weeks
Gram-negative rods	Third generation cephalosporin	6 weeks	6 weeks
Fungal endocarditis	Amphotericin B	6 weeks or more	An azole
Other organisms	According to sensitivity tests		
Culture negative	Initially benzylpenicillin with gentamicin, then according to most likely organism(s)		

[a]Patients hypersensitive to penicillins may be de-sensitised or treated with alternatives such as a cephalosporin or vancomycin.

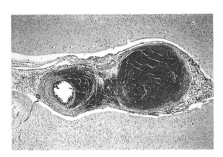

Fig. 4 **Histology of a mycotic aneurysm.**

Fig. 5 **Infected prosthetic valve.**

Infective endocarditis

- Endocarditis is infection of the cardiac valves or mural endocardium. It is usually bacterial, sometimes fungal.
- The two causative factors are 'soil' and 'seed': platelet–fibrin deposition on abnormal endothelium or valve; bacteraemia.
- The major causative organisms are oral streptococci, enterococci, staphylococci, Gram-negative rods and *Candida* spp.
- The clinical features result from four mechanisms: valve and heart damage, systemic bacteraemia (with fever and distant abscesses), emboli (causing distant infarcts and aneurysms) and immune complex deposition (causing e.g. glomerulonephritis).
- Endocarditis on prosthetic valves or in intravenous drug users has special additional features.
- Confirmation is mainly by blood cultures and echocardiogram.
- Management involves high-dose, long-term treatment with a suitable antibiotic(s).
- Surgery is needed for uncontrolled infection or cardiac failure.
- Prophylaxis, usually with amoxycillin, is given to those with valve disease before procedures likely to cause bacteraemia.

BACTERAEMIA, SEPTICAEMIA AND FUNGAEMIA

Bacteraemia or fungaemia are characterised by live organisms in the blood, shown by a positive blood culture. These two terms do not describe the clinical condition, which varies from asymptomatic to mildly unwell to seriously ill.

Septicaemia (called 'sepsis', or 'the septic syndrome' in some countries), in contrast, is a clinical description meaning severe infection with fever, rigors ('shaking chills'), tachycardia and severe systemic symptoms. Bacteraemia is usually found.

Septic shock means septicaemia plus low blood pressure vascular collapse and decreased organ perfusion; hence it is usually accompanied by oliguria with renal, cardiac and other organ failure, and high mortality. Septic shock occurs through widespread immune responses to endotoxins and other pathogen components (p. 26).

One useful classification (Table 1) divides bacteraemias into:

- focal source found on clinical examination

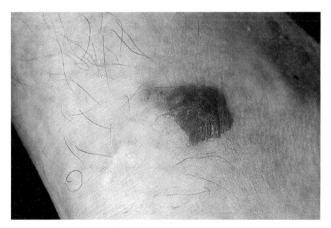

Fig. 1 **Septic embolus in *S. aureus* septicaemia.**

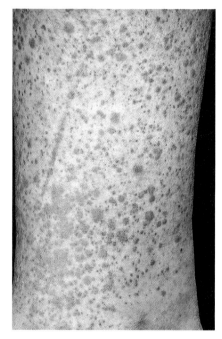

Fig. 2 **Vasculitis in *S. aureus* septicaemia.**

Table 1 **Types of septicaemias**

Type	Causative organism (commonest only)	Clinical features	Chemotherapy	Control
With focal source		**Specific**		
Skin	Staphylococci, S. pyogenes	Furuncles Erysipelas	Flucloxacillin Penicillin	Early treatment
Ear or sinus	S. pneumoniae H. influenzae	Otitis media Sinusitis	Penicillin Ceftriaxone	Early treatment
CNS	N. meningitidis S. pneumoniae	Meningitis Abscess (Fig. 5)	Penicillin Penicillin	Immunise some groups Immunise some groups
Heart	Streptococci Staphylococci	Endocarditis	Penicillin Fluclox + Gent	Chemoprophylaxis
Intra-vascular device	Staphylococci, resistant GNR	None, or local pus	Vanco + Gent	Asepsis
Lungs	S. pneumoniae H. influenzae	Pneumonia Abscess	Penicillin Ceftriaxone	Immunise some groups Immunise some groups
Abdomen	Aerobic GNR Anaerobes	Peritonitis Abscess	Ampi + Gent Ampi + Gent + Metro	Early treatment
Bone or joint	Staphylococci H. influenzae	Osteomyelitis Septic arthritis	Flucloxacillin Ceftriaxone	Early treatment
No focal source		**Non-specific**		
AIDS	S. pneumoniae	(Dyspnoea)	Penicillin +?	Immunise
Cardiac valve disease	Streptococci, Staphylococci	Murmur Emboli	Penicillin Fluclox + Gent	Chemoprophylaxis
Community acquired	GPC, GNR	None specific	Cefazolin + Gent	Immunise
Hospital acquired	GPC, GNR, P. aeruginosa	(Wound, UTI, pneumonia)	Timentin + Gent	Asepsis
Neonatal	Group B streptococci, GNR, Listeria sp.	Often none localising	Ampi + Gent	Asepsis
Neutropenia	GPC, resistant GNR	Often none localising	Timentin + Gent	Asepsis, chemoprophylaxis
Splenectomy	S. pneumoniae, H. influenzae	Often none localising	Penicillin, Ceftriaxone	Immunise

Ampi, ampicillin; Fluclox, flucloxacillin; Gent, gentamicin; Metro, metronidazole; Vanco, vancomycin; GNR, Gram-negative rods; GPC, Gram-positive cocci.

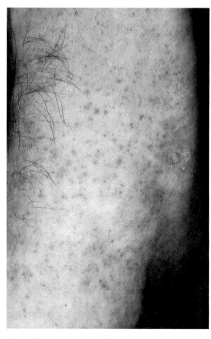

Fig. 3 **Haemorrhagic rash in meningococcal septicaemia.**

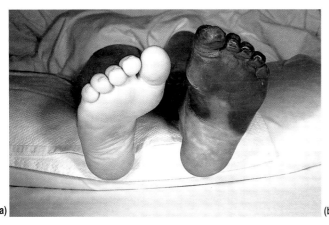 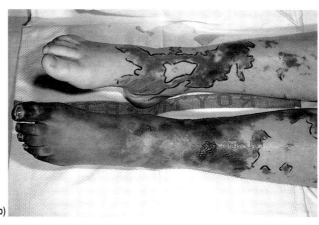

(a)   (b)

Fig. 4 **Infarcted toes and forefoot with severe haemorrhagic rash in meningococcal septicaemia.**

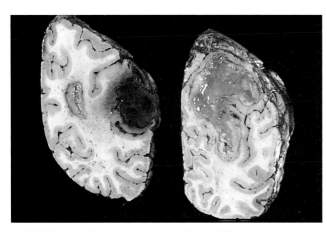

Fig. 5 **Multiple cerebral abscesses in cyanotic heart disease.**

CONFIRMATORY TESTS

Blood cultures, and microscopy and culture from likely sites, are always needed to find the organism. Chest X-ray, CT and other imaging are often needed to find the site. Serology is rarely useful in time.

MANAGEMENT

This obviously depends mainly on the pathogen and the severity of disease, and to a lesser degree on the source if known. Initial treatment may have to be broad spectrum, with modification when the pathogen and source are found (Table 1). Surgery to drain abscesses and remove necrotic tissue is essential. Circulatory, respiratory and renal support is essential in sepsis and septic shock.

Immunisation or chemoprophylaxis may be feasible to protect those with predisposing factors such as valve lesions, neutropenia or splenectomy. Aseptic technique is essential for hospital procedures, including intravascular device insertion and maintenance.

* no focal source, hence the epidemiologic setting is used to guide investigation and empiric therapy.

CAUSATIVE ORGANISMS

The pathogen depends on the source and the setting (Table 1). In general, staphylococci and streptococci are now more common than Gram-negative rods, while anaerobes, fungi and other organisms are uncommon.

CLINICAL FEATURES

Clinical features fall into four groups:

* general features of sepsis are found in all septicaemic patients: fever, headache, rigors ('shaking chills'), malaise, tachycardia and hypotension with pale, cold extremities
* specific clues are found in some patients:
  — a splenectomy scar
  — septic emboli, especially in staphylococcal sepsis (Fig. 1), which may also show vasculitis (Fig. 2)
  — meningococcal haemorrhagic rash with petechiae, (Fig. 3), larger haemorrhages, and even digital or forefoot infarction (Fig. 4).
* focal sources of infection, if present, show their own distinctive clinical features (Table 1)
* epidemiological setting may be vital (Table 1).

Bacteraemia, septicaemia and fungaemia

* Viable bacteria or fungi in the blood is known respectively as bacteraemia or fungaemia.
* Septicaemia is the clinical syndrome resulting from bacteraemia (or fungaemia). It can have a focal source or be without an obvious cause, when the epidemiological setting becomes important.
* Septic shock is septicaemia with hypotension and impaired tissue perfusion.
* Clinical features to be sought are:
  — general signs of sepsis: fever, rigors and tachycardia
  — specific signs giving clues to cause, e.g. meningococcal rash
  — specific focal source infections
  — epidemiologic setting.
* Chemotherapy initially may have to be broad spectrum, with modification when the pathogen and site are found.
* Corrective surgery is necessary for abscesses and infarcts.
* Circulatory, respiratory and renal support is essential in sepsis and septic shock.
* Chemoprophylaxis, immunisation and aseptic technique reduce the risk of bacteraemia in predisposed groups of patients.

TROPICAL SYSTEMIC INFECTIONS

Four major vector-borne systemic infections, mainly found in developing countries and the tropics, are described. Other systemic infections particularly found in the tropics are filariasis (p. 177), rickettsioses (pp. 60–1), yaws (p. 175) and the zoonoses (pp. 50, 184).

Table 1 **Leishmania spp. and their clinical diseases**

Major species	Distribution	Disease
L. donovani complex including L. infantum	India, Africa, Mediterranean	Kala azar (visceral)
L. tropica complex including L. major	India, Africa, Mediterranean	Cutaneous 'Old World'
L. mexicana complex	Central and South America	Cutaneous 'New World'
L. braziliensis complex	Central and South America	Cutaneous and mucocutaneous

Table 2 **Malaria**

Species	Areas	Asexual blood cycle	Relapses	Complications
P. falciparum	All malarious	48 hours (tertian)	None	Cerebral, renal, frequently fatal
P. vivax	Not W. Africa	48 hours (tertian)	Over 3 years	None
P. ovale	Tropical Africa	48 hours (tertian)	Over 20 years	None
P. malariae	Africa, South and Southeast Asia	72 hours (quartan)	Uncommon	Nephrotic syndrome

KALA AZAR

Leishmania (p. 76) are flagellated protozoa transmitted from animal reservoirs by biting flies or bugs. The genus is divided into four groups called complexes, and visceral leishmaniasis is caused by a member of the *L. donovani* complex, usually *L. donovani* itself. It is transmitted by *Phlebotomus* sandflies, either from canines or infected humans.

Leishmaniasis is classified into four types (Table 1) of which three are cutaneous (p. 177) and the fourth is kala azar or visceral leishmaniasis. In this form the parasite develops in the spleen and liver.

Clinical features

This chronic illness begins with fever and mild systemic symptoms, followed by wasting and, months or years later, by massive splenomegly (Fig. 1), with anaemia, grey-black skin pigmentation and hepatomegaly. The patient looks and feels surprisingly well until the terminal stage. Skin lesions can appear weeks, months or years after kala azar; this is known as post-kala azar leishmaniasis.

Confirmatory tests

The formol-gel test for very high serum immunoglobulins is simple but non-specific. Specific diagnosis is by seeing the amastigote stage of the parasite (Leishman–Donovan bodies) in bone marrow, splenic aspirate or liver biopsy. Culture and serology are less useful.

Chemotherapy

First choice chemotherapy are parenteral pentavalent antimony compounds such as sodium stibogluconate or meglumine antimonate, given slowly because of toxicity. Reserve drugs are amphotericin B and pentamidine.

Control and prevention

Bed nets with insecticide are protective. Unlike the cutaneous forms, immunity does not develop in kala azar, so a vaccine is unlikely.

MALARIA

Malaria is caused by a protozoan parasite, *Plasmodium* spp. (p. 72), which is transmitted by the bite of the female anopheline mosquito and is, therefore, restricted to areas where these can breed. Only the most dangerous, *P. falciparum*, does not have hypnozoites ('sleeping parasites') in the liver; these cause relapses even years later.

Clinical syndrome

Malaria is a very common, potentially fatal disease characterised by fever, headache and prostration, and often serious complications. It is classified by the interval between fever attacks (initiated by waves of release of parasite into the blood) and the causative parasite (Table 2). With each attack of fever (which may initially be daily with *P. falciparum*), the patient is first shivering cold, then hot and dry, then drenched with sweat. Headache is usually severe, with myalgia, leading at times to the tragic misdiagnosis of 'only influenza'. All malaria causes red cell destruction, with anaemia and splenomegaly in chronic disease. Falciparum malaria causes red cell sludging and cytokine release, causing cerebral malaria with convulsions and coma owing to capillary plugging. Death may follow.

Immune glomerulonephritis is common; nephrotic syndrome occurs especially in *P. malariae* infections; 'black water fever' with haematuria and renal failure is another feared complication, which can occur after quinine treatment.

Confirmatory tests

Thick and thin blood films will show the various forms of the parasites (Fig. 2). Occasionally bone marrow films are needed.

Chemotherapy

This is specialised and must change with drug resistance in various areas. Chloroquine is used for sensitive strains, and quinine or mefloquin for others, particularly for *P. falciparum*. Halofantrine and artemisinin derivatives are promising but resistance is already emerging. Primaquine is used to eradicate the liver 'hypnozoites'.

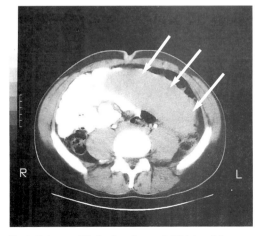

Fig. 1 **Kala azar: CT of spleen (upper right) into pelvis (shown by two white crescents of bone inferiorly).**

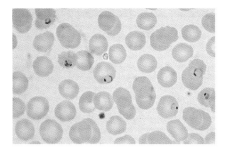

Fig. 2 **Malaria: blood film showing 'signet rings' of *P. falciparum*.**

Fig. 3 **Schistosomiasis with hepatosplenomegaly and ascites.**

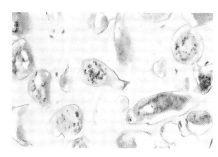

Fig. 4 *Schistosoma haematobium* **eggs in bladder biopsy.**

Control and prevention

Eradication is not currently possible because of resistance of mosquitoes to insecticides, and of many parasite strains to drugs. Control depends on insect repellants, bed-nets and imperfect chemoprophylaxis with chloroquine or mefloquine until the many difficulties in developing malaria vaccines are overcome.

SCHISTOSOMIASIS

Schistosomiasis is caused by *Schistosoma* spp. (p. 85). The vector is a snail, and the fluke larvae (cercariae) penetrate through human skin after they are released from the snail.

Clinical features

Schistosomiasis is characterised by urinary or bowel granuloma formation and haemorrhage, portal hypertension, and haemorrhage from oesophageal varices (thin-walled dilated veins).

It is classified into urinary schistosomiasis from *S. haematobium* (which causes haematuria), or intestinal from *S. mansoni* or *S. japonicum*. Infection causes symptoms that are dependent on the parasitic development stages:

1. Skin penetration by cercariae from the aquatic snail host, causes itch.
2. Migration causes allergic symptoms.
3. Maturation in the liver causes hepatitis.
4. Egg production in the vesical or mesenteric veins causes granuloma formation and haemorrhage.
5. Egg release causes portal hypertension, ascites, hepatosplenomegaly (Fig. 3) and oesophageal varices and haemorrhage.
6. Egg overflow into the systemic circulation causes pulmonary, cerebral and spinal cord granulomata.

Bladder polyps and cancer are frequent in chronic *S. haematobium* infection.

Confirmatory tests

The distinctive eggs are found by stool or urine microscopy, or by bladder (Fig. 4) or rectal biopsy.

Chemotherapy

Praziquantel is effective in early infections, but extensive tissue damage is irreversible.

Control and prevention

Control and prevention are difficult and depend on education, sanitation, mass treatment or the use of molluscicides against the snails in some circumstances. A vaccine is not yet developed.

TRYPANOSOMIASIS

Trypanosomiasis is caused by blood and tissue flagellated protozoa, the *Trypanosoma* spp. There are two forms.

- African: caused by variants of *T. brucei* and transmitted by the tsetse fly. Infection has a local, a systemic and an encephalitic (sleeping sickness) phase, and is finally fatal unless treated early (p. 93)
- South American (Chagas' disease): caused by *T. cruzi*. Infection is initially acute but major chronic cardiac and intestinal disease can occur years later as a result of continued parasitic activity (p. 124).

Confirmatory tests

Microscopy and specific staining (Giemsa) detects amastigotes in anti-coagulated blood, blood films or lymph node aspirates. Chronic Chagas' disease is confirmed serologically.

Chemotherapy

Suramin is the drug of choice for early *T. brucei* infections, the alternative being pentamidine. Toxic organic arsenicals such as melarsoprol are needed for CNS infection. Nifurtimox is most useful in the acute stage of *T. cruzi* infection, with benznidazole as a possible alternative.

Control and prevention

Control and prevention depend on tsetse fly/reduviid bug control, particularly around houses, protection from biting, and treatment of infected humans.

Tropical systemic infections

- Kala azar is visceral leishmaniasis; initial fever and wasting is followed by splenomegaly, anaemia and hepatomegaly.
- Malaria is caused by *Plasmodium* spp. in areas where anopheline mosquitoes breed. It causes fever, headache and prostration, with cerebral and renal complications as life-threatening sequelae.
- Schistosomiasis is caused by the flukes *Schistosoma* spp. and is characterised by urinary or liver damage.
- Trypanosomes are transmitted by tsetse flies (African sleeping sickness) and reduviid bugs (South American Chagas' disease). Both diseases begin with fever and local changes at the site of the bite. African trypanosomiasis progresses to encephalitis, while Chagas' disease can have an interval of years before chronic disease develops with cardiac and intestinal damage.

RARER SYSTEMIC INFECTIONS

A number of rarer systemic infections are discussed here; others are on pages 130, 175, 177, 184.

KAWASAKI DISEASE

Characteristics of this childhood disease are fever, rash and lymphadenopathy (hence the previous name of mucocutaneous lymph node syndrome, MCLS), with variable multisystem involvement. The causative organism is not known with certainty.

Clinical features are in three phases:

- acute phase of 1–2 weeks of fever, redness of conjunctivae, lips, oral mucosa with 'strawberry tongue', palms and soles, usually with oedema of hands and feet, and cervical lymphadenopathy in about 60%
- subacute phase of about 3 weeks of desquamation, arthralgia or arthritis in 40%, and cardiovascular disease (including myocarditis, arrhythmias, mitral incompetence or aneurysms) in 20%, with 2% mortality. There may be hepatic, renal, urethral, eye or meningeal involvement.
- convalescent phase until recovery after 6–10 weeks of illness.

Confirmatory tests are not diagnostic and may include neutrophilia, thrombocytosis, elevated ESR, pyuria, elevated transaminases and mild CSF lymphocytosis.

Chemotherapy is not apparently helpful; aspirin is used to suppress inflammation. **Control and prevention** is not yet feasible.

LISTERIOSIS

Listeriosis is caused by *Listeria monocytogenes*, a motile β-haemolytic (Fig. 1) Gram-positive rod (p. 32) which can grow at domestic refrigerator temperatures. It is found in many animals and plants, and infects adult humans directly or through unpasteurised milk or cheese, or uncooked vegetables or meat.

Clinical features vary:

- adult disease varies from asymptomatic carriage through mild 'influenza' to meningitis (particularly in the immunocompromised)
- intrauterine infection from mild bacteraemia in a pregnant woman causes abortion or fetal death
- neonatal infection leading to septicaemia, pneumonia or meningitis.

Confirmatory tests. Microscopy and culture of blood, CSF or other infected sites confirm listerial infection.

Chemotherapy uses penicillin or ampicillin.

Control and prevention are imperfect, with no vaccine. Pregnant women should avoid risky foods including unpasteurised milk or cheese, or uncooked vegetables or meat.

LYME DISEASE

Lyme disease is caused by *Borrelia burgdorferi* (p. 52). The reservoir hosts are deer and mice, and the vectors are hard-shelled ticks of the genus *Ixodes*. The disease has a patchy distribution on most continents.

Clinical features are in three phases:

- acute fever, systemic symptoms and the distinctive skin rash (erythema chronicum migrans, ECM); the rash has an

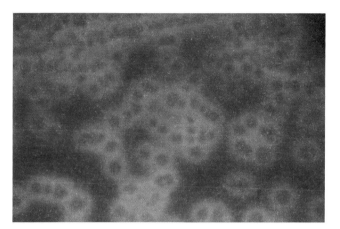

Fig. 1 *Listeria monocytogenes* colonies on blood agar showing beta haemolysis.

enlarging red border with central clearing and occurs initially at the site of tick bite, with later, transient lesions elsewhere
- subacute cardiac (myopericarditis, heart block, p. 124) and neurological (encephalitis, meningitis, peripheral neuritis, p. 89) lesions follow weeks or months later
- chronic arthralgias or arthritis, at times with neuropsychiatric manifestations, occur months or years later.

Confirmatory tests. These are usually serological, though the *Borrelia* may sometimes be seen or cultured from skin biopsies.

Chemotherapy for acute disease is doxycycline, or amoxycillin for children and pregnant women. In cardiac, neurological or joint disease, i.v. penicillin or ceftriaxone seem more effective.

Control and prevention in the absence of a vaccine depends on avoiding tick bites.

RELAPSING FEVER

Relapsing fever (RF), caused by spirochaetes of *Borrelia* spp. (p. 52), is characterised by afebrile intervals followed by febrile relapses. The repeated fevers result from antigenic variation in the infecting bacteria. There are two types of RF:

- epidemic louse-borne RF is caused by *B. recurrentis* and is spread person-to person by *Pediculosis humanus* (Fig. 2).
- endemic tick-borne RF is caused by numerous borrelia species; it is spread by soft *Ornithodorus* ticks from reservoirs in rodents.

Clinical features are similar in both endemic and epidemic RF: sudden fever, rigors, headache and myalgia, then hepatosplenomegaly. One week of bacteraemia alternates with 1 week without fever, repeated 3–13 times. Epidemic RF is similar to but usually more severe than endemic RF, with up to ten times higher mortality.

Confirmatory tests. Microscopy of Giemsa-stained blood films show the typical spiral bacteria; culture is used if films are negative. Serology is of little use.

Chemotherapy is by tetracycline, except for children or pregnant women. Chloramphenicol is an alternative.

Control and prevention are by avoidance of ticks and lice.

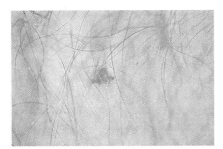

Fig. 2 **Louse and scratches.**

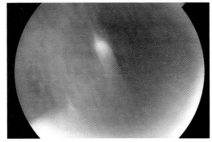

Fig. 3 *Toxocara* **eye infestation: scar and adhesion.**

TOXOCARIASIS AND VISCERAL LARVAL MIGRANS

Infection with the dog or cat nematode ascarid worms (*Toxocara canis, T. cati*, p. 81) occurs more often in children. The eggs are ingested accidentally from soil, and the resultant larva migrate via the blood stream to numerous tissues but cannot develop to the adult worm and so die. Their presence causes granulomas, tissue damage and necrosis. The most serious effects occur when the larvae infect the CNS or eye.

Clinical features include eosinophilia, cough and fever during the migration phase. Retinitis with tumour-like masses, fits or other organ damage, e.g. hepatitis, results from tissue invasion (Fig. 3).

Confirmatory tests. Eosinophilia is indicative and serology by fluorescent antibody detection confirms infection. Infected pets are found by stool tests.

Chemotherapy is by mebendazole, thiabendazole or diethylcarbamazine (DEC, Hetrazan).

Control is by protecting children from dog and cat faeces, by treatment of pets, and by disposal of animal faeces.

TYPHOID FEVER (ENTERIC FEVER)

Typhoid fever is caused by *Salmonella typhi* (p. 42). Paratyphoid fever is a milder illness caused by *S. paratyphi A, B* or *C*. Unlike other salmonellae, these four have no animal host, so transmission is person to person, usually via food or water.

Clinical features

Characteristic of this serious disease is high fever with systemic illness, constipation, increasing prostration, and sometimes a faint rash, intestinal haemorrhage or perforation.

Clinical features correlate with the three pathogenic stages:

* invasion of the small intestine through Peyer's patches, and multiplication within macrophages during the 10 day incubation period
* spread by bacteraemia to the reticulo-endothelial system

(liver, spleen, bone marrow), where further multiplication occurs, causing 1–2 weeks of fever, malaise, generalised aches, hepatosplenomegaly, neutropenia and sometimes faint red 'rose spots' on the skin
* re-invasion of the blood stream and infection of other organs including the kidney and Peyer's patches again, causing apparent relapse in the third week with higher fever, gut haemorrhage or perforation, 'toxaemia' with myocarditis, renal or hepatic impairment, and other organ infection including osteomyelitis or endocarditis.

A chronic asymptomatic carrier state persists in 1–2%.

Confirmatory tests. These follow from the above points. Positive blood cultures occur in the first 2 weeks, then positive urine and faecal cultures are obtained. Widal agglutination tests are most helpful in the unvaccinated, in non-endemic areas and when paired acute and convalescent samples are compared.

Chemotherapy is usually now by ciprofloxacin, unless contraindicated (children), not tolerated or unavailable, when ampicillin, chloramphenicol or cotrimoxazole are used for sensitive

strains. Haemorrhage, perforation and other complications obviously need specific management.

Control and prevention depend on personal hygiene, safe food and water, and good sewage disposal facilities to control endemic disease. Carrier detection and treatment help to control outbreaks, and killed or oral live vaccine will protect travellers.

YAWS

Yaws is caused by *Treponema pallidum* subspecies *pertenue*; this variant is visually and serologically identical with other human treponemes but is not yet cultured in vitro (p. 52). Infection follows contact of exudate from surface erosions with abraded skin.

Clinically, like the other treponematoses, there are three stages:

* a primary lesion which is a painless papule that becomes papillomatous with surface erosions, and then slowly heals
* a secondary stage with similar but multiple papillomata which heal but may relapse; lymphadenopathy and osteitis may occur
* a tertiary stage with multiple skin lesions of many types (plaques, nodules and especially ulcers), and gummata (chronic destructive ulcers) of bones, especially tibia, skull and nose (p. 175).

Confirmatory tests. Clinical diagnosis may be confirmed by dark ground microscopy (DGM) of exudates, or serology (VDRL or RPR, TPHA).

Chemotherapy is extraordinarily effective; one injection of benzathine penicillin is curative. Mass treatment with penicillin controls spread.

Rarer systemic infections

* Kawasaki disease is mucocutaneous lymph node syndrome, of unknown cause with no specific therapy or prophylaxis.
* Listeriosis is acquired from food and is particularly hazardous in the immunocompromised and the fetus; it causes meningitis, fetal death or neonatal sepsis.
* Lyme disease from *B. burgdorferi* transmitted by ticks has a distinctive rash and cardiac, neurological and joint sequelae.
* Relapsing fever is caused by borrelias; it is endemic tick-borne or epidemic louse-borne, and is marked by relapses of fever, with myalgia and hepatosplenomegaly.
* Toxocariasis occurs after the ingestion of larvae of *Toxocara* spp., migrating then dying in tissues to cause granulomas, particularly serious in the eye and brain.
* Typhoid fever is caused by *S. typhi* transmitted person to person, and is marked by fever, prostration and sometimes a 'rose spot' rash, gut haemorrhage or perforation.
* Yaws is a treponematosis caused by *T. pallidum* subspecies *pertenue*; it has a primary papule, secondary papillomata, and tertiary plaques and gummata.

PYREXIA OF UNKNOWN ORIGIN (PUO)

Classically, pyrexia of unknown origin (PUO) was defined as fever present for 3 weeks for which no cause had been found. Now, because of new diseases, new treatments and new prophylaxis, this 'classical PUO' is joined, particularly in hospital, by three new types of PUO: nosocomial PUO, neutropenic PUO and HIV-associated PUO. Further, outside hospitals, the general practitioner sees PUO as a 4–8 day fever (acute PUO) (p. 194). Fever is generally defined as a temperature greater than 38.3°C (101°F) on several occasions. In nosocomial neutropenic, HIV-associated PUO and in returned travellers or recent immigrants, diagnosis cannot wait 3 weeks. True fever must be distinguished from factitious fever, produced by the malingering or psychotic patient. Table 1 gives the main features of the different types of PUO. The mechanism of production of fever is described on page 26.

CAUSES

Acute PUO is caused by infections in 75% of patients; chronic PUO can have over 100 causes (Table 2). Classical PUO is associated with five major groups of causes:

- infections (20–40%)
- neoplasia, especially lymphomas and leukaemias (10–30%)
- collagen-vascular diseases (10–20%)
- miscellaneous (10–20%)
- undiagnosable, even after prolonged investigation (10–20%).

INVESTIGATION OF PUO

Because there is such a wide range of possible causes of PUO, investigation of the cause is best undertaken in steps.

Full history. This must be carefully taken. All types of PUO can be initiated by drugs; travel history is an important factor because many of the zoonoses or vector-borne diseases have a distinct geographic distribution.

- classical: hobbies, occupation, travel, contacts, animals, family history
- nosocomial: procedures, devices, anatomical factors, drugs
- neutropenic: underlying disease, chemotherapy, drugs
- HIV-associated: drug use, travel, contacts, duration and control of HIV, infections, lymphoma

Table 1 **Major types of PUO**

Classical PUO
Fever on several occasions or over 3 weeks duration
Diagnosis uncertain despite investigations as an outpatient or during at least 3 days in hospital

Nosocomial PUO
Fever on several occasions while receiving acute care
No sign of infection on admission
Diagnosis uncertain after 3 days despite investigations, including microbiological cultures incubating for 2 days

Neutropenic PUO
Fever on several occasions
Neutrophil levels in blood below 500 mm³ or expected to fall below this in 1–2 days
Diagnosis uncertain after 3 days despite investigation, including microbiological cultures incubating for 2 days

HIV-associated PUO
Fever on several occasions for 3 weeks as OP or 3 days as IP
Confirmed positive serology for HIV infection
Diagnosis uncertain after 3 days investigation, including microbiological cultures incubating for at least 2 days

Fever in general practice
4–8 day fever

Fever in returned traveller or in the tropics
Requires rapid diagnosis, particularly to exclude malaria

Table 2 **The ABC of PUO**

Classical PUO	
1. Infections	Abscesses (especially intra-abdominal), enteric fever, infective endocarditis, osteomyelitis, urinary tract infections
	Psittacosis, Q fever, relapsing and rat bite fevers, salmonellosis, tuberculosis
	Biliary tract infections, brucellosis, borreliosis (Lyme disease), chronic fatigue syndrome
2. Miscellaneous	CNS disease: pontine or hypothalamic
	Drugs
	Emboli, especially pulmonary
	Factitious fever
	Granulomatous diseases: Crohn's disease, hepatitis, sarcoidosis, temporal arteritis
	Habitual hyperthermia with no disease
	Inherited diseases: familial Mediterranean fever and others
3. Malignancies	Lymphomas: Hodgkin's and non-Hodgkin's; leukaemias; liver, kidney and other carcinomata
4. Collagen-vascular diseases	Mixed connective tissue disease, necrotising vasculitis, other vasculitis (allergic, drugs) polyarteritis nodosa, systemic lupus erythematosus, thromboses
5. Undiagnosed	Undiagnosable viral infections
Nosocomial PUO	
1. Underlying disease(s)	
2. Drugs	
3. Easy-to-diagnose infections	Pneumonia, urinary tract, wound
4. Concealed infections	Systemic candidiasis
	Acalculous cholecystitis or pancreatitis
	Vascular line-related infections
	Other device-related infections, e.g. sinusitis in intubated patient
	Transfusion-related infections, especially CMV, hepatitis C
Neutropenic PUO	
1. Underlying disease/condition	Organ transplants, leukaemia, lymphomas
2. Drugs	Numerous
3. Infection (causes 80–90% but confirmed in about 30%)	Early, bacterial: bacteraemia (including i.v. line infections), mouth infections, pneumonia, skin and soft tissue (e.g. peri-anal)
	Later, fungal and viral: candida, aspergillosis, CMV, HSV
HIV-associated PUO	
1. Underlying disease	Duration and control (viral load)
2. Drugs	Numerous
3. Infection	Mycobacteria: *M. avium* complex (MAC), TB, other
	Uncommon presentations of common infections in these patients, e.g. *P. carinii* pneumonia (PCP), cryptococcaemia or toxoplasmosis with fever alone
	CMV or other viral infections
	Intracellular infections: *Listeria*, *Salmonella*, *Histoplasma* spp., etc.
	Life-style related, e.g. endocarditis, syphilis
4. AIDS-associated tumours	Lymphomas etc.
Fever in travellers	Malaria, typhoid, dengue, etc.

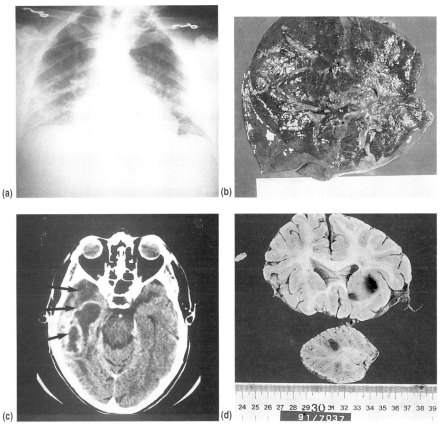

Fig. 1 **Aspergillosis. (a)** Chest X-ray showing widespread but non-specific opacities in a febrile patient with Hodgkin's disease; **(b)** autopsy showed disseminated aspergillosis; **(c)** CT showing brain abscesses; **(d)** autopsy confirming aspergillosis here also.

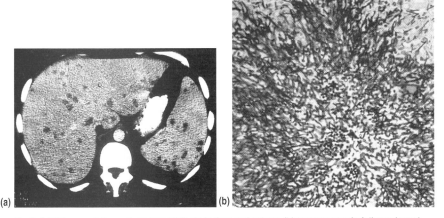

Fig. 2 **(a)** CT scan of disseminated candidiasis in liver and spleen; **(b) autopsy proof of disseminated candidiasis.**

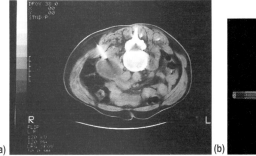

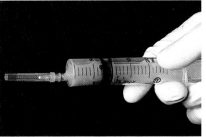

Fig. 3 **Tuberculous psoas abscess. (a)** CT scan showing psoas abscess, with needle; **(b)** the resultant pus, which was AFB smear positive and, later, culture positive for tuberculosis.

- travel associated (p. 188): malarial areas, zoonoses, vector-borne disease.

Full examination. The presence of a fever should be confirmed. The abdomen, liver, spleen, perianal area, lymph nodes, eyes, joints, muscles, lungs and heart should be particular areas of concern. Any wounds or i.v. devices should be checked.

Non-invasive investigations. Routine blood tests and chest X-rays (Fig. 1a) should be supplemented with other imaging techniques, e.g. CT scan (Fig. 1c, 2a) and aspiration of abscesses (Fig. 3)

Review history. Further questions may be indicated.

Repeat examination. Some physical signs may be transient, e.g. rashes.

Invasive investigations. Biopsies of liver, bone marrow and, possibly, skin, lymph nodes etc. may be indicated.

Therapeutic trial. If the diagnosis is probable but cannot be confirmed (e.g. culture-negative TB or endocarditis), empirical treatment can be started. If a non-infectious cause is suspected, corticosteroids or prostaglandin inhibitors may be used. In AIDS and neutropenic patients, infections can progress very rapidly.

MANAGEMENT

Treatment depends on the eventual diagnosis; empirical treatment in rapidly progressive disease is required while test results are awaited. Neutropenic PUO requires rapid diagnosis. Mortality is highest in HIV-associated PUO, lowest in neutropenic PUO.

Pyrexia of unknown origin

- Classical PUO is now separated from nosocomial PUO, neutropenic PUO and HIV-associated PUO in hospital practice, plus fever in general practice and returned travellers.
- Classical PUO is caused by infections, malignancies, collagen-vascular diseases and miscellaneous conditions, including factitious fever and undiagnosable illness.
- Nosocomial PUO, neutropenic PUO and HIV-associated PUO are usually the result of the underlying disease or are caused by infections or drugs (or AIDS-associated tumours).
- Diagnosis depends on history, examination, non-invasive and invasive investigations.
- Treatment depends on the likely or proven diagnosis.

DIARRHOEAL DISEASE I: GENERAL FEATURES

A number of clinical syndromes can arise from ingestion of pathogens. These can be confined to the gut or can spread to cause e.g. liver and intra-abdominal abscesses (p. 142). Diarrhoeal disease I discusses general features and special syndromes. Diarrhoeal diseases II describes bacterial, protozoal and worm infections in more detail.

Diarrhoea (frequent and/or loose bowel motions) is associated with a number of clinical syndromes.

- Gastroenteritis: this causes nausea, vomiting and diarrhoea. Abdominal discomfort, cramps and fever may occur. The commonest causes are *Campylobacter jejuni, E. coli, Salmonella* spp. and viruses. Rarer causes are *Bacillus cereus, Clostridium perfringens, Vibrio parahaemolyticus* and *Yersinia enterocolitica.*
- Food poisoning: often used loosely to mean any gastrointestinal illness following (dubious) food; strictly it means disease caused by food containing pre-formed toxin.
- Dysentery: diarrhoea with blood and mucus in the faeces. Abdominal pain, cramps and fever are common. It is usually caused by invasive large bowel infection, especially by *Entamoeba histolytica* or *Shigella* spp.
- Enterocolitis: inflammation of both the small and large bowel.
- Traveller's diarrhoea: any diarrhoeal illness associated with travel (p. 189).
- Cholera: caused by *Vibrio cholerae* and characterised by massive fluid loss through watery stools (p. 139).

PATHOGENESIS

There are several mechanisms by which gut pathogens cause disease.

- **Preformed toxin** causes short-incubation illness with marked vomiting. This is true food poisoning, for the bacteria produce the toxin while multiplying in the contaminated food; the bacteria may be destroyed by food preparation, the toxin is not (e.g. *C. botulinum* and *Staph. aureus*).
- **Toxin produced in the gut** by microbes multiplying there causes the symptoms, hence the longer incubation period and marked diarrhoea e.g. *V. cholerae,* enterotoxigenic *E. coli.* Toxin can also have distant effects.
- **Tissue invasion** by enteroinvasive pathogens (e.g. *Entamoeba histolytica, Salmonella* spp., enteroinvasive *E. coli*). The mucosal invasion often causes fever, cramps and blood and mucus in faeces. Deep invasion can lead to disseminated infection.
- **Parasitism** (frequently called infection) by protozoa and worms may cause diarrhoeal symptoms, from mild to chronic and may give rise to disseminated disease.
- **Perforation** can result from infection (or trauma) and gut microbes can then cause intra-abdominal disease (p. 142) or systemic sepsis.

Some pathogens can cause disease by more than one mechanism; for example *Campylobacter* spp. produce toxin in the gut and are tissue invasive (Table 1).

CLINICAL SYNDROMES

The clinical features can be grouped according to the pathogenesis; within these groups different pathogens show typical disease characteristics (Table 1).

SOURCES OF INFECTION

Infection occurs by the faecal-oral route from food, water or fingers. There are three major sources of infection:

- animal reservoirs: *Campylobacter* spp., *C. perfringens, Salmonella* spp. and *Y. enterocolitica*
- water: *Campylobacter* spp., enterotoxigenic *E. coli* (ETEC), *E. histolytica, Salmonella* spp. and *V. cholerae*
- food: all the causative organisms.

Prevention of infection thus depends on safe food handling, safe water supplies and proper sewage disposal.

CONFIRMATORY TESTS

Microscopy and cultures of food or stools are used to identify the causative organisms, important for individuals and in outbreaks of disease.

CAUSATIVE ORGANISMS

These may be grouped as:

- bacteria: 'true' gut pathogens
- protozoa
- worms
- viruses

Table 1 **Clinical features**

Classification/causative organism	Incubation period	Duration	Diarrhoea	Vomiting	Fever	Other
Preformed toxin (short incubation, vomiting prominent)						
B. cereus (emetic form)	1–6 hours	12–24 hours	+/–	Severe	0	Cramps
C. botulinum (adult form)	8–36 hours	Weeks to months	In 25%	In 50%	0	Constipation, paralysis
Staph. aureus	2–8 hours	12–24 hours	Frequent	Severe	Rare	–
Toxin produced in gut (medium incubation, marked diarrhoea)						
B. cereus (diarrhoeal form)	8–12 hours	12–24 hours	Marked	In 10%	0	Cramps ++
C. botulinum (infant)	8–48 hours	Weeks to months	Early	In 50%	0	Constipation, paralysis
C. perfringens (2 types)	8–24 hours	12–24 hours	Marked	In 10%	0	Cramps ++
E. coli (enterotoxigenic, ETEC)	12–48 hours	2–4 days	Mild to marked	+/–	0	–
V. cholerae	24–72 hours	1–7 days	Massive, watery	+/–	0	Severe dehydration
Tissue invasion (longer incubation, fever, pain, dysentery)						
Campylobacter spp. (especially *C. jejuni*)[a]	2–10 days	4–20 days	Marked	–	Yes	Cramps ++
E. histolytica	2–10 days	Days to weeks	Dysenteric	+/–	+/–	Cramps ++
E. coli (enteroinvasive, EIEC)	2–4 days	2–10 days	Some blood	–	+/–	Cramps ++
Salmonella spp.	1–3 days	2–7 days	Marked	Yes	Yes	Cramps +
Shigella spp.	1–4 days	2–4 days	Dysenteric	–	Yes	Cramps +++, tenesmus ++
V. parahaemolyticus[a]	1–2 days	2–3 days	Marked	Yes	Yes	Cramps
Y. enterocolitica[a]	3–7 days	1–2 weeks	Marked	30%	Yes	Cramps ++ and mesenteric adenitis

[a]Also produce toxin.

- unusual pathogens: e.g. those causing sexually transmitted diseases can cause bowel disease (see below)
- normal gut flora: antibiotic use can disturb normal bowel flora leading to disease (see below)
- bacterial overgrowth: occurs particularly in those with a predisposing abnormality.

Some pathogens have a distinct geographic distribution, some are found in animals and humans, others are only human pathogens. Some parasites are particularly common in AIDS patients. These differences are important for diagnosis, and for control and prevention. The individual pathogens are described in Diarrhoeal disease II.

SPECIAL SYNDROMES

Gay bowel syndrome
Homosexually active men can acquire gut infections in three ways:

- anal intercourse: sexually transmitted disease of distal bowel (proctocolitis)
- anal trauma: tears, ulcers, abscesses or fistulae
- oro–anal contact: ingestion of gastrointestinal pathogens.

Proctocolitis. The STD pathogens *N. gonorrhoeae*, *C. trachomatis*, *T. pallidum*, *H. ducreyi* and herpes simplex virus all cause ano–rectal pain and diarrhoea or dysentry. Ulceration may occur in syphilis, chancroid and herpes simplex.

Traumatic infections. Pyogenic infections from trauma usually involve Gram-negative rods and anaerobes. They cause ulcers, perianal and ischio-rectal abscesses and fistulae (Fig. 1) with haemoserous discharge, pain and fever.

Ingested gut pathogens. These cause diarrhoea, often with abdominal pain, cramps and nausea. A wide range of pathogens may be involved: *C. jejuni*, *Shigella* or *Salmonella* spp., *E. histolytica*, *Cryptosporidium* spp., *Giardia lamblia*, helminths, fungi or viruses, especially hepatitis A.

Confirmatory tests. Stool, ulcer or abscess microscopy and culture with special stains and cultures are needed for the range of likely pathogens. Syphilis, hepatitis and HIV serology is essential.

Chemotherapy depends on the specific pathogens found. Safer sex and better hygiene have decreased the incidence since the mid-1980s.

Antibiotic-associated disease
Antibiotic-associated diarrhoea is the general term for diarrhoea following antibiotic use (usually oral), while pseudomembranous enterocolitis (PME) describes the specific appearance on colonoscopy of severe cases with inflamed mucosa covered by typical fibrinous pseudomembrane. Symptoms vary from mild to severe diarrhoea to dysentery with blood loss, griping abdominal pain, fever and severe dehydration.

Many antibiotics alter bowel flora and allow the multiplication of *Clostridium difficile* an antibiotic-resistant member of normal bowel flora in many people; it can also be acquired by cross-infection from other patients. Its cytotoxin and enterotoxin both lead to diarrhoea. *Staph. aureus* and *Candida albicans* may contribute at times.

Confirmatory tests. These may be unnecessary in typical cases, but anaerobic stool culture on specific media detects *C. difficile* (so-named because difficult to grow), and toxin detection is by tissue culture.

Chemotherapy. Oral metronidazole is the treatment of choice. Oral vancomycin is also effective, but the emergence of vancomycin-resistant enterococci (VRE) should limit its use to relapses of pseudomembranous enterocolitis.

Control and prevention depend on limiting antibiotic use, and using narrow spectrum ones whenever possible.

Tropical sprue
The main characteristics are chronic diarrhoea, glossitis and malnutrition during or after residence in the tropics. The causative organisms are uncertain, though there is often coliform overgrowth in the small bowel.

Clinical features often begin with acute enteritis and diarrhoea, followed by chronic diarrhoea with abdominal discomfort, anorexia, weight loss, glossitis and malnutrition. Laboratory diagnosis is partly by excluding parasitic and other diseases, especially giardiasis. Small bowel biopsy shows characteristic abnormalities: broad villi with chronic inflammatory cell infiltrate.

Chemotherapy is usually with tetracycline, folic acid and vitamin B_{12}. **Control and prevention** are not yet feasible.

Bacterial overgrowth syndromes
Characteristics of these include chronic diarrhoea and malnutrition, from bacterial overgrowth in the upper small bowel. Predisposing structural or functional causes include achlorhydria, surgery causing a blind loop, or impaired motility from diabetes or scleroderma. Causative organisms are usually mixed aerobic enteric Gram-negative rods and anaerobes.

Clinical features are chronic bulky offensive fatty diarrhoea, fatigue, weakness and weight loss. Confirmatory diagnosis was by quantitative small bowel culture, now usually replaced by the C^{14}-glycocholic acid breath test for bile salt deconjugation by bacteria. Megaloblastic anaemia from vitamin B_{12} deficiency is usual.

Chemotherapy is usually with tetracycline, fat-soluble vitamins and vitamin B_{12}. **Control and prevention** is seldom feasible except by avoiding the surgical creation of blind loops.

Diarrhoeal disease I

- Diarrhoeal disease is produced by three main mechanisms:
  - preformed toxin is ingested in food and rapidly (hours) causes symptoms, include vomiting
  - toxin produced in the gut by proliferating bacteria causes marked diarrhoea with a longer incubation time (hours to days)
  - mucosal invasion causes disease after a number of days, often with fever, cramps and blood in faeces.
- Sources of infection are food, water and, occasionally, person to person or directly from animals.
- Gay bowel syndome results from ingestion or ano-rectal implantation of organisms including those causing sexually transmitted diseases.
- Antibiotic-associated diarrhoea and pseudomembranous colitis are caused by over-growth of *C. difficile* caused by suppression of normal gut flora by oral antibiotics.
- Bacterial overgrowth in the gut follows a structural or functional predisposition.

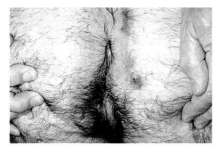

Fig. 1 **Fistulae from ischio-rectal abscesses.**

DIARRHOEAL DISEASE II: PATHOGENS I

BACTERIA

A number of bacterial pathogens cause intestinal disease (Table 1); pathogenesis includes all the mechanisms outlined in Diarrhoeal disease I.

Bacillus cereus

B. cereus causes either short-incubation emetic illness from ingestion of pre-formed toxin in food (usually rice, but also meats or vegetables) or longer-incubation diarrhoeal illness when ingested B. cereus produces toxin in the gut. In either case the illness is short in duration, and no specific treatment is necessary. Food or stool culture may show the cause.

Campylobacter spp.

These organisms, especially C. jejuni, are a common cause of gastroenteritis, often from chicken, other meats and milk. Toxin production and tissue invasion cause fever and marked cramps. Confirmatory diagnosis needs special stool culture, microaerophilic at 45°C. Early oral erythromycin may help the slow recovery.

Clostridium botulinum

This is the second organism causing two distinct syndromes, either the adult form from pre-formed toxin in incorrectly-preserved vegetables or fruits, or the infant form when ingested spores (usually in honey) produce toxin in the gut. In either case the toxin produces serious paralysis which may need respiratory support. Antitoxin treatment is unproven but usual.

C. perfringens

This is the third organism causing two distinct syndromes. The usual form occurs when food is cooked sufficiently to kill vegetative cells but not enough to kill spores which germinate and produce enterotoxin (α-toxin) which causes symptoms. The unusual 'pigbel' (necrotising enteritis) occurs when spores of type C organisms are ingested (usually in pork). These produce β-toxin which survives in people who are trypsin deficient in their gut as a result of eating a protein-deficient diet. It is traditionally associated in New Guinea with an occasional feast. The β-toxin produces diarrhoea, haemorrhage and perforation with 50% mortality.

Escherichia coli

E. coli is a member of the normal gut flora; some strains possess virulence factors that cause diarrhoea and other damage in the gut.

Enterohaemorrhagic E. coli (EHEC). These strains (also known as VTEC) produce verotoxin, which binds to receptors on gut mucosa and in the kidney leading to mucosal damage and haemorrhage. Serotype 0157 causes haemorrhagic colitis (HC) and the haemolytic–uraemic syndrome (HUS) of haemolytic anaemia, thrombocytopenia and acute renal failure.

Table 1 **Bacteria causing intestinal disease**

Organisms	Mechanism
Common in developed countries	
Campylobacter jejuni	Cytotoxin plus invasion
Clostridium perfringens	Enterotoxin plus β-toxin
Escherichia coli	Enterotoxin (ETEC) invasion (EIEC) verotoxin, (haemorrhagic, EHEC)
Salmonella spp.	Invasion
Staph. aureus	Preformed toxin
Common in developing countries	
Entamoeba histolytica	Invasion
Escherichia coli	Enterotoxin (ETEC) invasion (ELEC) verotoxin, (haemorrhagic, EHEC)
Salmonella spp.	Invasion
Shigella spp.	Invasion
Vibrio cholerae	Enterotoxin
Less common	
Clostridium botulinum	Preformed toxin; enterotoxin in infants
Vibrio parahaemolyticus	Enterotoxin and invasion
Yersinia enterocolitica	Enterotoxin and/or invasion
Bacillus cereus	Preformed toxin or enterotoxin

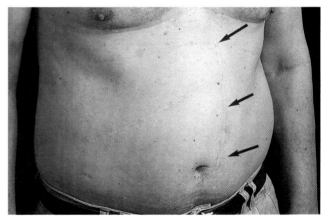

Fig. 1 **Two mouthfuls of take-away food resulted in severe salmonellal gastroenteritis and septicaemia in this man.** Note the gastrectomy scar. Other family members had only diarrhoea.

Enteroinvasive E. coli (EIEC). These strains invade the gut mucosa, multiply, spread and destroy the mucosa, causing bloody diarrhoea.

Enteropathogenic E. coli (EPEC). These organisms adhere to and destroy the gut microvilli.

Enterotoxigenic E. coli (ETEC). These strains produce a heat-labile enterotoxin (LT) similar to cholera toxin, and/or several heat-stable toxins (STs). They cause fluid secretion and hence diarrhoea: the commonest cause in children in developing countries.

Laboratory confirmation is specialised, though an ELISA test is available for LT in ETEC. Chemotherapy is usually unnecessary.

Salmonella spp.

Salmonellae (apart from S. typhi) were the commonest cause of gastroenteritis in the developed countries; now Campylobacter

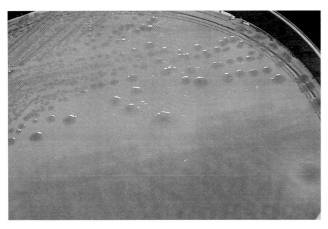

Fig. 2 *V. cholerae.* Golden yellow colonies on TCBS medium.

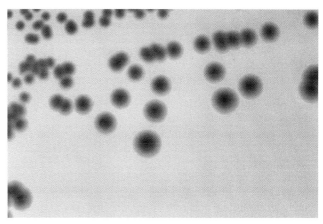

Fig. 3 *Vibrio parahaemolyticus* (non-cholera Vibrio) green colonies.

spp. are more common. Diarrhoea results from invasion of the small bowel mucosa, where inflammation occurs and fluid secretion increases. In certain patients (<1%) with predisposing conditions (e.g. sickle cell anaemia, gastrectomy (Fig. 1) or cancer) the organisms spread beyond the gut to cause septicaemia. Otherwise the illness is self-limiting and antibiotics will only prolong the period over which microbes are excreted. Diagnosis is by stool culture and serotyping; the latter is important when tracing the source of infection.

Control. There is a large animal reservoir of salmonellae, which are transmitted to humans by contaminated food (meat, eggs, dairy products) and can have secondary spread person to person. Control depends on food hygiene, and good water supplies and sewage disposal methods. People may continue to excrete organisms for several weeks after an infection so hand washing before handling food is essential, and those in occupations involving food should not work until faecal cultures for Salmonellae are negative.

Shigella spp.

Shigellae cause diarrhoeal disease of varying severity: mild (*S. sonnei*), more severe (*S. flexneri* and *S. boydii*), and severe bacillary dysentery with blood and mucus (*S. dysenteriae*). Shigellae cause diarrhoea by invasion of gut mucosa; they also produce enterotoxin but it may not contribute to the symptoms. Infection is common in children, in institutions and in developing countries because it is highly infectious (10–100 bacteria), so is spread person to person more commonly than by food or water. Diagnosis is by stool culture. Rehydration may be needed; antibiotics are only used for severe or systemic infections.

Staph. aureus

S. aureus causes sudden vomiting within 2–8 hours through preformed toxin, very like the emetic form of *B. cereus* disease. Culture of the causative food may, therefore be negative, while latex agglutination test for toxin is positive.

Vibrio cholerae

Characteristically, classical cholera causes profuse watery diarrhoea, but milder infections also occur.

The causative organism is *Vibrio cholerae*, usually serotype

01, which has two biotypes, el tor and cholerae, and three serologic subgroups. The organism is now widely spread in Asian, African and American rivers, estuaries and coastal waters, and infection is by drinking infected water. The symptoms of cholera result from production of an endotoxin. Virulence factors protect the organism and allow it to adhere to the gut mucosa.

Clinical features are watery diarrhoea (up to 1 litre hourly), fever and dehydration, with tachycardia, hypotension, hypokalaemia and bicarbonate loss. As a result metabolic acidosis and renal failure occur with death in a few hours in severe untreated disease.

Confirmatory tests must not delay treatment. An experienced microscopist can find motile vibrios by dark-ground microscopy of faeces. Culture is best on a specific medium containing thiosulphate, citrate, bile and sucrose (TCBS agar, Figs 2 and 3).

Chemotherapy is less important than prompt fluid and electrolyte replacement, which is essential; oral replacement solutions can minimise intravenous therapy. Antibiotics are not essential but tetracycline is often given to reduce the infective phase.

Control and prevention depend on water hygiene. There is no evidence for person-to-person spread. Current injectable vaccines give only partial protection for about 6 months, though new oral vaccines are promising.

Vibrio parahaemolyticus

This causes illness resembling *Salmonella* and *Shigella* spp. infections but has a longer incubation period, with cramps and fever from invasion and toxin production in the gut. The organism is unusual in being a salt-loving vibrio, hence illness is from infected seafood or fish, and special culture media are needed (Fig. 3). Chemotherapy is usually unnecessary.

Yersinia enterocolitica

This is a rare cause of longer incubation illness marked by fever, pain, enterocolitis and mesenteric adenitis mimicking acute appendicitis. The organism grows at low temperatures (4°C, but prefers 22–25°C), so infection can occur even from refrigerated food and is more common in colder climates. Special culture conditions are required in the laboratory. Chemotherapy is unproven, but gentamicin and/or ceftriaxone are used in septicaemic patients.

DIARRHOEAL DISEASE II: PATHOGENS II

PROTOZOA AND WORMS

Although a forbidding number of protozoa and worms enter the bowel and may cause abdominal symptoms including diarrhoea (Table 1) many are rare or of restricted geographic distribution.

Protozoa

Balantidium coli (p. 75). This is a large ciliate parasite of pigs and infects humans by the faecal–oral or person-to-person routes. It invades colonic mucosa and is a rare cause of dysentery.

Cryptosporidium parvum (p. 73). This rarely causes disease in immunocompetent people but is a very common cause of diarrhoea in AIDS patients (similar disease is rarely caused by *Isospora belli*, Microsporidia (including *Enterocytozoon*) and *Cyclospora* spp.). *Cryptosporidium* spp. are very resistant to disinfectants including chlorine, and infection is commonly water-borne, or person-to-person, including by sexual contact. Sometimes biliary tract infection causes cholangitis, but the major infection is in the small bowel: diarrhoea in AIDS is watery and profuse like cholera, causing dehydration and then malnutrition. Confirmation of diagnosis is by acid-fast stain of faeces (Fig. 1). Chemotherapy with paromomycin or azithromycin is seldom helpful; the hormone octreotide sometimes helps.

Dientamoeba fragilis (p. 75). This is not an amoeba but a motile flagellate. With no known cyst form, the delicate trophozoite is probably transported within *Enterobius vermicularis* (pin worm) eggs. It causes mild damage to colonic mucosa and rarely causes abdominal and diarrhoeal symptoms. Treatment is with iodoquine, tetracycline or paromomycin.

Entamoeba histolytica (p. 74). Infection occurs worldwide through ingestion of cysts in faecally contaminated food or water (or sexually). It causes acute amoebic dysentery with cramps, and diarrhoea with blood and mucus. It also causes relapsing chronic amoebic dysentery, and amoebic abscesses in the liver (Fig. 2), lung and brain. Confirmation of the diagnosis is by seeing the trophozoites in fresh warm stools. Metronidazole has replaced less effective drugs.

Giardia lamblia (p. 75). Trophozoites from ingested cysts of this flagellate adhere to small intestinal mucosa; giardiasis varies from asymptomatic infection through acute watery diarrhoea to chronic intermittent diarrhoea with malabsorption. Confirmation often needs small bowel content microscopy, as stool microscopy is often negative. Chemotherapy is by metronidazole, tinidazole or quinacrin; repeat courses may be needed. Cysts resist usual water chlorination levels.

Intestinal helminths

Ascaris lumbricoides (p. 79). Round worm infections (Fig. 3) are very common. They may be asymptomatic or may cause

Table 1 **Intestinal parasitic infections**

Causing diarrhoeal symptoms	Causing systemic symptoms
Protozoa	
*Giardia lamblia**	*Sarcocystis* spp.
*Cryptosporidium parvum**	*Toxoplasma gondii*
*Entamoeba histolytica**	
Isospora belli	
Dientamoeba fragilis	
Balantidium coli	
Nematodes	
*Ascaris lumbricoides**	*Dracunculus medinensis*
*Enterobius vermicularis**	*Toxocara canis*
*Trichuris trichiura**	*Trichinella spiralis*
Ancylostoma duodenale[a]*	
Necator americanus[a]*	
Strongyloides stercoralis[a]*	
Cestodes	
Taenia saginata and *T. solium**	
Diphyllobothrium latum	*Echinococcus* spp.
Hymenolepis nana and *H. diminuta*	*Taenia solium*
Trematodes	
Fasciolopsis buski	*Fasciola hepatica*
Heterophyes heterophyes	*Opisthorchis sinensis*
Metagonimus yokogawai	*Paragonimus westermani*

[a]Transmission by skin penetration not ingestion.
*Common

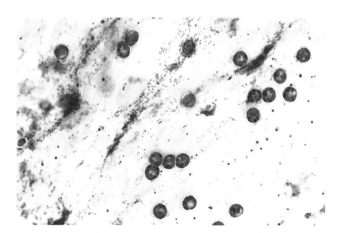

Fig. 1 *Cryptosporidium* spp. in faeces stained red with acid-fast stain.

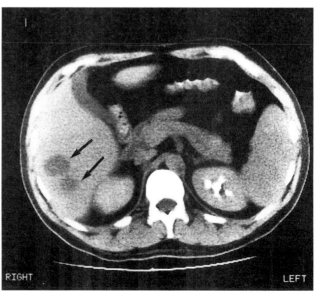

Fig. 2 **Multiple amoebic liver abscesses.**

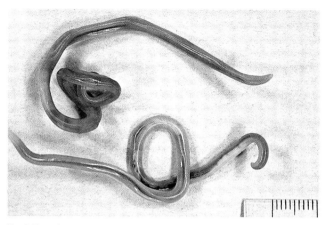

Fig. 3 **Roundworms.**

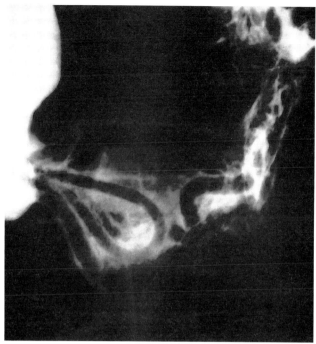

Fig. 4 **Round worms outlined by radioopaque dye in the stomach.**

Fig. 5 *Trichuris trichuria* **egg.**

diarrhoea; however, abdominal discomfort is more common, and even bowel obstruction (Fig. 4), biliary obstruction or peritonitis can occur. Diagnosis is by stool microscopy, and chemotherapy is by mebendazole or pyrantel.

Enterobius vermicularis (p. 78). Pin- (or thread) worm infection is also common and also rarely causes diarrhoea. Peri-anal itch is typical, and diagnosis is by microscopy of adhesive tape after peri-anal application. Chemotherapy is by mebendazole or pyrantel for the whole family.

Trichuris trichiura (p. 78 and Fig. 5). Whipworm infection occurs less commonly; symptoms range from none through mild abdominal discomfort and bloody diarrhoea to rectal prolapse. Luminal worms in appendicitis may not be causal. Chemotherapy is by mebendazole (better tolerated) or thiabendazole (more effective).

Ancylostoma duodenale, Necator americanus (**hookworms**) **and** *Strongyloides stercoralis* (p. 79). Unlike the three other helminths above, these infect by skin penetration. In the small intestine they can cause diarrhoea, but anaemia is the main effect of hookworm infection (Fig. 6); *S. stercoralis* causes systemic allergic symptoms, including rash and allergic pneumonitis. Treatment is by mebendazole or pyrantel for hookworm, while thiabendazole is more effective for *S. stercoralis*.

Cestodes. *Diphyllobothrium latum, Hymenolepis nana, Hymenolepis diminuta,* and *Taenia* spp. can all cause diarrhoea and other abdominal symptoms (p. 82).

Trematodes. *Fasciolopsis buski* and the rare *Heterophyes heterophyes* and *Metagonimus yokogawai* can also cause diarrhoea and other abdominal symptoms (pp. 84–85).

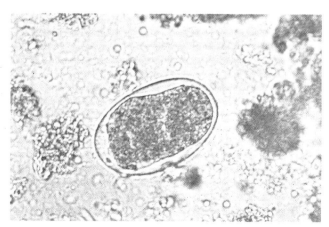

Fig. 6 **Hookworm egg.**

Diarrhoeal disease II: pathogens

- Many bacteria cause diarrhoeal disease; *Campylobacter* and *Salmonella* spp. and *E. coli* are frequent pathogens.
- Protozoa can live in the gut and may cause diarrhoeal disease, especially *E. histolytica, G. lamblia* and *Cryptosporidium* spp.
- Nematodes, cestodes and trematodes are gut parasites that may also cause diarrhoeal disease.
- Diarrhoeal disease is particularly threatening to the young, very old, malnourished and immunocompromised. Public health measures to prevent infection are vital to reduce disease incidence.

PERITONITIS AND INTRA-ABDOMINAL ABSCESSES

Peritonitis is *diffuse* infection in the peritoneal cavity, while intra-abdominal abscesses are *localised* collections of pus in the peritoneal cavity or abdominal organs. Peritonitis may localise to form one or more abscesses, while abscesses may rupture to cause peritonitis.

PERITONITIS

Peritonitis is characterised by diffuse inflammation of the peritoneum. It is chemical or infective in origin.

Chemical peritonitis. This occurs when acid gastric or duodenal contents are released into the abdominal cavity when a peptic ulcer perforates, or when bile is released from a perforated gallbladder.

Infective peritonitis. This may be primary, also called spontaneous, arising in individuals with predisposing conditions, e.g. cirrhosis or nephrosis, or may occur in childhood. Secondary peritonitis can follow intra-abdominal infection from various causes:

- perforation of a viscus by traumatic, ulcerative or ischaemic rupture
- surgical leaks
- pelvic inflammatory disease (PID)
- peritoneal dialysis (PD) including continuous ambulatory peritoneal dialysis (CAPD)
- intra-abdominal infections or abscesses.

Causative organisms

Primary peritonitis in children and nephrotic patients is almost always caused by pneumococci and other streptococci. In cirrhosis, it can also be caused by bowel flora: enteric Gram-negative rods (GNRs), enterococci and anaerobes, especially *Clostridium* spp.

Secondary peritonitis is usually caused by bowel flora, hence it is a mixed infection from enteric aerobic GNRs, enterococci and anaerobes, especially *Bacteroides fragilis*. In peritoneal dialysis, coagulase-negative staphylococci (CNS) are common. Rarely, abdominal tuberculosis or actinomycosis causes peritonitis.

Clinical features

In all cases, the features of peritonitis – fever, severe abdominal pain ('like the kick of a horse' on perforation, 'agonising' if diagnosed late), tenderness, rigidity and guarding on abdominal palpation, diminished or absent bowel sounds, plus systemic signs of sepsis – are superimposed on the features of the predisposing or underlying cause.

Confirmatory tests

Neutrophilia suggests infection; abnormal liver function tests suggest hepatic or perihepatic involvement. Gas under the diaphragm (without previous surgery) confirms a ruptured viscus, but only aspiration of peritoneal fluid (by needle, by peritoneal dialysis catheter or by laparoscopy) or surgical operation can prove peritonitis through subsequent microscopy and culture.

Management

Chemotherapy must cover as wide a range of organisms as possible, and hence is often triple therapy 'AGM' with ampicillin (for enterococci), gentamicin (for aerobic GNRs) and metronidazole (for anaerobes). Other possible drugs in severely ill patients include ticarcillin–clavulanate, piperacillin–tazobactam or imipenem. Supplementary surgery is essential for perforation, leaks or abscesses and may assist in cleansing the abdomen by lavage. In peritoneal dialysis peritonitis, intraperitoneal vancomycin is usual empiric initial therapy. In primary peritonitis in childhood or nephrosis, penicillin or a third-generation cephalosporin is used until the organism is known.

INTRA-ABDOMINAL ABSCESSES

Intra-abdominal abscesses are characterised by a localised collection of pus. They are classified by their site and by their mode of origin (Table 1 and Fig. 1) (as with abscesses in the chest; pp. 116–17). Infections adjacent to an organ are termed para- or peri-, e.g. para-appendiceal abscesses. Diverticula are out-pouchings from the lower large bowel (Fig. 2) by which infections often occur.

Table 1 **Types of intra-abdominal abscess**

Area affected	Type
Within viscera	Hepatic, pancreatic, splenic
Adjacent to bowel	Para-appendiceal, diverticular
Dependent peritoneal spaces	Subphrenic, paracolic, pelvic
Retroperitoneal	Perinephric, psoas

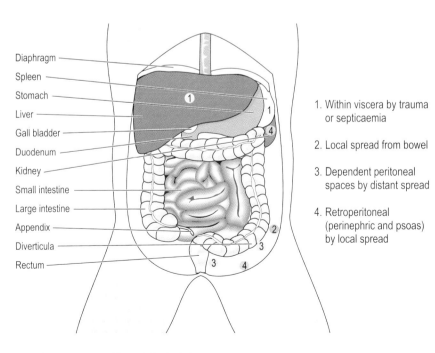

Diaphragm
Spleen
Stomach
Liver
Gall bladder
Duodenum
Kidney
Small intestine
Large intestine
Appendix
Diverticula
Rectum

1. Within viscera by trauma or septicaemia

2. Local spread from bowel

3. Dependent peritoneal spaces by distant spread

4. Retroperitoneal (perinephric and psoas) by local spread

Fig. 1 **Origin and types of intra-abdominal abscess.**

Fig. 2 **Diverticulitis, showing the dark opening of one large diverticulum on the mucosal surface.** The abscess on the outer surface is hidden in this view.

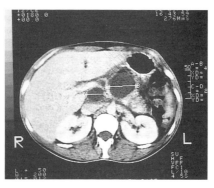

Fig. 3 **CT scan showing two large pancreatic abscesses (marked).**

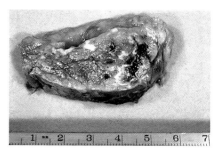

Fig. 4 **Acute pancreatitis: autopsy specimen showing necrosis and early abscess formation.**

Causative organisms

The causative organisms are usually bowel flora in bowel-related abscesses – appendiceal, diverticular and pancreatic abscesses (Fig. 3) – and in most subphrenic, paracolic and pelvic abscesses.

Hepatic abscesses (p. 145) are special, for they can arise from the upper bowel via the bile duct (cholangitis, p. 144), from the whole bowel by the portal vein (pylephlebitis, p. 145), from the blood stream by the hepatic artery, or from trauma. As a result bowel flora, amoebae, hydatids, septicaemic organisms and skin or soil organisms can be involved (pp. 144–6).

Psoas and retroperitoneal abscesses usually arise either from the kidney,

hence contain uro-pathogens as does a perinephric abscess (p. 150), or from the spine, hence often caused by *S. aureus* or *M. tuberculosis* (pp. 180–1).

Splenic abscesses usually result from septicaemia, caused by enteric GNRs, staphylococci, streptococci, salmonellae or anaerobes (about 20% each).

Clinical features

Though modified by the site and the underlying pathology, abscesses in general are marked by a swinging temperature, continuing longer than expected after, for example, appendicitis, diverticulitis or pancreatitis (Fig. 4). Paralytic ileus with constipation and diminished or absent bowel sounds is common. In contrast to peritonitis, pain is not prominent early because the abscess is often separated from the sensitive peritoneum by the 'abdominal policeman', the omentum. A mass usually occurs late.

Confirmatory tests

While plain X-ray may be helpful, especially for gas in subphrenic abscesses

(Fig. 5a), the CT scan has revolutionised the diagnosis of many abscesses (Fig. 5b), though small or multiple abscesses may not be visualised separately from bowel. Indium scans using the patient's labelled white cells are quicker and more specific than gallium scans. The causative organisms are diagnosed by Gram stain with aerobic and anaerobic culture of the abscess pus.

Management

Once visualised, the abscess (es) must be drained, under CT scan (Fig. 5b) or by surgery. For bowel-related abscesses, triple therapy 'AGM' (as for peritonitis, above) is still appropriate, but other regimens of similar broad spectrum can be used. Psoas, retroperitoneal and splenic abscesses need specific chemotherapy for the pathogen.

Control and prevention

This depends on early surgery for appendicitis or diverticulitis, and early recognition and treatment of septicaemia and bowel, urinary and spinal infections.

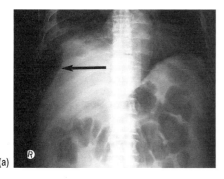

(a)

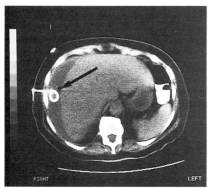

(b)

Fig. 5 **Subphrenic abscess. (a)** On abdominal X-ray (often seen best on chest X-ray); **(b)** on CT scan with 'pig-tail' drain inserted.

Peritonitis and intra-abdominal abscesses

- Chemical peritonitis is caused by leakage of gut contents or bile through trauma or perforation.
- Primary peritonitis occurs in childhood or in nephrosis (pneumococcal or streptococcal) or cirrhosis (bowel flora).
- Secondary peritonitis is usually caused by bowel flora following damage to the guts. Coagulase-negative staphylococci are common in peritonitis secondary to peritoneal dialysis.
- Clinical features are pain, guarding and rigidity, plus systemic evidence of sepsis.
- Initial chemotherapy for peritonitis must cover all likely possibilities, e.g. ampicillin, gentamicin, and metronidazole. Perforation and other anatomic defects need surgical cure.
- Intra-abdominal abscesses are collections of pus that arise by traumatic implantation, local spread, distant spread or by septicaemia.
- Clinically, fever and systemic features are more prominent than pain or a mass until late.
- Confirmatory tests are CT or other imaging, and microscopy and culture when essential drainage is done. Chemotherapy is initially empiric against likely organisms, then specific.

BILIARY AND HEPATIC INFECTIONS

BILIARY INFECTIONS

CHOLECYSTITIS

Cholecystitis is inflammation of the gall-bladder; it is classified into acute (with or without calculi) and chronic. The latter usually follows acute attacks and infection plays a lesser part.

Acute cholecystitis

Acute cholecystitis (Fig. 1) in over 85% of patients results from gallstones, usually obstructing the cystic duct, which leads to distension and inflammation. Complications include:

* arterial damage causing ischaemia, necrosis, gangrene, perforation and peritonitis, with *Clostridium perfringens* producing gas in the wall (emphysematous cholecystitis)
* bacterial complications
  – local infection with pus in the gallbladder ('empyaema', Fig. 2)
  – local spread to a pericholecystic collection or to the common bile duct, causing cholangitis (see below)
  – distant spread causing septicaemia
* contiguous complications: lymphadenitis, pancreatitis (often with cholangitis) and peritonitis.

Acute acalculous ('without calculi') cholecystitis in the other 15% of patients is found particularly in debilitated hospitalised patients. Acute microvascular damage is probably mediated by endotoxin and activated Hageman factor XII.

Causative organisms. These are bowel flora, including enterococci, aerobic Gram-negative rods and anaerobes, including *Bacteroides fragilis* and *C. perfringens*.

Clinical features. Pain is usual in the right upper quadrant (RUQ), commonly in the back. Nausea, vomiting, fever and tenderness over the gallbladder are usual, but rigors, jaundice or hypotension suggest complications such as ascending cholangitis.

Confirmatory tests. Ultrasound (Fig. 3) shows a dilated gallbladder and may show a dilated common bile duct. The HIDA scan shows no gallbladder in acute infection if the cystic duct is blocked; non-visualisation also occurs in early scans in over 50% of chronic cholecystitis patients. Any operative specimens must be cultured.

Management. Debate continues whether chemotherapy should be given to all, or only to elderly or severely ill patients or those with complications (including cholangitis, emphysematous cholecystitis, perforation and peritonitis). Empiric therapy is against bowel flora, often triple 'AGM' (ampicillin, gentamicin and metronidazole). Surgery is often delayed in uncomplicated patients but is imperative in complicated disease.

CHOLANGITIS

Cholangitis is infection in the bile ducts, a feared complication of biliary, pancreatic or bowel disease. A stone in the common bile duct is the most common cause. Cholangitis is usually caused by bowel flora, but it may result from parasites: *Ascaris lumbricoides*, *Clonorchis sinensis*, *Cryptosporidium* spp. and *Fasciola hepatica*.

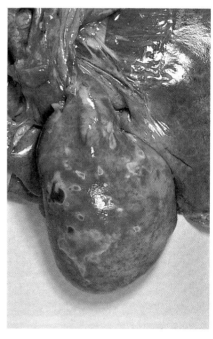

Fig. 1 **Acute cholecystitis, showing red, inflamed gallbladder with patchy necrosis.**

Clinical syndrome. In 85% of patients, Charcot's triad – fever, rigors and jaundice – occurs, which progresses to shock and liver failure.

Confirmatory tests. Ultrasound and HIDA scans have diminished the need for percutaneous transhepatic cholangiography (Fig. 4). Blood and bile cultures are essential.

Management. Triple 'AGM' therapy as above is usual, but surgery to remove obstruction is essential. Control and prevention are by early treatment of precipitating factors, e.g. gallstones.

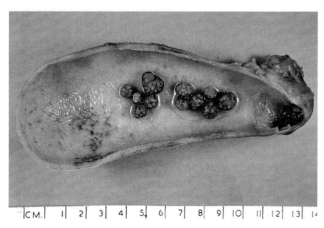

Fig. 2 **Empyema of the gall bladder.**

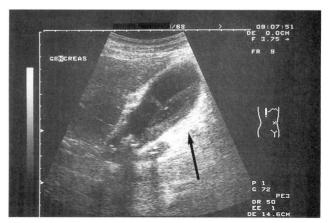

Fig. 3 **Ultrasound showing thick-walled gall bladder.**

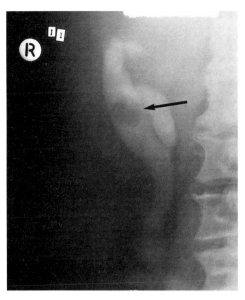

Fig. 4 **Dilated common bile duct (white dye) with large obstructing stone (dark gap).**

PORTAL PYLEPHLEBITIS

Portal pylephlebitis (p. 143) is septic thrombophlebitis (venous infection with thrombosis) of the portal vein. It arises from intraabdominal, often appendiceal, infection. It can spread locally to the liver or distantly by bacteraemia to the lung. As expected, bowel flora are the dominant pathogens.

Clinical features. Persistent high fever and signs of sepsis follow appendicitis or other intra-abdominal infection. If not diagnosed and treated, septic emboli lead to liver abscesses, with right upper quadrant pain and enlarging liver.

Confirmatory tests. CT scanning can show the intra-abdominal cause and the resultant liver abscesses; it rarely shows the portal pylephlebitis. Blood cultures are essential. Chest X-ray may show lung abscesses and pleural fluid. Abscess pus must be aspirated or drained, and cultured aerobically and anaerobically.

Management. Pylephlebitis needs appropriate antibiotics against gut flora, such as 'AGM', plus removal of the cause and drainage of any abscesses.

LIVER ABSCESSES AND CYSTS

Diffuse liver infections occur in kala-azar, malaria and schistosomiasis (pp. 128–9). There are a number of viral hepatic infections (hepatitis A–G, yellow fever). The liver is damaged in cholangitis (see above), which can be caused by parasites in the bile duct. Localised liver abscesses and cysts occur in a number of infections; three other important ones follow.

Amoebic liver abscess

Amoebic abscesses are a complication of *Entamoeba histolytica* infection that can occur long after intestinal amoebiasis. Over 60% of patients are unaware of the preceding bowel infection. The abscesses arise by spread of the trophozoites up the portal vein and are often multiple and in the right lobe, but may be single or left-sided. They may spread to serous spaces (pleura, peritoneum or pericardium) or the lung, then

brain. The *lysis* of liver tissue actually produces necrosis, 'like anchovy sauce', not a true pus-containing abscess.

Clinical features. Right upper quadrant pain and tenderness with fever are usual; only 25% have active amoebic dysentery. Systemic features including weight loss are prominent. Referred shoulder tip pain, pleural fluid and a raised right hemidiaphragm occur with abscesses near the diaphragm.

Confirmatory tests. Neutrophilia is usual, not eosinophilia. Ultrasound (Fig. 5) or CT scan show the abscess(es) (Fig. 2, p. 141). Stool microscopy for amoebae is usually negative unless diarrhoea is present. Drainage is less common now with better chemotherapy. Amoebae are not often seen in the aspirate, being found chiefly in the wall of the 'abscess'. Conversely, serology is usually positive.

Chemotherapy. Metronidazole has replaced emetine or chloroquine, and pain relief is usually rapid. Diloxanide furoate is used to treat intestinal infection. Emetine is sometimes used with metronidazole for complications such as peritonitis or pericarditis after rupture of an abscess.

Control and prevention depend on food hygiene and effective treatment of intestinal amoebiasis.

Pyogenic liver abscess

Hepatic abscesses arise in four ways:

- from the upper bowel via the bile duct (cholangitis, p. 144)
- from the whole bowel via the portal vein (pylephlebitis, see above)
- from the blood stream via the hepatic artery
- from outside by a wound.

The causative organisms depend on the source and route: bowel flora, septicaemic organisms, or skin or soil organisms.

Clinical features arise from:

- the liver abscess itself: fever, right upper quadrant pain and tenderness (not with deep abscesses) and enlarging liver
- the source: rigors and jaundice with cholangitis, right lower abdominal pain and tenderness with appendicitis, rigors and shock with septicaemia.

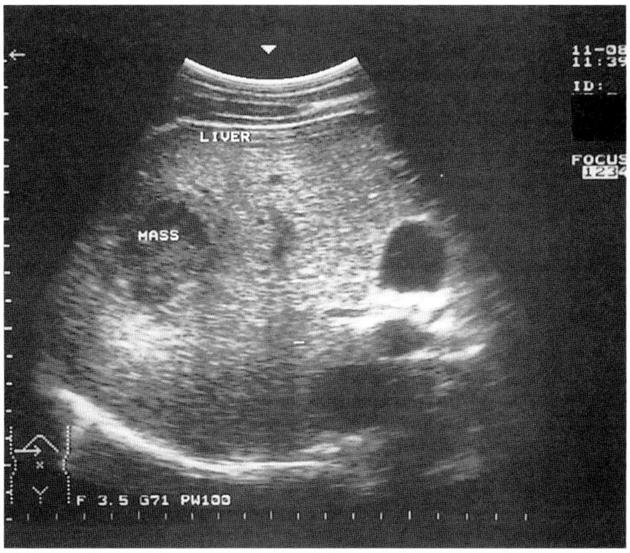

Fig. 5 **Amoebic liver abscess: ultrasound shows large single abscess ('mass').**

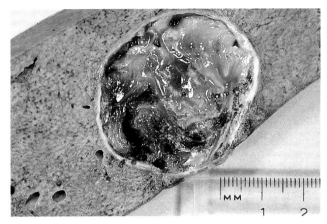

Fig. 6 **Hydatid cyst of liver: dead and inspissated.**

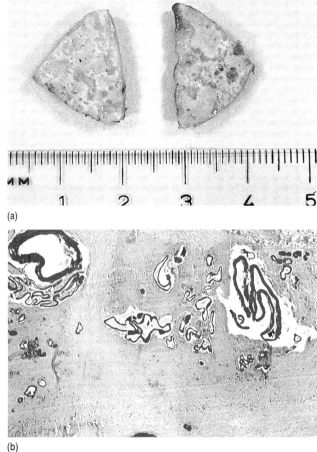

(a)

(b)

Fig. 7 *E. multilocularis* infection, with many locules infiltrating the liver: (a) liver specimen; (b) microscopy.

Confirmatory tests. Imaging by ultrasound or CT shows the abscess (Fig. 6) and, often, the source. Blood cultures and culture of the abscess pus show the causative organism(s).

Chemotherapy. Therapy depends on the cause; broad-spectrum empiric therapy such as 'AGM' (ampicillin plus gentamicin plus metronidazole) is used initially for bowel sources or septicaemia.

Hydatid cysts

Hydatid disease of the liver is usually cystic resulting from *Echinococcus granulosus* (pp. 82–3), but rarely it is invasive caused by *E. multilocularis* spreading like a malignancy (Fig. 7). Cysts are usually single and in the right lobe, but may be multiple or in any part of the liver. Humans are infected with the encysted larvae of *Echinococcus* spp. in canine faeces. The intermediate hosts are herbivores especially sheep and cattle (p. 82).

Clinical features. Often the cyst is found because of pressure on local structures, particularly the bile ducts, causing jaundice or cholangitis. Rarely spontaneous rupture or secondary infection occurs. Cysts elsewhere may be found in life or at autopsy, especially in lung or brain (pp. 120, 91). Old liver cysts calcify.

Confirmatory tests. Serology, particularly for arc 5 in a gel immunodiffusion test, is highly specific; other serology including latex agglutination is used for screening or supplementary testing. X-rays, CT and liver and brain scans are used for localisation.

Management. Albendazole is more active than mebendazole for treatment. Cysts can be aspirated under CT control in experienced units, but surgery is still often necessary, being very careful not to spill the infective cyst contents.

Control and prevention include treating adult worms in farm herbivores, preventing dogs from eating raw infected animal (e.g. sheep) viscera, and hand washing after dog or soil contact.

Biliary and hepatic infections

Biliary tract infections

- Cholecystitis, inflammation of the gallbladder, usually results from obstruction by gallstones. It causes fever and RUQ pain. Infection is usually by bowel flora, which are treated by broad-spectrum antibiotics, often 'AGM', with or without surgery.
- Cholangitis is infection of the bile ducts, usually by bowel flora. It causes fever, rigors and jaundice, often with shock. It is diagnosed clinically and by imaging, and is treated with broad-spectrum antibiotics and urgent surgery, because it is lethal.
- Portal pylephlebitis is infection of the portal vein, usually by bowel flora from appendicitis or other abdominal infection. It causes liver abscesses, septicaemia and lung abscesses. It is treated by broad-spectrum antibiotics, removal of the cause, and drainage of abscesses.

Liver abscesses and cysts

- Amoebic liver abscess is characterised by past or present residence in an endemic area for *E. histolytica*, and by pain and fever, seldom by simultaneous dysentery. It is diagnosed by ultrasound or CT scan and treated by metronidazole.
- Pyogenic liver abscess follows cholangitis, portal vein bacteraemia from abdominal sepsis, septicaemia or an external wound. Symptoms are local pain, fever and tenderness, plus symptoms from the source. Surgery may be needed for both the source and the abscess, and chemotherapy is essential.
- Hydatid cysts occur in endemic areas and often cause few symptoms. They are shown by imaging, may be confirmed by serology, and may respond to albendazole, although drainage or surgery are still often needed.

TROPICAL AND RARE ABDOMINAL INFECTIONS

Cholera is described on page 137. Worms and parasites causing abdominal disease in the tropics are described on pages 138 and 139.

FITZ–HUGH CURTIS SYNDROME

Fitz–Hugh Curtis syndrome is a rare condition involving inflammation of the serous membrane covering the liver: perihepatitis. Causative organisms are either *N. gonorrhoeae* or *C. trachomatis*. Clinical features are right upper quadrant pain, tenderness and guarding, plus fever. Symptoms of general peritonitis, salpingitis, cervicitis or urethritis may be present. Confirmatory tests are microscopy and culture of cervix and urethra swabs. Adhesions like violin strings can be visualised at laparoscopy (Fig. 1). Chemotherapy is with ceftriaxone and/or doxycycline. Division of the 'violin strings' can abolish chronic pain. Control and prevention is by safer sex to decrease gonorrhoea and genital chlamydial infection.

GRANULOMATOUS HEPATITIS

Granulomatous hepatitis is an uncommon condition which can have infectious and non-infectious causes; idiopathic cases also

Fig. 1 **Fitz-Hugh–Curtis perihepatitis showing 'violin string' adhesions at laparoscopy.**

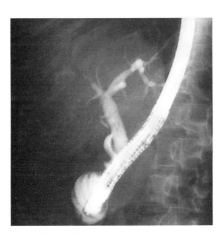

Fig. 2 **Clonorchiasis.**

occur where no cause can be found. Hepatic granulomas are collections of macrophages transformed into epithelioid cells, often with multinucleate giant cells, and sometimes with central 'cheesy' necrosis (caseation).

Causative organisms include: bacteria, e.g. *M. tuberculosis, M. leprae, M. avium–intracellulare complex, Brucella abortus* and *T. pallidum*; fungi, including *Histoplasma*; viruses including CMV; parasites, e.g. *Schistosoma* spp. and rickettsia (including *C. burneti*: Q fever).

Sarcoidosis, hypersensitivity and allergic diseases, and lymphomas are the commonest non-infectious causes. Clinical features vary with the cause, but fever, malaise and weight loss are usual. Confirmatory tests are liver biopsy to confirm granulomatous hepatitis, and Mantoux test, chest X-ray, microscopy with special stains, cultures and serology to find the cause. Chemotherapy depends on the cause, but anti-tuberculosis therapy should be given before steroids if no cause can be found. Control and prevention in contacts depends on the cause.

YERSINOSIS (ABDOMINAL)

Gastrointestinal infection with *Yersinia enterocolitica* or *Y. pseudotuberculosis* can uncommonly cause enterocolitis or mesenteric adenitis (infected mesenteric lymph glands). Yersiniae are found in many domestic and wild animals (i.e., these infections are zoonoses). *Y. enterocolitica* infection gives enterocolitis (p. 137) with diarrhoea, pain and fever; it also causes mesenteric adenitis, which mimics appendicitis, and, rarely, arthritis or erythema nodosum. *Y. pseudotuberculosis* causes pseudotuberculosis of the abdomen, i.e. mesenteric adenitis. Confirmatory tests are faecal cultures after 28 days of cold (4°C) enrichment, or cultures from lymph nodes or blood at 25°C. Serology is of limited use. Chemotherapy is not well established, but gentamicin or third-generation cephalosporins are active against both species. Control and prevention depends on food and water hygiene, and avoiding infected animals.

FLUKE INFECTIONS

Fasciolopsiasis occurs commonly in Southeast Asia. It is caused by *Fasciolopsis buski*, which attaches to the intestine causing diarrhoea with possible bleeding and ulceration. Later, the face, legs and abdomen may swell. The fluke *Clonorchis sinensis* also occurs in Southeast Asia. It attaches in the bile duct (Fig. 2) and can cause cholangitis (p. 144). Flukes can be treated with praziquantel.

Tropical and rare abdominal infections

- Cholera is water-borne and widespread; it causes water and electrolyte loss through mild to overwhelming diarrhoea.
- Fitz–Hugh Curtis syndrome is a perihepatitis caused by gonorrhoeal or chlamydial infection.
- Granulomatous hepatitis has both non-infectious and infectious causes, especially mycobacterial infection.
- *Yersinia enterocolitica* and *Y. pseudotuberculosis* both cause mesenteric adenitis mimicking appendicitis.
- Fluke infections cause gastrointestinal symptoms and cholangitis. *F. buski* and *C. sinensis* are found in Southeast Asia.

URINARY TRACT INFECTIONS: CYSTITIS AND PYELONEPHRITIS

Urinary tract infections (UTI) are classified into lower and upper tract infections.

Lower urinary tract infections include:

- urethritis: infection of the urethra, usually a sexually transmitted disease (p. 154)
- cystitis: infection of the bladder, often loosely called lower UTI, because it is so common
- trigonitis: localised cystitis of the triangle between the urethral and two ureteric orifices
- the urethral syndrome: dysuria and frequency without cystitis
- prostatitis: infection of the prostate (p. 150).

Upper urinary tract infections include:

- ureteritis: infection of the ureter; this is rare, usually caused by renal tuberculosis and is not discussed further
- pyelitis: infection of the pelvis of the kidney; it probably does not occur alone, without some renal infection, hence pyelonephritis
- acute pyelonephritis: infection of the renal pelvis and some of the renal tissue: it is the commonest upper UTI
- chronic pyelonephritis: diffuse interstitial nephritis with inflammatory changes. Infection is difficult or impossible to prove, therefore it is not considered further.

Bacterial infection

Infection is usually acquired by the ascending route so faecal flora are common pathogens.

Bacteriuria is the presence of bacteria in urine. It is found in symptomatic UTI but it may also be asymptomatic.

Significant bacteriuria, the criterion of UTI, has been considered to be 10^5 or more bacteria per ml of voided mid-stream urine. Numbers below 10^5 usually mean a contaminated specimen, not infection. However, with dysuria and pyuria, 10^3 to 10^4 bacteria per ml can signify infection.

Uncomplicated UTI is cystitis in the adult non-pregnant woman.

Complicated UTIs are infections in pregnancy, in children, in men, in sites other than the bladder, or with structural or functional defects. Complicated infections may be mixed, with two or more bacterial genera.

Recurrence is re-appearance of symptoms or infected urine after treatment or spontaneous improvement. It is either a **relapse** with the same organism, or a **re-infection** with a different organism. Re-infection means poor perineal hygiene or a persisting structural or functional defect, while relapse means such a defect, or inappropriate anti-microbial treatment. The fundamental determining factors are the virulence and numbers of the infecting organism balanced against normal host defences and pre-disposing factors.

Pathogenesis

There are a number of factors that predispose to infection, all of which either disrupt the flow of urine or make the access of pathogens easier:

- urethra length: shorter in females
- presence or increased numbers of pathogens: poor perineal hygiene, sexual activity
- obstruction to complete bladder emptying: pregnancy, prostatic hypertrophy (Fig. 1), congenital, tumours etc.
- foreign bodies: calculi, catheters, etc.
- vesicoureteric reflux: congenital, pregnancy.

Obstruction, foreign bodies or reflux lead to stasis, residual bladder urine and loss of the normal flushing defences.

CYSTITIS

The bladder is usually infected from the perineum and urethra, especially when sexual activity or poor perineal hygiene cause faecal contamination.

Causative organisms

As most uro-pathogens come from faecal flora, the commonest is *E. coli*, causing 80% of community-acquired and 40% of hospital-acquired infections; pathogens in descending order of frequency are:

- *E. coli*
- faecal Gram-negative rods, *Klebsiella, Enterobacter, Serratia, Proteus* spp. and *Pseudomonas aeruginosa* (particularly in hospital-acquired UTI)
- coagulase-negative *Staphylococcus saprophyticus* (especially in sexually active young women)
- Gram-positive cocci including enterococci
- other organisms including *M. tuberculosis* and *Candida* spp.

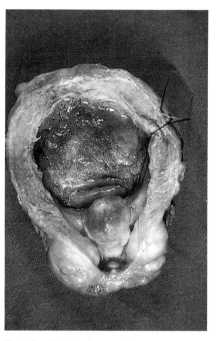

Fig. 1 **Cystitis, showing red, inflamed mucosa and greatly thickened wall, from prostatic obstruction.**

Fig. 2 **Pyelonephritis with intra-renal abscess.**

Clinical syndromes

Dysuria (pain on passing urine), urinary urgency, frequency and nocturia, suprapubic pain and tenderness, and low-grade fever are common. High fever with loin and renal angle pain and tenderness may occur, particularly with obstruction. Because there is much overlap in symptoms and signs between lower and upper UTIs, it is impossible to distinguish them clinically with certainty. Particularly in the elderly, UTI can be present with no symptoms. The prostate must be examined in men with cystitis. The **urethral syndrome** is dysuria and frequency without cystitis; it can be caused by urethritis, vaginitis, prostatitis, herpes genitalis or pathogens that are difficult to detect, e.g. Chlamydia. Complications of cystitis are rare.

Confirmatory tests

Microscopy and culture of a mid-stream urine specimen confirm a UTI, while ultrasound and/or IVP (intra venous pyelography) are needed to localise the site in women with recurrent infections and in men. Cystoscopy, biopsy and special stains and cultures are rarely needed. Micturating cysto-urethrography (MCU) is used to show reflux.

Chemotherapy

Similar cure rates in uncomplicated cystitis are found with numerous oral anti-microbials, e.g., amoxycillin–clavulanate, cephalexin or trimethoprim. Nitrofurantoin is an alternative to trimethoprim in pregnancy. Norfloxacin should be reserved for resistant infections.

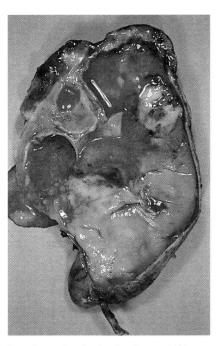

Fig. 3 **Pyonephrosis, showing destroyed kidney filled with pus.**

Single-dose or short course (3 day) therapy is effective in uncomplicated infections in non-pregnant women. Complicated or recurrent infections should be treated according to sensitivity test, for 7–10 days. Intravenous therapy is rarely needed. Predisposing factors should be removed if possible.

Control and prevention

Education in perineal and sexual hygiene, and removal of anatomic or functional defects are necessary; nightly or post-coital antibacterials are needed if the above fail or are not feasible. Triple micturition (urinating three times in succession with 1–3 minute intervals) minimises residual urine in patients with vesicoureteric reflux.

PYELONEPHRITIS

Only acute pyelonephritis is discussed here. The routes are ascending infection as in cystitis, or (rarely) haematogenous in endocarditis, bacteraemia or septicaemia, especially caused by *Staph. aureus.*

Determining factors are the same as in cystitis: virulence and numbers of the organism versus host defences and predisposing factors.

Causative organisms

These are similar to those in cystitis. *Proteus mirabilis* is particularly associated with calculi (urinary stones), probably because its urease produces ammonia and alkaline urine, favouring stone formation.

Clinical features

Dysuria (pain on passing urine), urinary urgency, frequency and nocturia, and high fever with vomiting, loin and renal angle pain and tenderness are common.

As with cystitis there is much overlap in symptoms and signs between lower and upper UTIs, so it is impossible to distinguish them clinically with certainty.

Complications of pyelonephritis are:

- local destruction, causing renal abscesses (p. 150), or papillary necrosis (particularly in diabetes, sickle cell disease or analgesic abuse), which can produce further obstruction (Fig. 2) or pyonephrosis (Fig. 3)
- local spread, causing perinephric abscess (p. 150)
- distant spread, causing bacteraemia (p. 128).

Confirmatory tests

As with cystitis, microscopy and culture of a mid-stream urine specimen confirms a UTI, while ultrasound and/or IVP are needed to confirm the renal site and show renal size, calculi and other abnormalities, e.g. staghorn calculus with pyonephrosis (Fig. 4). Cystoscopy, biopsy and special stains and cultures are rarely needed in acute infection.

Chemotherapy

Intravenous therapy is often needed initially, for example ampicillin and gentamicin. This may be replaced by oral amoxycillin or cephalexin according to sensitivity tests in community-acquired infections, but severe disease and/or more resistant hospital-acquired organisms may need continuing intravenous reserve drugs. Rehydration and pain relief are important.

Control and prevention

Careful catheter and operative techniques minimise hospital-acquired infection, while removal of underlying urologic abnormalities prevents recurrence.

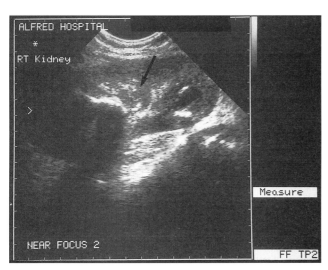

Fig. 4 **Pyonephrosis resulting from staghorn calculus (ultrasound).**

Cystitis and pyelonephritis

- Infection is usually ascending up the urethra by faecal flora.
- Infection occurs when the virulence and numbers of organisms exceed host defences.
- Host defences are impaired by poor perineal hygiene or anatomical or functional abnormalities, especially foreign bodies, obstruction or vesicoureteric reflux.
- Dysuria, frequency, fever, suprapubic and loin pain and tenderness are common in both lower and upper UTI.
- Urine microscopy and culture confirm infection, while renal ultrasound or IVP show renal abnormalities.
- Empiric treatment is followed by specific treatment after antibiotic sensitivity tests.
- Prevention depends on education and correction of defects more than on anti-bacterials.

RENAL AND PERINEPHRIC ABSCESSES/PROSTATITIS

These three infections, adjacent to the urinary stream, are alike because they:

- may cause dysuria and frequency like a UTI
- often have negative or ambiguous urine culture and microscopy results
- are undiagnosed without specific investigation
- are serious if undiagnosed
- need specific treatment, which may be operative.

INTRARENAL ABSCESS

Renal abscesses are characterised by collections of pus within the renal parenchyma. They can be classified by their causative mechanisms:

- ascending infection in pyelonephritis, especially with obstruction; this is the commonest mechanism, usually causing multiple abscesses (Fig. 1)
- haematogenous infection in bacteraemia, also usually multiple
- infection of a renal cyst, by bacteraemia or through an operative procedure; usually single.

Causative organisms

Ascending infection is with the usual uro-pathogens: *E. coli, Klebsiella* spp. and other faecal pathogens. Haematogenous infection is often with *Staph. aureus.*

Clinical features

Dysuria and frequency may occur if abscesses communicate with the collecting system (usually those from pyelonephritis), but otherwise are absent. The important features are fever with loin or flank pain and tenderness which *persist* for 3 days or more after treatment starts for a UTI or bacteraemia. A palpable mass is rare.

Confirmatory tests

Urine biochemistry, microscopy and culture show proteinuria, haematuria, pyuria and bacteriuria in abscesses which communicate with the collecting system, but these tests are often negative or show only low counts in abscesses from bacteraemia or infected cysts. Blood cultures are positive in one third. The diagnostic tests are ultrasound and/or CT scanning (Fig. 2) to confirm the diagnosis and show the number and position of the abscesses.

Management

Initial broad-spectrum empiric therapy with, for example, ampicillin and gentamicin, or an anti-staphylococcal beta-lactam, is replaced by specific therapy after positive urine or blood cultures, usually continued for 6 weeks or more. Percutaneous drainage under CT control has largely replaced open surgery when drainage is needed for large or unresponsive abscesses. Therapy is needed for the underlying defect (e.g. obstructive uropathy, endocarditis or papillary necrosis (Fig. 3)).

Control and prevention depend on early recognition and treatment of pyelonephritis, urinary obstruction and bacteraemia.

PERINEPHRIC ABSCESS

Perinephric abscesses are collections of pus next to the kidney (Fig. 4). They can be classified by their cause:

- local spread in pyelonephritis or renal abscess, especially with obstruction
- haematogenous infection in bacteraemia; this is rare.

Causative organisms

Spread from pyelonephritis involves the usual uro-pathogens. Haematogenous infection is often with *Staph. aureus.*

Clinical features

Dysuria and frequency occur from pyelonephritis but otherwise are absent. Like a renal abscess, the important features are fever with loin or flank pain and tenderness which *persist* for 3 days or more after treatment starts for a UTI or bacteraemia. A mass develops if diagnosis is delayed.

Confirmatory tests

The tests used are the same as those described for intrarenal abscesses. Blood cultures are often positive. The diagnostic tests are ultrasound and/or CT scanning.

Management

Treatment for pyelonephritis or bacteraemia is initiated empirically and is then directed by urine or blood cultures. Drainage is essential, usually percutaneous under CT control (Fig. 5). Surgery is needed only for very large or unresponsive abscesses.

Control and prevention, as in renal abscesses, depend on early recognition and treatment of pyelonephritis, urinary obstruction and bacteraemia.

PROSTATITIS

Prostatitis is characterised by diffuse infection of the prostate, which may progress to prostatic abscess. Prostatitis is classified as acute bacterial, chronic bacterial or chronic 'non-bacterial' ('prostatosis'). Prostatitis is probably caused by ascending infection from the urinary tract, either from a UTI (cystitis or pyelonephritis) or from a sexually

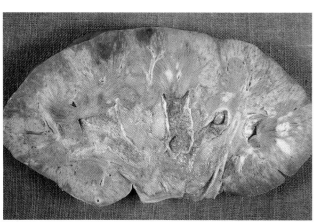

Fig. 1 **Intrarenal abscesses from acute pyelonephritis.**

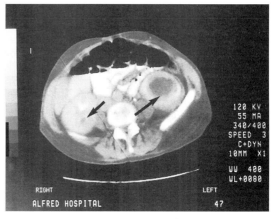

Fig. 2 **Intrarenal abscesses on CT.**

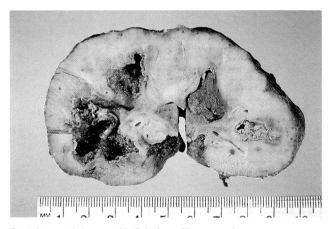

Fig. 3 **Intrarenal abscess with diabetic papillary necrosis.**

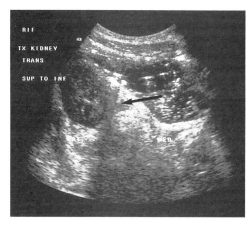

Fig. 4 **Peri-nephric abscess on ultrasound.**

transmitted disease (usually urethritis). Haematogenous infection, or lymphogenous spread from the rectum may occur.

Causative organisms

Pathogens causing prostatitis are usually:

- uro-pathogens: *E. coli, Klebsiella* spp. or other faecal Gram-negative rods or enterococci, from a UTI
- sexually transmitted pathogens: *N. gonorrhoeae* or *C. trachomatis*.

By definition, routine cultures are negative in 'non-bacterial' prostatitis, which may be caused by *C. trachomatis* or ureaplasma.

Clinical syndromes

In acute prostatitis, fever and perineal pain are usual, and symptoms of a lower UTI (dysuria and frequency) or of urethritis (dysuria and discharge) are frequent.

In chronic bacterial prostatitis, symptoms range from none (asymptomatic bacteriuria), to perineal or low back pain, to recurrent UTIs only temporarily responsive to anti-bacterials.

In chronic 'non-bacterial' prostatitis, symptoms again range from none to perineal or low back pain to recurrent UTIs.

Confirmatory tests

In acute prostatitis, urethral swabs, urine microscopy and culture or direct immunofluorescence (DIF) for *Chlamydia* usually show the pathogen, and acute prostatic tenderness is present rectally.

In chronic bacterial prostatitis the above tests are often negative, and prostatic localisation studies are needed: (urethral) urine is voided first (VB1), then a mid-stream (bladder) urine

(VB2), then expressed prostatic secretions (EPS) during rectal prostatic massage, then a third voided urine (VB3). Positive microscopy and culture on VB1 indicates urethritis, on VB2 indicates cystitis, while positive EPS and/or VB3 indicate prostatitis. Seminal ejaculate is more often culture positive than is EPS. Ultrasound may show an abnormal prostate. Biopsy is the final arbiter.

In chronic 'non-bacterial' prostatitis' all the above tests are negative except EPS show more than 10 white blood cells per HPF and ultrasound may be positive. Routine cultures, by definition, are negative.

Chemotherapy

Acute prostatitis usually responds well to oral norfloxacin or cotrimoxazole for 2–4 weeks. Initially, intravenous broad-spectrum beta-lactams and/or gentamicin may be needed for severe cases.

Chronic bacterial prostatitis usually only responds to a few drugs which penetrate into the non-acutely inflamed prostate: norfloxacin or ciprofloxacin alone or with rifampicin, or cotrimoxazole for 6–12 weeks.

Chronic non-bacterial prostatitis responds poorly, but doxycycline or erythromycin for 2–4 weeks are worth a trial, and aspirin helps pain relief.

Control and prevention

These depend on early diagnosis and treatment of UTIs and urethritis.

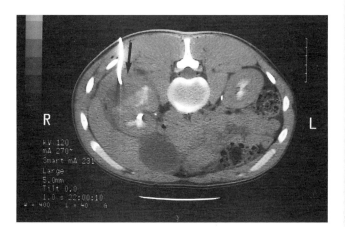

Fig. 5 **Peri-nephric abscess on CT, with catheter drain.**

Renal and perinephric abscesses/prostatitis

- These three infections, adjacent to the urinary stream, all may cause dysuria and frequency like a UTI, but usually have negative or ambiguous urine test results and, therefore, may be undiagnosed without specific investigation.
- Renal and perinephric abscesses typically show persistent fever and loin or flank pain that persists after apparently appropriate treatment for a UTI has started. They need urine and blood cultures, ultrasound or CT scans and continuing antibiotics. Perinephric abscesses usually need CT-directed drainage, rarely needed for renal abscesses.
- Prostatitis ranges from acute symptomatic to chronic asymptomatic with recurrent UTIs. Treatment may have to be empiric with norfloxacin or cotrimoxazole, or with doxycycline if *C. trachomatis* is suspected.

TROPICAL AND RARE URINARY INFECTIONS

ACUTE GLOMERULONEPHRITIS (AGN)

Unlike acute pyelonephritis, bacteria are *not* found in the urine and kidney when infection causes acute glomerulonephritis. The disease is caused by deposition of immune complexes (antibody with bacterial antigens or bacteria) onto the glomerular basement membrane. These complexes arise as a result of a distant infection (e.g. streptococcal impetigo). The immune complexes activate the complement and coagulation systems (pp. 21–4) causing inflammation, with clotting, increased permeability, endothelial cell damage and mesangial proliferation (Fig. 1). Numerous other types of glomerulonephritis, not caused by infection, are not considered here.

Causative organisms

The most common cause is *Streptococcus pyogenes*; acute glomerulonephritis is more common after a skin infection than after a streptococcal sore throat. The nephritogenic types are four or five of the 65 M types of *S. pyogenes*. Other organisms giving rise to acute glomerulonephritic immune complexes are:

- *Plasmodium species*, especially *P. malariae*, then *P. falciparum*
- streptococci or coagulase-negative staphylococci (rarely other organisms) in bacteraemia resulting from endocarditis or infected intravascular shunts (Fig. 1b).
- viruses, including Epstein–Barr and hepatitis B.

Clinical features

Poststreptococcal AGN follows 10–14 days after streptococcal pharyngitis (especially with group A, serotype M12) or after pyoderma or infected scabies (especially with group A, serotype M49). Rarely group C streptococci are responsible. Inflammation from the immune complexes causes haematuria and proteinuria, oedema and hypertension with headache. Nephritis from *P. malariae* often progresses to the nephrotic syndrome with persisting oedema and massive proteinuria. Focal or diffuse AGN in bacteraemia causes haematuria, proteinuria and renal impairment.

Confirmatory tests

In all cases, urine examination shows haematuria, proteinuria and hyaline casts, urine culture is sterile, and serum electrolytes, creatinine and urea show renal impairment. In streptococcal infection, throat or skin swabs are usually positive before therapy, and antistreptolysin O and DNAase B tests usually become positive in 1–2 weeks. In acute malaria with nephritis, thick and thin blood films are positive, but they are usually negative in malarial nephrotic syndrome, though antibody is positive.

Blood cultures and other tests in endocarditis are discussed on page 127.

Management

Penicillin for at least 10 days is given to eradicate causative streptococci. Malaria and endocarditis are treated as described (pp. 130, 127). Because only 6–8% of *S. pyogenes* M types cause AGN, repeat attacks are rare (in contrast to rheumatic fever), so penicillin prophylaxis is not given after poststreptococcal AGN.

HYDATID DISEASE

Larvae of the tapeworms *Echinococcus* spp. may be carried to any organ where they encyst, forming a hydatid cyst. They occasionally lodge in the kidney, producing a space-occupying lesion (Fig. 2). Drugs often do not penetrate well, so treatment is operative removal of the cyst only, if possible.

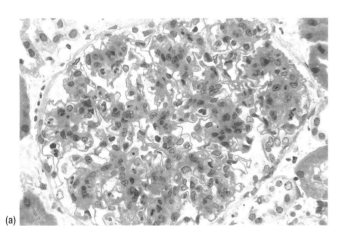

(a)

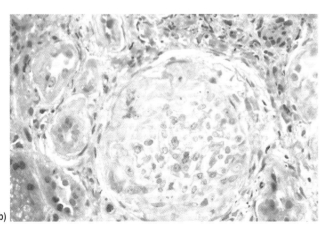

(b)

Fig. 1 **Glomerulonephritis. (a)** Acute poststreptococcal glomerulonephritis; **(b)** 'shunt' glomerulonephritis.

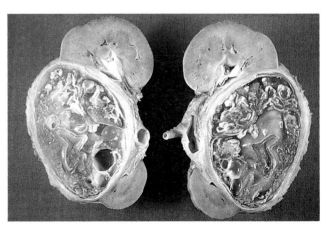

Fig. 2 **Hydatid of the kidney.**

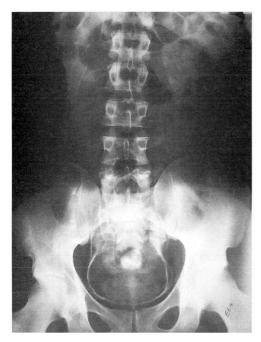

Fig. 3 **Schistosomiasis causing calcified bladder wall.**

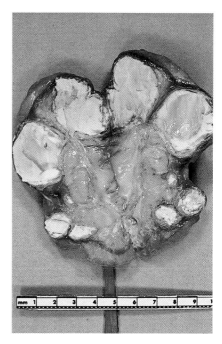

Fig. 4 **Renal TB showing massive destruction with caseation.**

SCHISTOSOMIASIS

The eggs of the blood flukes (*Schistosoma* spp., usually *S. haematobium*, p. 85) can deposit in the bladder. Early there is cystitis with haematuria, then hepatosplenomegaly with portal hypertension (p. 131). Later there may be bladder calcification (Fig. 3). Treatment is with praziquantel.

TUBERCULOSIS

Urinary tract tuberculosis, usually renal (Fig. 4) is secondary to haematogenous spread from a primary focus elsewhere, usually pulmonary. Spread may also have involved bone, joint, adrenal (Fig. 5), epididymis or other uro-genital organs. It is characterised by caseating destruction and heals with fibrosis, which often causes obstruction and bacterial UTIs.

Causative organism

This is *M. tuberculosis*, usually a human strain by inhalation causing primary pulmonary TB, occasionally bovine *M. tuberculosis* by ingestion, causing gastrointestinal TB (p. 54).

Clinical features

UTI symptoms of dysuria, frequency and loin or back pain are common, but fever and systemic symptoms are surprisingly uncommon.

Confirmatory tests

Urine examination often shows haematuria and proteinuria, but characteristically shows 'sterile pyuria' i.e. pyuria with usual cultures sterile. Urine cultures are positive in 80–90% when three successive early morning urine (EMU) specimens are cultured on specific mycobacterial media. Urine microscopy for acid-fast bacilli is unreliable because false positives occur from saprophytic mycobacteria (e.g. *M. smegmatis*) in healthy people. Chest X-ray shows pulmonary TB, often inactive, in over 70%. Renal ultrasound is often suggestive, while intravenous pyelography or CT are usually diagnostic.

Management

Triple therapy including isoniazid and rifampicin is usually needed for 9–12 months, longer in complicated or relapsed patients. Surgery is now used only for complications, including stricture or failed medical therapy. Control and prevention depend on BCG, and early, effective treatment of TB elsewhere.

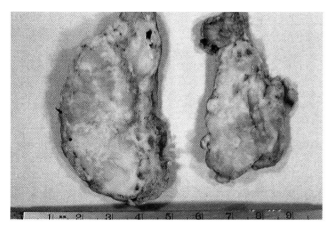

Fig. 5 **Adrenal TB showing typical tubercles.**

Tropical and rare urinary infections

- Acute glomerulonephritis related to infection is an immune complex disease, with haematuria, proteinuria, oedema and hypertension. Treatment depends on the cause, usually streptococcal infection, malaria or endocarditis.
- Renal tuberculosis is blood borne, usually from pulmonary TB. Sterile pyuria is usual, and special urine cultures are essential. Specific long-term anti-tuberculous therapy including isoniazid and rifampicin is needed.
- Helminth parasites (tapeworms and flukes), can deposit in the pelvic veins, bladder and kidney causing lesions.

URETHRITIS

Urethritis is inflammation of the male or female urethra. It may be:

- chemical, e.g. from local disinfectants or anaesthetics
- mechanical, e.g. from catheters or other foreign bodies, or from a worried man 'stripping', i.e. digitally milking, his urethra on several days for signs of sexually transmitted disease (STD; also called venereal disease or VD).
- infective, the most common cause, often as a STD.

The two major types of infective urethritis are **gonorrhoea** and so-called **non-specific urethritis (NSU)** or **non-gonorrhoeal urethritis (NGU).**

CAUSATIVE ORGANISMS

Gonorrhoea is caused by infection with *Neisseria gonorrhoeae* (p. 36). This pathogen only infects humans and is spread person to person, usually by sexual contact. It survives poorly outside the human host.

NGU has two major causes: *Chlamydia trachomatis* and *Ureaplasma urealyticum* (pp. 58, 57). It can also be caused, less commonly, by *Trichomonas vaginalis* and herpes simplex and, rarely, by *Gardnerella vaginalis* and yeasts.

In addition to these sexually transmitted diseases, **bacterial urethritis** (*Escherichia coli, Klebsiella* spp., *Staphylococcus aureus,* etc.) occurs with catheters, strictures and urinary infections including prostatitis (Table 1).

The relative incidence of these causes varies with:

- **gender**: *C. trachomatis* in women causes cervicitis (p. 156) more commonly than symptomatic urethritis
- **sexual orientation**: gonorrhoea predominates in homosexual men, *C. trachomatis* in heterosexual men
- **clinical presentation**: infections vary in their usual form of presentation (Table 2).

CLINICAL FEATURES

The two cardinal symptoms are discharge and dysuria. Discharge from the urethra varies from a scanty, clear or mucopurulent discharge, particularly in NGU, to copious yellow or yellow-green pus, particularly in gonorrhoea.

Dysuria means pain on passing urine and results directly from urethral infection and inflammation. It varies from mild (particularly in NGU) to extremely severe, 'like passing razor blades', particularly in gonococcal infections. It is often accompanied by urgency and frequency of micturition and by nocturia, i.e. the (unusual) passage of urine during the night.

Systemic symptoms including fever and malaise may occur in gonorrhoea but are absent in NGU.

Signs include urethral tenderness and the discharge, which may only be visible after 'milking' the urethra from penile base towards the glans. Inguinal lymph nodes may be enlarged in gonorrhoea. The testes, epididymis, prostate, rectum and throat or, in females, the cervix, Fallopian tubes and pelvis (cervicitis and pelvic inflammatory disease, pp. 156–59), should be examined for local spread or co-existent infection. The skin and joints may show disseminated infection.

Complications of gonorrhoea

Complications of gonorrhoea are infrequent unless treatment is delayed or inadequate.

Local spread. This causes periurethral abscesses, urethral stricture, epididymitis or prostatitis, salpingitis and PID.

Distant spread. This causes gonococcaemia with skin or joint infection. It is rare in men, but occurs in 1–3% of women, particularly if there is unrecognised, asymptomatic cervicitis (p. 156).

Co-existent infection. The rectum and pharynx can be infected by direct contact with infectious discharge, usually urethral, and the pharyngitis is often asymptomatic. The proctitis (infection of the rectum) may be asymptomatic but more often causes pain and anal discharge.

Postgonococcal urethritis (PGU). This is the persistence of symptoms of urethritis after treatment for gonorrhoea. It is caused by gonococci resistant to the antibiotic chosen, re-infection, co-existing chlamydial infection (the commonest cause) or co-existing ureaplasmal infection.

Complications of NGU

Local spread. Acute epididymitis (Fig. 1) and prostatitis are not uncommon with chlamydial infection but are rare in ureaplasmal infection.

Reiter's syndrome. This is the triad of urethritis, arthritis and uveitis. Additionally, 50% of patients have unusual skin or mucous membrane lesions (with even more unusual names): keratodermia blenorrhagica (hard skin papules with a waxy

Table 2 **Clinical presentation of non-gonorrhoea urethritis**

	Acute	NGU (%) Persistent	Recurrent
C. trachomatis	50	0	5
Ureaplasma	30	50	20
Other infections	20	50	75

Table 1 **Characteristics of urethritis (infective)**

Classification	Causative organism	Clinical features Discharge	Dysuria	Confirmatory tests	Chemotherapy	Control
Gonorrhoea	*N. gonorrhoeae*	Purulent, severe	Moderate to gross	Gram-negative diplococci in pus cells	Ceftriaxone or amoxycillin	Treat partners
Non-gonococcal urethritis (NGU)	*Chlamydia trachomatis*	Mucoid to thin pus	Mild to moderate	Direct immunofluorescence or culture	Doxycycline	Treat partners
	Ureaplasma urealyticum	Mucoid to thin pus	Mild to moderate	Special culture	Doxycycline (erythromycin)	Treat partners
	Trichomonas vaginalis	Minimal or none	Minimal or none	Wet mount microscopy	Metronidazole	Treat partners
	Herpes simplex	Minimal or none	Minimal or none	Viral culture	Acyclovir	Treat partners
	Rarer causes	Minimal or none	Minimal or none	M&C	Specific	Treat partners
Bacterial urethritis	*S. aureus* or enteric GNR	Purulent, moderate	Moderate to severe	Gram stain and culture	Flucloxacillin or gentamicin	Remove cause

M&C, microscopy and culture; GNR, Gram-negative rods.

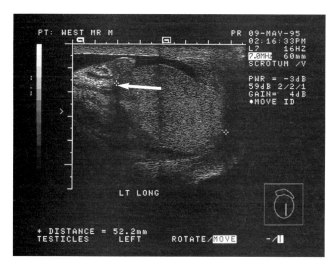

Fig. 1 **Chlamidial epididymitis with hydrocoele on ultrasound.**

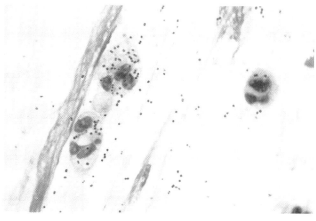

Fig. 2 *N. gonorrhoea* on Gram stain.

yellow centre), circinate balanitis (rash with circular outline on penile skin), circinate or ulcerative vulvitis, pharyngitis or glossitis. Reiter's syndrome is most often caused by genital chlamydial infection but may also be post-dysenteric, i.e. following bacterial gastroenteritis caused by *Salmonella, Shigella, Campylobacter* or *Yersinia* spp.

CONFUSING CONDITIONS

In men, while prostatitis, cystitis or an upper urinary tract infection may need to be considered, the symptoms of urethritis are usually clear-cut, and only the causative organism is unclear.

In women, however, dysuria, urgency, frequency, and nocturia, with or without urethral discharge, may also be caused by:

- bacterial cystitis: infection of the urinary bladder, with pyuria (i.e. polymorphonuclear leucocytes (PMNs) in the centrifuged deposit) and >100 000 bacteria per ml
- 'the urethral syndrome': most often urethritis caused by gonococcal or chlamydial infection but may be an atypical cystitis, or urethral trichomoniasis
- vulvo-vaginitis, usually candidial, less often associated with *Gardnerella vaginalis*.

CONFIRMATORY TESTS

Microscopy of the smear from a urethral swab showing increased numbers of pus cells, some containing Gram-negative intracellular diplococci is almost pathognomonic ('certainly diagnostic') for gonorrhoea (Fig. 2). A special small calcium alginate swab must be used, as routine swabs are too large, and cotton inhibits fastidious pathogens like the gonococcus.

Direct immunofluorescent (DIF) microscopy using one of the commercially available special kits is necessary to diagnose chlamydial infection.

Culture by direct 'bed-side' inoculation on special non-selective media (chocolate agar) and selective media (e.g. modified Thayer–Martin) in CO_2 is necessary to grow the fastidious gonococcus.

Identification is by a positive oxidase reaction and by sugar fermentation.

Culture for chlamydia in tissue culture is a specialised technique.

Antibiotic sensitivity testing is necessary where the incidence of penicillinase-producing *N. gonorrhoeae* (PPNG) is increasing or unknown. Serology is useless for gonorrhoea, and seldom used to diagnose chlamydial infection, though an IgM titre above 128 is highly suggestive of acute infection.

CHEMOTHERAPY

As negative investigations do not exclude infection, and double infections are so common, all patients with urethritis should be treated for **both** gonorrhoea and NGU.

Procaine penicillin by a single intramuscular injection or 3.0 g amoxycillin with probenicid cures uncomplicated gonorrhoeal urethritis caused by sensitive strains but has been replaced by ceftriaxone where PPNG or chromosomally mediated resistance is common. Spectinomycin or ciprofloxacin are also usually effective.

A tetracycline for 10 days, usually doxycycline for convenience, is used for chlamydial infection, and a second course may be needed. Erythromycin is used in pregnancy, and for *Ureaplasma* infections resistant to tetracycline.

CONTROL AND PREVENTION

Asymptomatic infected people, particularly women, form a reservoir for the gonococcus and for chlamydiae. As there are no vaccines and chemoprophylaxis is not feasible, control depends on education, safer sexual practices, and rapid exact diagnosis with effective treatment of patients and their sexual partners.

Urethritis

- Urethritis may be chemical or mechanical but is usually infective, particularly as a STD.
- Gonorrhoea and non-gonorrhoeal urethritis (NGU, or NSU), usually caused by *C. trachomatis,* are most common.
- Usual symptoms are discharge and dysuria. Co-existent cervical, rectal or pharyngeal disease is common and often asymptomatic.
- Spread locally causes epididymitis, prostatitis, salpingitis and PID. Disseminated disease is rare. Reiter's syndrome can be a sequel of genital chlamydial infection.
- Diagnosis depends on the urethral smear, Gram stain, DIF and culture.
- Treatment is usually ceftriaxone for gonorrhoea and doxycycline for chlamydial infection.
- Control depends on education, safer sex, and rapid exact diagnosis with effective treatment of patients and their sexual contacts.

CERVICITIS

Cervicitis means inflammation of the cervix of the uterus. It is usually caused by infection. The important causes are:

- *Neisseria gonorrhoeae*
- *Chlamydia trachomatis*
- herpes simplex.

Less commonly, cervicitis may be:

- granulomatous: resulting from TB, anaerobes, schistosomiasis (Fig. 1)
- non-infectious.

CAUSATIVE ORGANISMS

Gonorrhoea is caused by the Gram-negative diplococcus *N. gonorrhoeae* (p. 36), while *C. trachomatis* (p. 58) is the other common cause of endocervical infection. These two infections commonly co-exist.

Herpes simplex virus causes infection and ulceration of the ectocervix. Granulomatous cervicitis caused by *Mycobacterium tuberculosis,* anaerobes such as *Bacteroides* and *Fusobacterium* spp., and schistosomal infections are all rare and not further discussed (Table 1).

CELLULAR DAMAGE

The gonococcus infects epithelial cells and multiplies in vacuoles. These vacuoles fuse with the basement membrane allowing the bacteria access to the connective tissues. Here inflammatory responses cause the tissue damage. There is no clear exotoxin production. Strains of gonococcus that cause disseminated disease are resistant to the bactericidal action of serum.

Chlamydiae are obligate intracellular parasites that enter through minute abrasions in the mucosa. After binding to specific cell receptors, they enter the cells and begin to replicate (p. 58). Damage results from cell destruction and the host inflammatory response.

CLINICAL FEATURES

Many, perhaps most, infections are asymptomatic and only discovered when a sexual partner is found to be infected.

When symptoms occur they include:

- vaginal discharge
- dysuria if urethritis (p. 154) is present
- abdominal pain and fever if there is spread to salpingitis or pelvic inflammatory disease (PID)
- further symptoms from distant spread (see below).

The diagnostic sign is cervical discharge, but this varies from purulent to clear, and from profuse to inconspicuous.

Differential diagnosis of uncomplicated cervicitis is principally from vaginitis (p. 161).

Complications of gonorrhoeal cervicitis

Local spread. The gonococcus moves like the knight in chess, infecting urethra not vagina, cervix not uterus, fallopian tube not fimbriae (primarily), ovary not posterior abdominal wall. Salpingitis (Fallopian tube infection) leads to tubal blockage, tubo-ovarian abscesses (Fig. 2) and PID in about 15% of patients.

Distant spread. This causes gonococcaemia, with skin or joint infections. It occurs in 1–3% of women because of unrecognised asymptomatic cervicitis. Disseminated infection is characterised by fever, headache and prostration (which may even progress to septic shock, p. 26), a pustular skin rash, migratory arthralgias and then suppurative ('septic') arthritis in large joints, especially knees, wrists and ankles.

Congenital infection. Infection can be transmitted vertically to the child during childbirth (Fig. 3).

Co-existent infection. The rectum and pharynx can be infected by direct contact with infectious discharge, usually urethral. The pharyngitis is often asymptomatic. The proctitis (infection of the

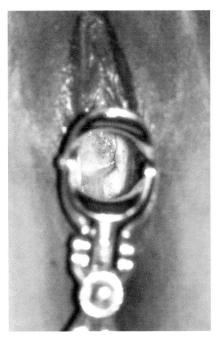

Fig. 1 **Cervicitis associated with schistosomiasis.**

rectum) may be asymptomatic but can cause severe anal pain and discharge.

Postgonococcal cervicitis (PGC). This is the persistence of symptoms of cervicitis after treatment for gonorrhoea. It is caused by gonococci resistant to the antibiotic chosen or reinfection or co-existent chlamydial infection (the most common cause).

Complications of chlamydial cervicitis

Local spread. Bartholinitis (infection of the glands of Bartholin on each side of the vagina at the base of the labia) is troublesome, but acute salpingitis (in 10%), and consequent tubal obstruction, ectopic pregnancy, infertility and PID are serious consequences.

Distant spread. This may cause peritonitis and, rarely, perihepatitis with peritoneal adhesions like violin strings: the Fitz–Hugh Curtis syndrome (p. 147).

Table 1 **Characteristics of cervicitis (infective)**

| Classification | Causative organism | Clinical features | | Confirmatory tests | Chemotherapy |
		Discharge	Dysuria		
Gonorrhoeal	*N. gonorrhoeae*	Purulent, severe	Moderate to gross	Gram-negative diplococci in pus cells	Ceftriaxone or amoxycillin
Chlamydial	*Chlamydia trachomatis*	Mucoid to thin pus	Mild to moderate	Direct immunofluorescence or culture	Doxycycline
Herpetic	Herpes simplex	Mucoid to thin pus	Nil to mild	Viral culture	Acyclovir
Granulomatous	*M. tuberculosis*	Mucoid to thin pus	Nil to mild	Microscopy and culture	Specific
	Anaerobes	Mucoid to thin pus	Nil to mild	Anaerobic culture	Specific
	Schistosomes	Mucoid to thin pus	Nil to mild	Microscopy, biopsy	Specific

In all cases, treat sexual partners.

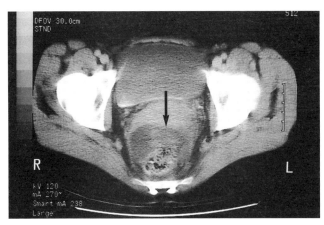

Fig. 2 **Infective cervicitis and salpingitis led to this pelvic tubo-ovarian abscess.**

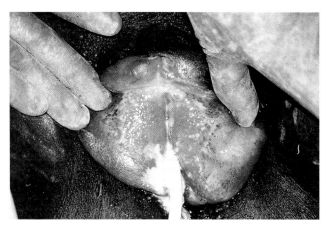

Fig. 3 **Cervical gonorrhoea in pregnancy.**

Co-existent infection. Chlamydial urethritis is present in over 50% of women with chlamydial cervicitis, and gonorrhoea also commonly co-exists at either site.

Cervical cytologic metaplasia. This often regresses after treatment of chlamydial infection. While cervical cytologic atypia and dysplasia (intraepithelial neoplasia) are statistically associated with chlamydial infection, the human papilloma virus (HPV) is considered to be the causal agent.

Congenital infection. Infants born through an infected cervix usually become infected and often develop neonatal chlamydial pneumonia.

CONFIRMATORY TESTS

Microscopy of the smear from an endocervical swab showing increased numbers of pus cells, some containing Gram-negative intracellular diplococci, is pathognomonic for gonorrhoea. A special calcium alginate swab must be used, as cotton inhibits fastidious pathogens like the gonococcus.

Direct immunofluorescent microscopy using one of the commercially available special kits is necessary to diagnose chlamydial infection.

Culture by direct 'bed-side' inoculation on special non-selective media (chocolate agar) and selective media (e.g. modified Thayer–Martin) in CO_2 is necessary to grow the fastidious gonococcus (Fig. 4). Identification is by a positive oxidase reaction and sugar fermentation. Culture for chlamydia in tissue culture is a specialised technique.

Antibiotic sensitivity testing is necessary where the incidence of penicillinase-producing *N. gonorrhoeae* (PPNG) is increasing or unknown.

Serology is useless for gonorrhoea and is seldom used to diagnose chlamydial infection, though an IgM titre above 128 is highly suggestive of acute infection.

CHEMOTHERAPY

As negative investigations do not exclude infection and double infections are so common, all patients with cervicitis should be treated for **both** gonorrhoea and NGU.

Procaine penicillin by a single intramuscular injection or 3.0 g of amoxycillin with probenicid cures uncomplicated gonorrhoeal cervicitis resulting from sensitive strains. This has been replaced by ceftriaxone where PPNG or chromosomally mediated resistance is common. Spectinomycin or ciprofloxacin are also usually effective.

A tetracycline, usually twice-daily doxycycline for convenience, for 10 days is used for chlamydial infection, and a second course may be needed. Erythromycin is used in pregnancy.

CONTROL AND PREVENTION

As there are no vaccines, and chemoprophylaxis is not feasible, control depends on education, safer sexual practices, and rapid exact diagnosis and effective treatment of patients and their sexual contacts.

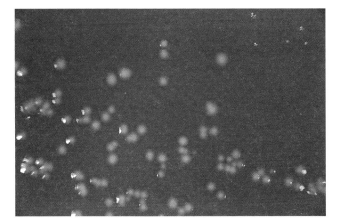

Fig. 4 **Culture plate of *N. gonorrhoea*.**

Cervicitis

- Cervicitis is usually infective, particularly as a STD.
- The most common types are gonorrhoeal caused by *N. gonorrhoeae*, and chlamydial caused by *C. trachomatis*.
- Vaginal discharge, abdominal pain and dysuria are the usual symptoms. Co-existent urethral, rectal or pharyngeal disease is common and often asymptomatic.
- Spread locally causes bartholinitis, salpingitis and PID (p. 158). Peritonitis, perihepatitis, disseminated disease and Reiter's syndrome are all rare. Infants are infected by vaginal delivery through an infected cervix.
- Diagnosis depends on the endocervical smear, Gram stain, DIF and culture.
- Treatment is usually ceftriaxone for gonorrhoea, and doxycycline for chlamydial infection.
- Control depends on education, safer sex, and rapid exact diagnosis with effective treatment of patients and their sexual contacts.

SALPINGITIS AND PELVIC INFLAMMATORY DISEASE

The term 'pelvic inflammatory disease' (PID) is unfortunately used in two different ways. An exact (sensu strictu) use, especially by gynaecologists, means only acute salpingitis (Fallopian tube infection). A more broadly defining use (sensu latu) includes endometritis, salpingitis, salpingo-oophoritis, inflammation of the adnexal ('adjacent') tissues and inflammatory 'pelvic mass', pelvic abscess, tubo-ovarian abscess and peritonitis. This broader usage is convenient, for these conditions are a continuum, extending into each other, and clinically it is often impossible to tell either the exact anatomical extent or the pathological stage (Table 1).

CAUSATIVE ORGANISMS

These conditions are usually polymicrobial, i.e. caused by a number of microbial species in a mixture, not by a single species. The most important causes are:

- *Neisseria gonorrhoeae* (p. 36), as a STD, particularly causing salpingitis and ascending infection
- *Chlamydia trachomatis* (p. 58) as a STD, also causing salpingitis and ascending infection
- anaerobic or mixed anaerobic and aerobic bacteria, particularly in endogenous postabortal, postpartum or postoperative endometritis and endomyometritis, when uterine muscle is infected
- *Clostridium perfringens* (p. 34), as a life-threatening pathogen that can infect where illegal abortions are done without aseptic technique
- *Actinomyces* spp. (p. 56), important in IUD-related infections
- *Mycoplasma hominis* (p. 57) is less important; *Ureaplasma* spp. and viruses, including HSV and CMV, are unimportant.

CLINICAL SYNDROMES

The clinical features vary in severity, depending on the cause, organism, site and extent of disease.

Endometritis. PV bleeding and fever are prominent, often with some pelvic pain and uterine tenderness.

Salpingitis, salpingo-oophoritis and adjacent tissue infection. The classic triad of vaginal discharge, pelvic pain and fever is only present in about 25% of patients. Pain on moving ('rocking') the cervix and adnexal tenderness in the fornices are usual on vaginal examination. A pelvic mass may be present from swollen infected tissues and omentum without actual abscess formation. Systemic symptoms develop.

Local peritonitis. If present, this produces abdominal pain and tenderness.

Table 1 **Characteristics of salpingitis and PID**

Classification	Causative organisms	Clinical features	Confirmatory tests	Chemotherapy*
Endometritis	MAA, CP	PV bleeding, fever	Cervical swab, M&C	Cefoxitin +/- gentamicin
Salpingitis and salpingo-oophoritis	MAA, NG, CT	Vaginal discharge, pelvic pain, fever, cervical tenderness	Cervical swab, culdocentesis	Ceftriaxone + doxycycline
Tubal and tubo-ovarian abscess	MAA, NG, CT	As salpingitis, plus tubal mass and systemic symptoms	Culdocentesis then operate	Cefoxitin, doxycycline, gentamicin
Pelvic abscess	MAA, NG, CT	As salpingitis, plus pelvic mass and systemic symptoms	Culdocentesis then drain	Cefoxitin, doxycycline, gentamicin
Peritonitis	MAA, NG, CT	Abdominal pain and guarding and rigidity, shock	Laparotomy, M&C	Cefoxitin, doxycycline, gentamicin

MAA, mixed anaerobic and aerobic organisms; NG, *N. gonorrhoeae*; CP, *Cl. perfringens*; CT, *C. trachomatis*; M&C, microscopy and culture
* In all STDs, treat sexual partners

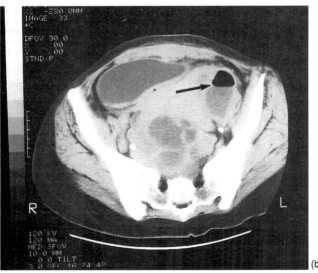
(a)

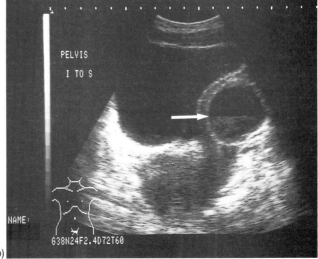
(b)

Fig. 1 **Tubo-ovarian abscess. (a)** CT scan; **(b)** ultrasound. Both showing fluid levels (different patients).

Urethritis. If present (p. 154), it causes dysuria.

Tubo-ovarian abscess. A tubal mass develops (Fig. 1) and systemic symptoms with fever usually increase.

Pelvic abscess. A tender pelvic mass is present, and systemic symptoms with fever are prominent.

Generalised peritonitis from rupture of a tubal or pelvic abscess. Both abdominal pain and fever increase, and abdominal muscle guarding and rebound tenderness appear, with paralysed bowel, called 'paralytic ileus', vomiting and severe systemic symptoms and shock.

Differential diagnosis

Differential diagnosis in mild cases includes vaginitis (p. 163) or cervicitis (p. 156), while more severe disease must be distinguished from acute appendicitis, endometriosis, ectopic pregnancy, or ovarian cyst rupture or haemorrhage.

CONFIRMATORY TESTS

Useful specimens include endocervical swabs, pus from culdocentesis (aspiration through the posterior vaginal fornix for salpingitis or tubal abscess), and pus from laparotomy.

Laboratory tests for gonorrhoeal and chlamydial infections are outlined below:

- **Microscopy** of the smear from an endocervical swab showing increased numbers of pus cells, some containing Gram-negative intra-cellular diplococci, is almost pathognomonic ('certainly diagnostic') for gonorrhoea. A special calcium alginate swab must be used, as cotton inhibits fastidious pathogens like the gonococcus.
- **Direct immunofluorescent microscopy** using one of the commercially available special kits is necessary to diagnose chlamydial infection.
- **Culture** by direct 'bed-side' inoculation on special non-selective media (Chocolate agar) and selective media (e.g. modified Thayer-Martin) in CO_2 is necessary to grow the fastidious gonococcus. Identification is by a positive oxidase reaction, and sugar fermentation. Culture for chlamydia in tissue culture is a specialised technique. Blood culture may be positive in peritonitis.
- **Antibiotic sensitivity testing** is necessary where the incidence of penicillinase-producing *N. gonorrhoea* (PPNG) is increasing or unknown.
- **Serology** is useless for gonorrhoea, and seldom used to diagnose chlamydial infection, though an IgM titre above 128 is highly suggestive of acute infection.

- **Imaging** by ultrasound and/or CT scan can show the site and extent of infection, including abscesses (Fig. 1).

CHEMOTHERAPY

Empiric chemotherapy usually combines a beta-lactam active against *N. gonorrhoeae* and anaerobes, with a tetracycline for chlamydial infection, or aminoglycoside for Gram-negative aerobes. For infections probably acquired sexually, give ceftriaxone (or cefoxitin) plus doxycycline for gonorrhoea and chlamydial infection (or amoxycillin/clavulanate plus doxycycline orally for mild infection). For postabortal, postpartum and postoperative infections, cefoxitin plus gentamicin, clindamycin plus gentamicin, or ampicillin plus metronidazole plus gentamicin are all effective. Benzyl penicillin or metronidazole must be used for suspected or proved clostridial infections

Mild cases may be treated orally as out-patients, but intravenous therapy for at least 4 days in hospital is needed in pregnancy, presence of IUD, presence of abscess, peritonitis, possible other diagnosis, or previous failed oral treatment.

Chemotherapy alone is insufficient in five situations:

- foreign bodies such as IUDs or retained products of conception must be removed
- unruptured abscesses must be drained
- intra-abdominal ruptured abscess with peritonitis needs immediate abdominal surgery
- clostridial infections need surgery and often hyperbaric oxygen
- extensive infection with destroyed organ function needs excision (Fig. 2).

CONTROL AND PREVENTION

For sexually acquired infections control depends on education, safer sexual practices and rapid exact diagnosis and effective treatment of patients and their sexual contacts. For postabortal, postpartum and postoperative infections, control depends on chemoprophylaxis when relevant, aseptic procedural technique and rapid exact diagnosis and treatment.

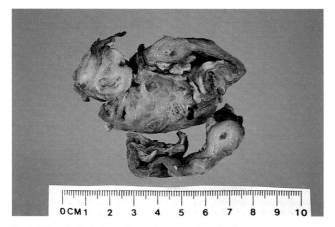

Fig. 2 **PID pathological specimen of uterus and both tubes.**

Salpingitis and PID

- PID can include endometritis, salpingitis, salpingo-oophoritis, inflammation of the adnexal ('adjacent') tissues and inflammatory 'pelvic mass', pelvic abscess, tubo-ovarian abscess and/or peritonitis.
- PID is either sexually acquired or else follows abortion, birth, an IUD or gynaecologic surgery.
- Most infections are polymicrobial: the commonest organisms are *N. gonorrhoeae, C. trachomatis*, or mixed anaerobic and aerobic organisms.
- Vaginal discharge, pelvic pain and fever are the usual symptoms, but spread of infection causes abdominal pain, tubal and pelvic masses, signs of peritonitis, paralytic ileus and shock.
- Diagnosis depends on the endocervical smear and special stains and cultures on aspirated pus. Serology is of little use. Imaging shows the site and extent of abscess formation.

EPIDIDYMITIS, ORCHITIS AND BALANITIS

EPIDIDYMITIS

Epididymitis is inflammation of the epididymis, usually caused by infection, rarely by trauma. Infection from a STD or from the urethra, bladder or prostate ascends via the vas deferens. It can also result from septicaemic or local spread in systemic disease.

Causative organisms

Ascending infection from the bladder occurs in children, infancy and older men. In the latter, prostatic hypertrophy, catheterisation or instrumentation can cause partial obstruction which predisposes to ascending infection. The usual organisms are *E. coli* and *Klebsiella* spp.

In sexually active men infection is usually sexually acquired – gonococcal or chlamydial – but it can involve *E. coli* from an unrecognised structural abnormality or anal intercourse.

In children, septicaemia can cause epididymitis, involving, for example, *Neisseria meningitidis* or *Haemophilus influenzae*.

Rarely epididymitis can occur as the result of systemic disease, such as TB (Fig. 1), blastomycosis, coccidioidomycosis, brucellosis or schistosomiasis.

Fig. 1 **Tuberculous epididymitis.**

Clinical features

From the urinary and epididymal infection there is urethral discharge (which may be minimal), dysuria, frequency and nocturia. In addition, the scrotum is painful and swollen with redness and tenderness of the epididymis, progressing to hydrocoele, orchitis and fever.

Complications. Complications include chronic epididymitis, with or without a chronic sinus, and epididymo-orchitis, which may progress to testicular abscess, testicular infarction or infertility.

Differential diagnosis. This includes torsion, infarction, abscess, traumatic rupture, tumour of the testis or torsion of the appendages.

Confirmatory tests

These are special microscopy and culture of urine and urethral discharge for specific pathogens. Ultrasound is useful to exclude testicular torsion. Aspiration of the epididymis is used in complicated cases. Cystoscopy, intravenous pyelography and micturating cysto-urethrography are used to define structural abnormalities of the urinary tract.

Chemotherapy

For sexually acquired infections, amoxycillin/clavulanate or ceftriaxone (depending on the presence of penicillinase-producing *N. gonorrhoeae* (PPNG), p. 36) plus tetracycline or doxycycline are given. For ascending infections from bacteriuria, a broad-spectrum penicillin or cephalosporin is usual until sensitivity results are available. Scrotal support and analgesia are helpful, and surgery may be needed for complications.

Control and prevention

This depends on the origin of the infection. Sexual partners should be traced if necessary and treated. For ascending infections from bacteriuria, predisposing obstruction or stasis must be remedied.

ORCHITIS

Orchitis is inflammation of the testis, commonly caused by viral disease such as mumps, less commonly by spread from epididymitis (see above) and, rarely, from metastatic, blood-borne spread. Causative organisms, therefore, are gonococci, chlamydiae, uro-pathogens such as *E. coli* or *Klebsiella* spp., or systemic infections including TB (Fig. 2). Clinical features are high fever, nausea and vomiting with acute testicular pain, swelling and tenderness. Confirmatory laboratory tests are urethral and urine microscopy and culture. Ultrasound differentiates from epididymitis. Orchitis is managed in a similar manner to epididymitis.

BALANITIS

Balanitis is inflammation of the glans, corona or shaft of the penis. It is usually caused by a *Candida albicans* infection acquired from an infected sexual partner. Small vesicles develop into thrush-like patches with itching and burning. It may spread to scrotum, groins and perineum. Microscopy and culture confirm. It is cured with local nystatin or clotrimazole creams, and treatment of the sexual partner(s).

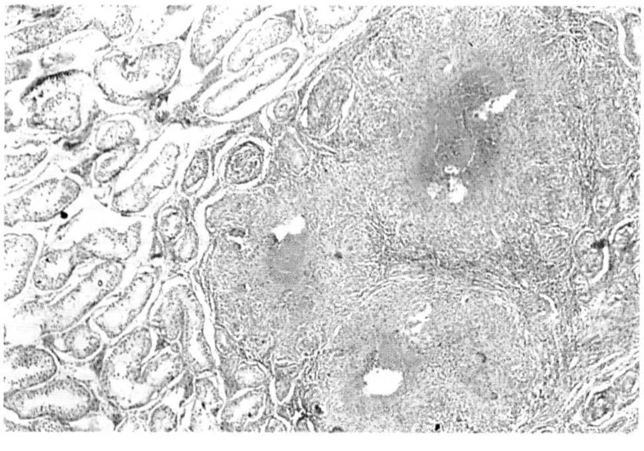

Fig. 2 **Tuberculous orchitis.**

Epididymitis, orchitis and balanitis

- **Epididymitis** is usually sexually acquired or an ascending infection from the urinary tract, but rarely it can arise from systemic disease.
- The common causes are gonorrhoea and chlamydial infections, or uropathogens such as *E. coli*.
- Local pain, swelling, redness and tenderness, urethral discharge and dysuria are the common symptoms.
- Spread locally causes hydrocoele and **orchitis**, which can progress to abscess and sinus formation.
- Orchitis (inflammation of the testes) is usually due to mumps, less often to spread from the epididymis. Severe pain and tenderness need analgesia and support.
- **Balanitis**, inflammation of the skin of the penis, is usually caused by *C. albicans* and treated by local creams.

VAGINITIS AND VULVO-VAGINITIS

Vaginitis is inflammation of the vagina; co-existent inflammation of the vagina and exterior vulva (of any cause) is called vulvo-vaginitis. Infection in a Bartholin's gland causes a vulval abscess (Fig. 1).

Differential diagnosis is from excessive physiological discharge, non-infectious vaginitis (e.g. chemical), and from cervicitis (p. 156).

There are three common infectious causes: candidiasis, trichomoniasis and bacterial vaginosis.

Candidiasis

Candida infection is usually endogenous and is favoured by increased glycogen, loss of normal bacterial flora (p. 16) and poor cell-mediated immunity (CMI) (p. 23), so predisposing causes include pregnancy or the 'pill', systemic antibiotics, diabetes, treatment with steroids, or excessive local warmth and moisture from tight (nylon) underwear.

Candidiasis is marked by pruritis and a thick, white, cheesy discharge, which adheres to the vaginal wall in white spots, looking like the breast of a thrush (Fig. 2). Dyspareunia (pain on intercourse) and dysuria may occur.

The yeasts and pseudohyphae of candidiasis are diagnosed with a wet mount in 10% KOH (which incidentally destroys trichomonads).

Culture for *Candida* is easy on specific media but does not confirm the diagnosis as it is found in 40% of normal women.

Candidiasis is treated with local clotrimazole or miconazole. Oral ketoconazole, because of toxicity, is reserved for chronic relapsing cases. Male sexual partners with balanitis should be treated also (p. 160).

Predisposing factors should be controlled where possible.

Trichomoniasis

Trichomoniasis is caused by the protozoan flagellate *Trichomonas vaginalis,* and infection is acquired sexually.

Trichomoniasis may be asymptomatic, but often causes severe itching with a copious, purulent, frothy, foul discharge. The erythematous vaginal wall is reddened, but a friable cervix with punctate haemorrhages, the distinctive 'strawberry cervix', is uncommon.

Fresh vaginal discharge is examined in a saline wet mount to show motile trichomonads.

Trichomoniasis is a STD, so patients and their sexual partners should be treated with metronidazole. It is controlled like other STDs, by education, safer sex, and rapid diagnosis and treatment of patients and partners.

Bacterial vaginosis

Bacterial vaginosis is associated with *Gardnerella vaginalis,* probably acting

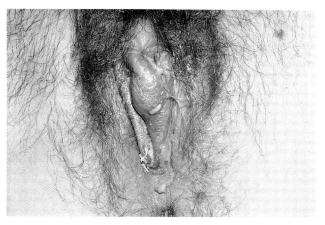

Fig. 1 **Vulval abscess.**

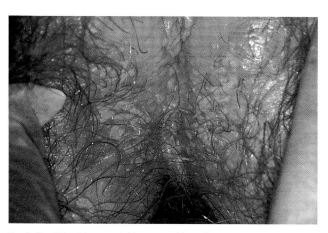

Fig. 2 *Candida albicans* vaginitis: aceto-white stain.

with various anaerobes, including *Bacteroides* spp. and the curious curved *Mobiluncus* (Fig. 3). Infection is probably endogenous.

Bacterial vaginosis is marked by an absence of inflammation, so the previous name of 'non-specific vaginitis' is not now used. Symptoms are absent or mild, with thin, frothy discharge and some vaginal odour.

Fresh vaginal discharge in a saline wet mount may show 'clue cells', which are squamous epithelial cells with their borders obscured by innumerable tiny coccobacilli.

Bacterial vaginosis is also treated with metronidazole. In pregnancy, ampicillin or clindamycin can be used.

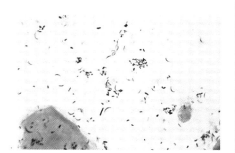

Fig. 3 *Mobiluncus* sp. in bacterial vaginosis.

Vaginitis and vulvo-vaginitis

- Vaginitis is usually caused by candidiasis (from *C. albicans*), trichomoniasis (from *T. vaginalis*) or bacterial vaginosis (from *G. vaginalis* and mixed anaerobes).
- Clinical features are commonly vaginal discharge and vulval itch, sometimes with dysuria or dyspareunia.
- Diagnosis is by microscopy of a wet mount of fresh vaginal discharge, with culture if negative.
- Treatment is an imidazole such as clotrimazole for candidiasis, and metronidazole for trichomoniasis or bacterial vaginosis.
- Control is by reducing the predisposing factors, by education and safer sex, and early diagnosis and treatment of sexual partners.

TROPICAL AND RARE SEXUALLY TRANSMITTED DISEASES

Four sexually transmitted diseases (STDs) characterised by genital ulceration are discussed here. Except for syphilis, they are rare in developed countries, but common in many developing countries. Table 1 compares their major features with herpes genitalis caused by herpes simplex virus infection, for although virus infections (including HIV-AIDS and genital warts) are outside the scope of this book, herpes genitalis is the most common ulcerative STD.

Differential diagnosis includes fixed drug eruption and traumatic ulcers. Other STDs discussed in this book include gonorrhoea and chlamydial infections, urethritis, cervicitis, salpingitis and pelvic inflammatory disease (PID), epididymitis, orchitis, balanitis and vaginitis (pp. 154–161), lice and scabies (p. 71) and the so-called gay bowel syndrome (p. 137).

Confirmatory tests should be done whenever available, for the clinical features are sufficiently variable that clinical diagnosis even by experienced clinicians is wrong in 40% of patients!

Control of all STDs depends on education, safer sex and rapid exact diagnosis and treatment of patients and their sexual partners.

CHANCROID

Chancroid ('like a syphilitic chancre') is an ulcerating venereal disease also known as 'soft sore'. This describes two major differences from syphilis: the ulcer is soft, and it is sore. In Africa and other tropical areas, it is frequently associated with HIV infection. The causative organism is *Haemophilus ducreyi* (p. 38), a small Gram-negative cocco-bacillus.

Chancroid is a painful ulcer with ragged edges and a necrotic base. The clinical features, summarised in Table 1, help in differential diagnosis although clinical diagnosis can be unreliable. Auto-inoculation, by contact of adjacent surfaces, can produce 'kissing' ulcers, while scarring during healing can produce a 'saxophone penis' (Fig. 1).

Confirmatory testing by culture uses special media of sheep blood with vancomycin. Gram stain is unreliable because of numerous similar bacteria.

Chemotherapy is by erythromycin for 10 days, or cotrimoxazole.

GRANULOMA INGUINALE

This condition has many synonyms, of which 'granuloma venereum' is more accurate than granuloma inguinale (as this is the one ulcerative STD in which inguinal lymphadenopathy is **not** present) and 'ulcerating granuloma of the pudenda' is the most descriptive.

It is spread by non-sexual trauma as well as sexually, and as vaginal and rectal lesions are often inconspicuous,

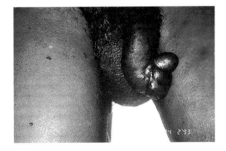

Fig. 1 **Chancroid: 'saxophone penis'.**

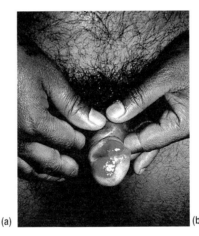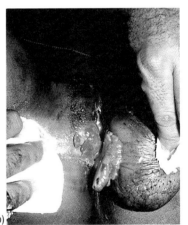

(a) (b)

Fig. 2 **Granuloma inguinale. (a)** Penile ulcer; **(b)** vulval elephantiasis.

Table 1 **Comparison of rare and 'tropical' ulcerating STDs with herpes simplex**

	Chancroid ('soft sore')	Granuloma inguinale	Lymphogranuloma venereum	Syphilis	Herpes simplex
Causative organism	*Haemophilus ducreyi*	*Calymmatobacterium donovani*	*Chlamydia trachomatis*	*Treponema pallidum*	Herpes simplex
Clinical features					
Ulcer	Marked, 1–10	Marked, 1 or more	Inconspicuous	Single	Clusters
Border	Flat, ragged	Elevated, pearly	Flat	Rolled	Flat
Induration	Soft	Firm	Nil	Firm–hard	Soft
Pain	Marked	Minimal	None	None	Marked
Depth	Deep	Moderate	Shallow	Shallow	Shallow
Base	Necrotic, yellow	Granulomatous	Pink-red	Clean	Red
Secretion	Blood or pus	Sero-sanguineous	Serous	Serous	Serous
Progression	Progresses, coalesces	Progresses, coalesces	Heals	Heals	Heals
Lymph nodes	**Painful, enlarged**	Nodes not enlarged Inguinal granuloma (pseudo-bubo)	**Painful, many enlarged, form sinuses**	**Painless, enlarged, 'rubbery'**	Mild pain
Systemic	Minimal	Usually none	Marked	Minimal	Minimal
Confirmatory tests					
Preferred	Special culture	Crushed tissue	Serology	Dark-field microscopy	Culture
Alternative	Gram stain	Histology	Culture	RPR/TPHA	Tzanck
Chemotherapy					
Preferred	Erythromycin	Tetracycline	Tetracycline	Penicillin	Acyclovir
Alternative	Cotrimoxazole	Cotrimoxazole	Erythromycin	Tetracycline/Erythromicin	Famciclovir or valaciclovir

sexual partners may appear uninfected. The causative organism is *Calymmato-bacterium granulomatis*, a Gram-negative bacterium of uncertain affiliation (p. 47).

The clinical features of this painless destructive genital ulcer (Fig. 2) with pearly everted edges are summarised in Table 1.

Confirmatory tests are Giemsa stain of ulcer tissue crushed between two microscope slides or histology, as culture is not yet feasible. Bipolar-staining rods (Donovan bodies) are seen within macrophages.

Chemotherapy is by tetracycline or cotrimoxazole for 3 weeks or until healing is achieved.

LYMPHOGRANULOMA VENEREUM (LGV)

As the name describes, LGV is a STD characterised by inguinal lymphadenopathy; genital ulceration is inconspicuous.

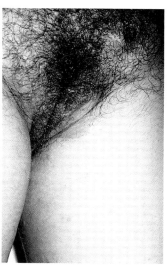

Fig. 3 **Secondary syphilis, mucous patch.**

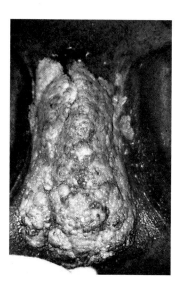

Fig. 5 **Massive vulval warts.**

The causative organism is *Chlamydia trachomatis*, serotypes L1, L2 and L3 (p. 58), whereas urethritis, cervicitis and other chlamydial genital disease (pp. 154–160), is caused by serotypes D–K.

Clinical syndromes

The clinical features are a primary phase, which is often undiagnosed, with a self-healing genital ulcer, a secondary phase with systemic symptoms and marked inguinal and other lymphadenopathy that progresses to abscess, sinus and fistula formation, and a third phase of fibrosis, strictures and lymphatic obstruction. By contrast, if the initial lesion is ano-rectal, the primary phase is marked by severe local symptoms (diarrhoea, anorectal pain and discharge) and systemic upset with fever. Confirmatory tests are serology (immunofluorescence being preferable to complement fixation testing) or culture of bubo (lymph node) aspirate for chlamydia.

Management

Chemotherapy is by tetracycline for at least 14 days, or erythromycin. Abscesses need aspiration through adjacent normal skin, and surgery may be needed for fistulae, strictures or elephantiasis from lymphatic obstruction.

Fig. 4 **Secondary syphilis: condylomata lata.**

SYPHILIS

A painless, indurated, infectious, self-healing genital ulcer called a chancre is the characteristic of the primary stage of syphilis. The second stage has systemic symptoms such as fever and myalgia, and may include lymphadenopathy, patchy alopecia, a highly variable but usually symmetrical rash of the palms and soles, with painless, shallow, infectious ulcers ('mucous patches', Fig. 3) of mucous membranes and genitalia, and infectious condylomata lata ('flat warts') around the anus (Fig. 4). There is then a latent stage of many years before the third stage of cardiovascular, CNS, bone, joint or skin involvement. The causative organism is *Treponema pallidum*, a spirochaete that does not stain with Gram's or other routine stains and cannot be cultured on media (p. 52).

The differential diagnosis of genital syphilis includes chancroid, granuloma inguinale and sometimes LGV, as well as TB, herpes simplex, venereal warts (Fig. 5) and traumatic ulcers.

Confirmatory tests are dark-field microscopy for spirochaetes, and serology, such as rapid plasma reagin (RPR) and *T. pallidum* haemagglutination (TPHA) tests, which are often negative in primary syphilis, but almost always positive in secondary syphilis. HIV antibody testing, after counselling and consent, is usually indicated also.

Single-dose benzathine penicillin has been widely used for chemotherapy, but a 10 day course of procaine penicillin is probably superior. Tetracycline is used in penicillin hypersensitivity, and erythromycin in hypersensitivity in pregnancy. All sexual contacts within 90 days must be treated.

Tropical and rare sexually transmitted diseases

- **Chancroid** is a soft ragged necrotic painful ulcer with painful inguinal lymphadenopathy; it is caused by *Haemophilus ducreyi* and is treated with erythromycin or cotrimoxazole.
- **Granuloma inguinale** has an ulcer with a firm, elevated pearly edge and granulomatous base, with little pain and some inguinal infiltration. Causative organism is *Calymmatobacterium donovani*. Treatment is tetracycline or cotrimoxazole.
- **Lymphogranuloma venereum** has an inconspicuous ulcer but marked painful lymphadenopathy, which forms abscesses, sinuses, fistulae, scars and strictures. Causative organism is *Chlamydia trachomatis*, serotypes L1, L2 and L3. Treatment is by tetracycline or erythromycin.
- **Syphilis** has a primary chancre which is firm, painless, infectious and self-healing, and a secondary stage which includes infectious genital mucous patches and condylomata lata. Treatment is by penicillin if possible, otherwise tetracycline or erythromycin.
- Confirmatory tests of all are special stains, special culture and/or specific serology.
- Control of all STD involves education, safer sex, exact diagnosis and rapid treatment of patients and their sexual partners.

STREPTOCOCCAL SKIN AND SOFT TISSUE INFECTIONS

Local streptococcal infections of the skin and soft tissues occur at all levels (Fig. 1).

In addition, *distant* streptococcal infections affect the skin:

- erythema nodosum
- erythema marginata
- purpura fulminans
- scarlet fever.

Pyoderma is a confusing term used either generally to mean all skin infections or, sometimes, specifically for impetigo.

The appearance of skin lesions in disease are described clinically in several ways:

- colour
  — erythema: redness caused by vasodilatation
  — petechiae: tiny purplish discolorations from bleeding
  — purpura: large purplish discoloration from bleeding
- texture
  — macules: flat spots
  — papules: small raised lumps
  — nodules: larger firm lumps
  — plaques: large flattish lesions with little height
- contents
  — vesicles: small watery blisters
  — bullae: large watery blisters (> 0.5 cm)
  — pustules: blisters containing pus
- skin integrity
  — erosion: superficial loss of skin
  — ulcer: deeper hole, may bleed
  — crust: leakage of blood or serum drying on skin.

LOCAL INFECTIONS

Impetigo

Impetigo (Fig. 2) is a superficial infection caused by *Streptococcus pyogenes* (group A) (p. 30), and may follow nasopharyngeal carriage or infection. Some strains are nephritogenic, so acute glomerulonephritis follows. In addition, *Staphylococcus aureus* (p. 28) often secondarily infects impetigo.

Clinical features. Impetigo begins, often on the face, with small papules or vesicles that ulcerate and ooze highly infectious, thin, sero-purulent fluid which dries in typical crusts that last up to 2 weeks and heal without scars. Usually there is no systemic illness. School children and other groups in close contact are commonly affected. Spread is typical, both to other parts of the body, and to other people (contacts). It must be distinguished from bullous impetigo,

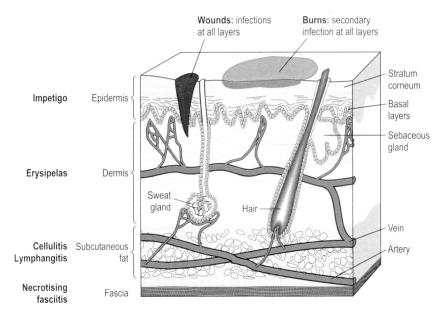

Fig. 1 **Section of skin and subcutaneous tissues, showing the levels affected by different streptococcal infections.**

with big (> 1 cm) bullae, caused by staphylococci.

Confirmatory tests. Diagnosis is usually clinical, but swabs of the oozing fluid grow the streptococcus and show any secondary infection.

Chemotherapy. Systemic penicillin is effective, making the lesions non-infectious within 48 hours; obviously a penicillinase-resistant type is used if staphylococci are present. Local antiseptics may be used in conjunction with penicillin to diminish infectivity but are ineffective alone.

Erysipelas

Unlike impetigo, which is troublesome but trivial, erysipelas is a serious and even life-threatening disease, with rapidly spreading infection in the dermis (Fig. 3). Erysipelas is also caused by *Strep. pyogenes* (group A), but sometimes by group B, C or G streptococci. Some patients suffer many episodes.

Clinical features. A bright red area, tingling more than painful, appears and spreads rapidly, with a raised sharply defined edge. The face is the classical site, so the diagnosis may be missed when a leg or other site is involved. Lymphangitis follows. Predisposing factors, both to the initial attack and to recurrence, are lymphatic obstruction, skin disease (which may be occult, e.g. tinea between the toes), injury or surgery.

Confirmatory tests. These are often unnecessary, and skin swabs or aspirate are usually negative without other skin

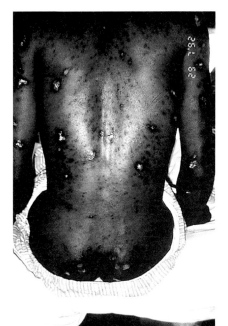

Fig. 2 **Impetigo.** Note superficiality, scabs or crusts, and serous discharge.

disease. Throat cultures are positive in about 30% of patients, blood cultures in only 5%.

Chemotherapy. Systemic penicillin must be given immediately (unless the patient is hypersensitive). Cephalothin or erythromycin are less satisfactory alternatives.

Control and prevention. Recurrent attacks can be prevented or minimised by attention to the predisposing factor(s).

If these cannot be removed, long-term penicillin can be used.

Wounds and burns

These may be infected by streptococci (pp. 170–1).

Cellulitis and lymphangitis

While these can still be caused by *Strep. pyogenes* infection, they are now much more commonly caused by *Staph. aureus* or other organisms (p. 168).

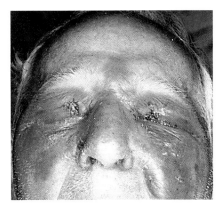

Fig. 3 **Erysipelas.** Note fiery red colour and sharply demarcated edge.

Fig. 4 **Erythema marginata.** Note faint red–brown rash with distinctive irregular margin and central paler area.

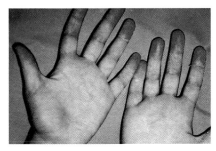

Fig. 5 **Scarlet fever.**

Necrotising fasciitis

This is now rarely caused by *Strep. pyogenes* (see gangrenous infections, p. 169).

SKIN MANIFESTATIONS OF DISTANT INFECTIONS

Erythema nodosum (EN)

Erythema nodosum is a condition characterised by red or red-purple lumps, often tender and often on the legs. It can be caused by a number of infections:

- mycobacterial infections (TB or leprosy) (p. 92)
- streptococcal infections
- *Chlamydia trachomatis* (p. 163) (lymphogranuloma venereum)
- *Yersinia* infections (p. 51)
- systemic fungal infections (pp. 62–7).

There are also numerous causes which cannot be proved to be infective. These include Crohn's disease, ulcerative colitis, sarcoidosis and systemic lupus erythematosus. Drugs (e.g. sulphonamides) also cause EN.

The lumps appear at the onset or during the course of the conditions listed above. They seldom ulcerate.

Confirmatory tests. Tests are directed at the underlying disease.

Management. This is directed at the underlying condition. Corticosteroids, if not otherwise contraindicated, may be used to hasten resolution and relieve pain.

Erythema marginata

This is a faint, widespread, migrating skin rash, with an irregular red-brown margin and a paler centre (Fig. 4). It is very rare but is pathognomonic of (found only in) rheumatic fever.

Purpura fulminans

This is usually defined as symmetric peripheral gangrene, with distal ischaemic necrosis of two or more extremities without large vessel occlusion. It is an extremely serious sign of disseminated intravascular coagulation (DIC), occurring in streptococcal (and other) septicaemia, often heralding death.

Scarlet fever

Scarlet fever follows a streptococcal infection. Certain strains of *Strep. pyogenes* contain a lysogenic phage which codes for an erythrogenic toxin which has three effects:

- erythrogenic, giving the typical rash
- pyrogenic, giving fever
- endotoxic, giving shock.

Clinical syndrome. Scarlet fever has a number of features:

- the primary streptococcal infection, usually pharyngitis and tonsillitis, rarely a skin or wound infection
- diffuse red rash (Fig. 5), spreading from the chest over the whole body except for the face, palms and soles; it feels rough like sandpaper, then desquamates as the rash fades
- circum-oral pallor
- a coated 'white strawberry' tongue which becomes a beefy 'red strawberry tongue' when the coating disappears
- an enanthem of small red spots on the palate
- fever and prostration
- rarely, septicaemia, arthritis, jaundice, death.

The incubation period is about 3 days and the disease is most commonly seen in children under 10 years.

Confirmatory tests. Swabs from the portal of entry (throat, skin, wound or even uterus) grow the organism. Skin tests are no longer used.

Chemotherapy. Penicillin is the drug of choice, with supportive measures.

Control. Widespread use of penicillin to treat throat infections, and improved socioeconomic conditions, including less crowding and better ventilation, have greatly decreased the incidence of scarlet fever, rheumatic fever and acute glomerulonephritis in developed countries.

Streptococcal skin and soft tissue infections

- Local streptococcal infections include impetigo, erysipelas, cellulitis, fasciitis (depending on the level of tissue infected) and wound and secondary infections of burns.
- Distant streptococcal infections can cause erythema nodosum, erythema marginata, purpura fulminans and scarlet fever.
- Impetigo is superficial and has a serous ooze that is very infectious.
- Erysipelas is serious, with systemic symptoms, and needs urgent systemic penicillin.
- Cellulitis and fasciitis are now less commonly streptococcal.
- Erythema nodosum, erythema marginata and purpura fulminans can be important but rare signs of distant streptococcal infection.
- Scarlet fever with scarlet rash and strawberry tongue is caused by erythrogenic toxin-producing *Strep. pyogenes.*

STAPHYLOCOCCAL SKIN AND SOFT TISSUE INFECTIONS

Local staphylococcal infections, beginning with the most superficial, are:

1. bullous impetigo (in the epidermis)
2. paronychia (around nails)
3. folliculitis (in hair follicles)
4. furuncles (in epidermis and dermis)
5. carbuncles (in subcutaneous tissues)
6. cellulitis (in subcutaneous tissues).

Staphylococcal infections here, as elsewhere in the body, are characterised by:

- suppuration (pus formation)
- necrosis of local tissues
- abscess formation.

In addition, distant staphylococcal infections affect the skin.

LOCAL INFECTIONS

Bullous impetigo

Bullous impetigo is caused by particular strains of *Staph. aureus* and is distinguished from streptococcal impetigo (p. 164) by large bullae (blisters) containing yellow pus and, of course, many staphylococci on Gram stain and culture.

Although these strains, like those causing the scalded skin syndrome (SSS), produce exfoliatin (the epidermolytic toxin), bullous impetigo differs from SSS in three ways: blister culture is positive, erythema is only local around the bullae and Nikolsky's sign is negative. Bullous impetigo particularly infects infants and young children, is highly infectious by direct spread, and is controlled by stopping this spread and giving a penicillinase-resistant penicillin, e.g. flucloxacillin.

Paronychia

An acute paronychia is an infection next to a fingernail or toenail, characterised by quite severe pain and tenderness, redness, heat, swelling and pus formation. Surgical incision is usually necessary, plus an anti-staphylococcal antibiotic such as flucloxacillin.

Chronic paronychiae differ in the following ways:

- usually affect several fingers
- associated with repeated immersion in water
- less pain, swelling and pus formation
- causes include *Candida* spp. and other organisms
- medical therapy is difficult, but surgery is unnecessary.

Folliculitis

Folliculitis (Fig. 1) is a common but usually trivial infection localised to a hair

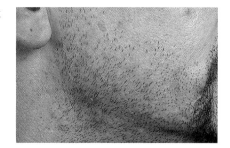

Fig. 1 **Folliculitis.** Note small yellow pustules around hair follicles, with little surrounding inflammation.

follicle, with a small bead of pus and some adjacent erythema. It often heals spontaneously but otherwise is cured by needling the apex and using a dab of antiseptic. When it affects an eyelash it is called a **stye** or **hordeolum**, and removal of the lash plus local antibiotic are usually necessary. When it affects the beard area, it is called **sycosis barbae**, is more troublesome, and oral antibiotics and a dermatologist's care are usually necessary.

Furuncles (boils)

Boils are a common and more severe form of folliculitis, often affecting sebaceous and sweat glands. A boil is a painful tender elevated red-rimmed pustule (Fig. 2) that grows until it discharges or is opened with a sterile needle or blade.

There is usually no host abnormality apart from nasal carriage of the causative strain of *S. aureus*.

In contrast, there is often some host abnormality such as acne, diabetes or abnormal white cell function in:

- **furunculosis:** multiple boils
- **chronic furunculosis:** persistent or recurrent boils
- **hidradenitis suppurativa**, a deep suppurative infection of the sweat glands, usually in axilla or groin, for which skilled antibiotic therapy is often necessary.

Fever or chills, however, are uncommon with any of the above.

Carbuncles

Fusion of a number of furuncles causes a carbuncle, a very unpleasant but now uncommon condition, with extension deeply and widely in the subcutaneous tissues, and multiple but ineffective sinuses to the skin (see Fig. 4, p. 29); the back of the neck was a common site. Fever, chills and even bacteraemia can follow. High-dose parenteral anti-staphylococcal

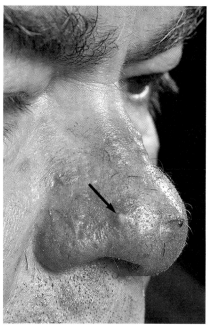

Fig. 2 **Furuncle.** Note larger collection of yellow pus, raised from skin surface with definite red inflamed edge.

antibiotics are essential, and surgical removal of the central core of necrotic infected tissue is usually needed.

Cellulitis

What may be called 'simple' cellulitis is an infection of the skin and subcutaneous tissues, characterised by redness, swelling and pain, with little or no skin or tissue necrosis. Complicated cellulitis (and deeper infections) characterised by necrosis (gangrene) and/or gas formation are discussed on pages 168–9.

Causative organisms. *S. aureus* and *Strep. pyogenes* are easily the most common causes of cellulitis; less often, group B, C or G streptococci, *Aeromonas hydrophila* from fresh water (p. 170), *Erysipelothrix rhusiopathiae* from meat or fish (p. 33), enteric Gram-negative rods or other organisms are responsible.

The clinical features are somewhat variable, depending on the organism, the host response and the route of infection, which can be from a skin abrasion, a traumatic or surgical wound, a furuncle or subcutaneous infection, or even blood-borne from septicaemia.

- Streptococcal cellulitis has the brilliant red, hot, shiny skin of erysipelas but lacks the sharp raised edge, and there is greater swelling from the subcutaneous infection.

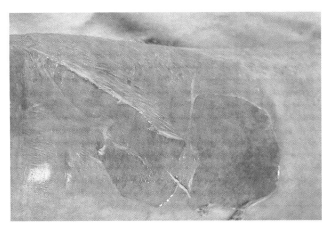

Fig. 3 **Scalded skin syndrome.**

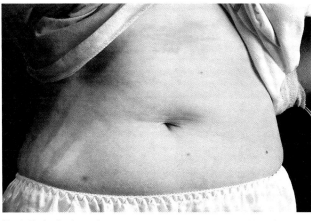

Fig. 4 **Toxic shock syndrome.** Note erythema with no local skin lesion.

Lymphangitis and lymphadenopathy are prominent, but there is minimal pus without a prior wound. High fever and tachycardia characterise this serious infection, in which bacteraemia and shock can develop in 6 to 12 hours.

- Staphylococcal cellulitis has less brilliant redness, but pain, tenderness and thick yellow pus are more prominent. Fever, malaise and regional lymphadenopathy usually develop more slowly than in streptococcal cellulitis. Local abscesses, limited skin necrosis, thrombophlebitis, bacteraemia and distant ('metastatic') abscesses follow if treatment is delayed or inadequate.

Confirmatory tests are made on pus using Gram stain and cultures; blood cultures are also used if bacteraemia is possible. Swabs from unbroken skin are useless.

Chemotherapy. Intravenous penicillin G is the treatment of choice for streptococcal cellulitis, but if *Staph. aureus* is the known or probable cause, a penicillinase-resistant penicillin such as flucloxacillin must be given. Pain relief and immobilisation in the acute stage are important, and pus must be drained.

Control and prevention. Cellulitis may recur because of predisposing factors including:

- lymphatic, venous or arterial insufficiency
- scars or retained foreign bodies from injury or surgical operation
- skin disease
- immune deficiency or immunosuppression.

The predisposing factor(s) obviously should be treated or removed if possible. Otherwise, suppression with the relevant oral penicillin or erythromycin may be necessary long term or even life long, as relapses have occurred when penicillin

was stopped after 5 years apparently successful control.

SKIN MANIFESTATIONS OF DISTANT INFECTIONS

Scalded skin syndrome

Staphylococcal scalded skin syndrome (SSS) is also known in neonates as Ritter's disease or pemphigus neonatorum. It is caused by strains of *Staph. aureus* of phage group II; these produce toxins called exfoliatins which cause intra-epidermal cleavage planes. Staphylococcal scarlet fever is a rare, milder form with erythema but no exfoliation.

Clinical features. The three clinical characteristics are erythema, enormous bullae and extensive exfoliation (desquamation). Fever and erythema develop, then bullae, then exfoliation, often in sheets and occurring initially in the face, axillae and groins. Nikolsky's sign is removal of the upper layers of the epidermis by gentle sliding movement of the examining finger (Fig. 3).

Confirmatory tests. The distant causative infection (e.g. conjunctival, umbilical) is swabbed for Gram stain and culture, as the dramatic skin lesions do not contain the staphylococcus.

Chemotherapy. A penicillinase-resistant penicillin such as flucloxacillin is given, and fluid replacement is usually necessary.

Control depends on early diagnosis and active treatment of all staphylococcal infections, especially in neonates and infants.

Toxic shock syndrome (TSS)

TSS is a serious systemic infection caused by *Staph. aureus* strains producing toxic shock syndrome toxin 1 (TSST-1). Predisposing factors include multiplication of the organisms and magnesium binding in certain high-absorbency tampons and a genetically ineffective antibody response. However, skin or other tissues may be the primary site in men and non-menstruating women.

Clinical features. High fever, myalgia, vomiting and diarrhoea develop quickly. This is followed by a skin rash (Fig. 4) (which later exfoliates at a deeper level than in SSS), then shock, hypotension, and renal and often hepatic impairment within 36–48 hours.

Confirmatory tests. Treatment must precede laboratory results, but the primary site must be sampled for Gram stain and culture. Toxin detection is usually only available in special laboratories.

Chemotherapy. The primary site is removed or treated, and high-dose intravenous anti-staphylococcal penicillin is given, while shock and organ failure are treated appropriately. TSS is much less common since certain high-absorbency tampons were withdrawn from sale.

Staphylococcal skin and soft tissue infections

- Local staphylococcal infections (*Staph. aureus*) are characterised by pus formation, necrosis of local tissues and abscess formation. They range from simple boils to cellulitis.
- Management is with a penicillinase-resistant penicillin and possibly surgical drainage.
- Distant infections affect the skin through toxins produced by *Staph. aureus.*
- Scalded skin syndrome is characterised by erythema, enormous bullae and extensive exfoliation (desquamation).
- Toxic shock syndrome (TSS) is characterised by temperature, skin rash and shock.

GAS-FORMING AND GANGRENOUS INFECTIONS

Gas-forming and gangrenous (necrotising) infections are limb- and life-threatening conditions that can occur at various tissue levels. Three necessary conditions are:

- portal of entry: traumatic or surgical wound
- anaerobic area: foreign body, ischaemic tissue
- pathogen entry: faecal or soil contamination usually.

The following conditions are classified by tissue level, necrosis or gas formation and immune status.

CHANCRIFORM ULCERS

The major feature is a punched-out ulcer rather like a syphilitic chancre, with skin and subcutaneous tissue necrosis, but little adjacent cellulitis initially.

Infectious causes include chancroid, cutaneous diphtheria, mycobacterial infections, syphilis and tularaemia (p. 185), ecthyma (see below) and anthrax.

Anthrax progresses from a painless papule through a vesicular 'malignant pustule' to a black necrotic ulcer with extensive red non-pitting oedema. The patient becomes very ill with high fever and bacteraemia. Gram stain of the lesion and blood culture show distinctive 'bamboo-like' chains of Gram-positive rods. High-dose intravenous penicillin is the drug of choice. Now very rare in developed countries, the patient's occupation or exposure to infected animals or their products, and lack of pain, are important diagnostic clues.

Pyoderma gangrenosum is a complication of ulcerative colitis or rheumatoid arthritis in which one or more nodules break down to deep necrotic coalescing ulcers, colonised by *Staphylococcus aureus*, streptococci and Gram-negative bacilli.

SYNERGISTIC GANGRENE

Synergistic gangrene (Fig. 1) is characterised by rapidly progressive destruction of skin and subcutaneous tissue caused by different bacteria acting synergistically. It is also called symbiotic gangrene and progressive bacterial synergistic gangrene.

The causative organisms are always multiple; the classical pair are *S. aureus* and a micro-aerophilic or anaerobic streptococcus in cultures from the advancing edge, but other anaerobes and *Proteus* spp. are often found in the ulcer.

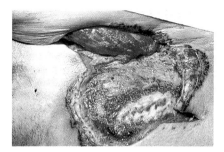

Fig. 1 **Synergistic gangrene.**

Clinically, this unusual lesion typically complicates a contaminated wound, with expanding massive ulceration, and destruction of all tissue down to muscle (Fig. 1). There is often surprisingly little 'systemic toxicity'.

Meleney's gangrene is a related disease characterised by synergistic gangrene *plus* subcutaneous necrotic tracks to distant sinuses.

Confirmatory diagnosis is secondary to clinical diagnosis; ulcer cultures are misleading, but excised tissue should be cultured.

Management involves broad-spectrum antibiotics, e.g. flucloxacillin, gentamicin and metronidazole, but radical surgical excision of the advancing edge is almost always necessary, and at times an old-fashioned diathermy 'firebreak' must be cut.

CREPITANT ANAEROBIC CELLULITIS

Crepitant anaerobic cellulitis is classified into clostridial and non-clostridial types. The former must be distinguished from clostridial **myonecrosis** (see below).

The causative organisms are *Clostridium perfringens* or other Clostridia, or mixed anaerobes, including *Bacteroides* spp. and *Peptostreptococci*, rarely *Escherichia coli* or *Klebsiella* spp. *Cl. septicum* is often associated with colonic cancer.

Clinically, there is gradual onset of a predominantly subcutaneous tissue infection around a wound, with bubbly crepitus on palpation, but little pain, tenderness, pus, skin involvement or toxaemia.

Confirmatory tests are Gram stain, and aerobic and anaerobic cultures, which are important to determine the causative organisms. X-ray shows bubbles of gas in connective tissue, *not* muscle.

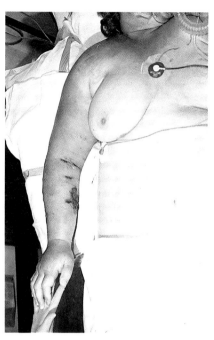

Fig. 2 **Fatal necrotising fasciitis (streptococcal) after elbow operation.** Note relatively small area of skin necrosis initially, yet even amputation did not save her life.

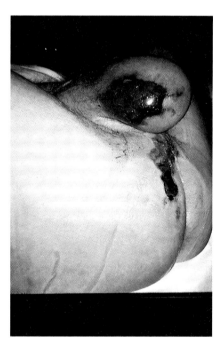

Fig. 3 **Fournier's gangrene.**

Chemotherapy must usually begin before culture results are available and should include penicillin and metronidazole.

Surgical exploration is necessary to determine the extent of disease, drain any pus and remove necrotic subcutaneous tissue (not uninvolved muscle).

NECROTISING FASCIITIS

Necrotising fasciitis is caused by *Strep. pyogenes* group A (now rare, Fig. 2) or a mixed anaerobic infection. If it affects the male genitalia it is called **Fournier's gangrene** (Fig. 3).

Clinically, the illness is acute, with severe systemic toxicity and fever, but initially little skin involvement; at fascial level there is extensive destruction followed by skin necrosis; the muscle is surprisingly spared.

Tests using skin swabs are ineffective, but pus or tissue grow the causative organisms.

Intravenous high-dose penicillin plus metronidazole should be appropriate chemotherapy, but surgery is essential.

NON-CLOSTRIDIAL MYONECROSIS

Non-clostridial myonecrosis is characterised by muscle necrosis, with less subcutaneous and lesser skin involvement. Gas formation is not prominent.

It may be divided into four subtypes:

Anaerobic streptococcal myositis. This rare condition is caused by anaerobic streptococci, often with Group A streptococci or *Staph. aureus*. This is an acute disease with red swollen skin and a 'sour' exudate. This is followed by pain and gas in the muscles, then gangrene, shock and death unless urgent treatment is commenced with high-dose penicillin and surgical debridement.

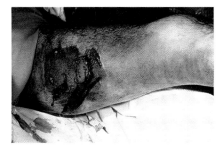

Fig. 4 **Gas gangrene.**

Synergistic necrotising cellulitis. This rare disease is caused by mixed anaerobes. It is an aggressive infection with marked local tenderness from muscle and subcutaneous tissue destruction, with some gas production, but little skin change except drainage of thin 'dishwater' pus from small ulcers. Systemic toxicity is marked, and urgent treatment with antibiotics, for example penicillin and metronidazole, and surgery is essential.

Infected vascular gangrene. This is caused by mixed anaerobes (e.g. *Bacteroides* spp., anaerobic streptococci) and enteric aerobes in patients with arterial impairment, especially diabetics. There is gas and foul pus, but this less aggressive infection is usually controlled by appropriate antibiotics and local excision of dead muscle.

Fulminant myonecrosis. See immunocompromised host below.

CLOSTRIDIAL MYONECROSIS (GAS GANGRENE)

Gas gangrene (Fig. 4) is a rapidly fatal disease characterised by muscle necrosis *and* gas formation with consequent skin and subcutaneous tissue destruction and multisystem failure. The causative organism is usually *Cl. perfringens*, rarely other *Clostridia*, e.g. *Cl. septicum*.

Clinically the onset is abrupt after an incubation period of 6–72 hours, with severe local pain, fever, pallor and sweats. There is local oedema, then skin discoloration, blistering and necrosis, with crepitus and dirty blood-stained pus with a musty or mouse-like odour. If treatment is delayed, then shock, haemolytic anaemia, jaundice, renal failure and death follow from the action of clostridial toxins particularly on muscle and blood.

Confirmatory laboratory diagnosis by urgent Gram stain must not delay treatment. If X-rayed on the way to theatre, there are bubbles and linear gas shadows in the muscles and soft tissues.

Urgent treatment combines high-dose intravenous penicillin and metronidazole or imipenem with surgical removal of all dead or dying muscle, and, if available, hyperbaric oxygen.

Prevention depends on antibiotic prophylaxis and removal of ischaemic tissue and foreign bodies from pre-disposed wounds.

IMMUNOCOMPROMISED HOST

In addition to the forms described above, unusual necrotising infections can occur modified by impaired host response:

- *Aspergillus* spp. and the Phycomycetes (*Rhizopus*, *Mucor* and *Absidia* spp.) cause serious, often untreatable and hence fatal gangrenous cellulitis, with a black anaesthetic ulcer and a relentlessly advancing purplish oedematous edge.
- *Aeromonas hydrophila* causes a plaque, followed by a necrotic ulcer. It can also cause crepitant cellulitis.
- *Bacillus cereus* causes crepitant cellulitis.
- *Cryptococcus neoformans* causes a plaque, then a necrotic ulcer.
- *Klebsiella* spp. or *Nocardia* spp. cause fulminant myonecrosis.
- *Pseudomonas aeruginosa* can cause a gangrenous cellulitis, or **ecthyma gangrenosum**, a black necrotic ulcer with surrounding erythema.

DIABETIC FOOT INFECTIONS

Foot infections in diabetics are more likely to lead to gangrene, particularly digital, because of the impaired host response and vascular disease (Fig. 5). Mixed infections with staphylococci, streptococci and anaerobes are common, with enteric Gram-negative colonisation. Pus and excised tissue should be cultured and appropriate chemotherapy given. Surgery should be conservative whenever possible, but underlying osteomyelitis should be sought (p. 179).

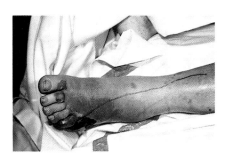

Fig. 5 **Diabetic foot infection with gangrene.**

Gas-forming and gangrenous infections

- Predisposing factors are a wound, an anaerobic area and contamination by pathogens.
- Infections are often mixed; anaerobes including Clostridia are prominent.
- Clinical classification depends on tissue level involved and whether necrosis or gas formation predominates.
- Immunocompromised patients have special pathogens and unusual infections.
- Treatment with chemotherapy and surgery is urgent for rapidly progressive infections; identification of pathogens should occur as fast as possible.

WOUND, BITE AND BURN INFECTIONS

TRAUMATIC WOUND INFECTIONS

Wound infections are best classified into those infected from:

- skin flora
- a perforated viscus
- water or animals
- soil.

Wounds infected by skin flora

The causative organisms will usually be *Staphylococcus aureus* or *Streptococcus pyogenes*, and the resultant infections, depending on the level infected, will be erysipelas (uncommonly; p. 164), cellulitis (commonly; p. 166), or one of the rare infections described on pages 168–9.

Immobility particularly in severely ill patients can lead to bed sores, which can become infected (Fig. 1).

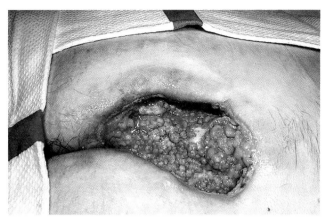

Fig. 1 **Infected bed sore.**

Wounds perforating a hollow viscus

The causative organisms obviously depend on the viscus perforated: most feared is colonic flora, with huge numbers of pathogenic aerobic and anaerobic Gram-negative and Gram-positive organisms (p. 142). These cause peritonitis and local abscesses, followed by septicaemia and distant infections if untreated.

Small bowel, gastric and biliary wounds cause fewer infective problems initially, though chemical peritonitis from gastric acid or bile becomes secondarily infected if untreated.

Respiratory flora causes few infections at skin or subcutaneous levels, though the pleural cavity can become infected, called an empyema (p. 117).

In all cases, management includes empiric antibiotics aimed at the likely flora, operative cultures, cleansing and closure of the perforation, drainage as necessary, maintenance of vital organ functions, and modification of antibiotic therapy if culture results and the clinical course so indicate.

Wounds infected from water or animals

A variety of organisms can be responsible:
Aeromonas hydrophila. This Gram-negative rod causes acute cellulitis around traumatic wounds sustained while swimming. The organism is sensitive to gentamicin, tetracycline and cotrimoxazole.

Erysipelothrix rhusiopathiae. This Gram-positive rod causes erysipeloid, an uncommon occupational skin infection of those handling raw meat or fish. Usually confined to the fingers or hands, the lesion has a purplish-red raised edge and a centre fading as the lesion enlarges. Pus is uncommon, and there is burning rather than pain. Microbiologic confirmation of the diagnosis usually needs Gram stain and culture of biopsy of the edge of the lesion, surface swabs being

negative (p. 31). Healing is accelerated by penicillin or cephalosporin treatment. Septicaemia is uncommon, but when it occurs, endocarditis often follows and is fatal in about 30% of patients.

Mycobacterium marinum. This atypical mycobacterium (p. 53) growing optimally at 25–32°C in water causes 'fish tank granuloma' and 'swimming pool granuloma' at the site of abrasions. The lesions are chronic, nodular or ulcerating, on the hands or over bony prominences, usually single but rarely ascending like sporotrichosis. Biopsy is often necessary for diagnosis. Specialist referral is essential, for treatment is difficult, as the organism is relatively resistant. Early lesions may respond to long-term cotrimoxazole, tetracycline or combined rifampicin–ethambutol, but excisional surgery and skin grafting may be necessary for advanced lesions.

Soil-contaminated wounds

The major causative organisms are:

- *Clostridium perfringens* and related clostridia (p. 32)
- *Cl. tetani*: tetanus (p. 37)
- *Cl. botulinum*: wound botulism (p. 33)
- *Mycobacterium ulcerans* (Buruli or Bairnsdale ulcer) (p. 174)
- *Cladosporium* spp. and *Phialophora* spp.: subcutaneous mycoses (p. 176)
- *Sporothrix schenckii*: sporotrichosis (p. 176).

SURGICAL WOUND INFECTIONS

These are classified into *superficial* if above the deep fascia, and *deep* if below (Fig. 2). Extension beyond skin and soft tissues is considered on pages 168–9.

Organisms causing infection in surgical wounds include:

- *Staph. aureus* commonly and *S. epidermidis* rarely
- Streptococci, especially group A, and enterococci
- enteric Gram-negative rods, e.g. *Escherichia coli, Klebsiella* spp.
- anaerobes, usually mixed
- *Pseudomonas aeruginosa*
- *Cl. botulinum* or other clostridia, very rarely.

Infection can occur from a variety of sources: the patient's own normal flora (usual), the theatre staff (uncommon but infamous) and the environment or instruments (unacceptable).

Clinical features. Redness, swelling, local pain and heat develop, and progress to a purulent discharge from the wound or around a suture ('stitch abscess') if untreated or if the infection is deep.

Confirmatory tests. Swabs of the discharge should always be Gram stained and cultured to isolate the pathogen, determine the antibiotic sensitivities and convince the surgeon.

Management

Pus must always be drained and foreign bodies removed. If cellulitis is established or spreading, antibiotics are needed. Initial empiric chemotherapy is aimed at the likely pathogens, e.g. flucloxacillin against *Staph. aureus* for many superficial operations; cephalosporin or gentamicin plus metronidazole against bowel flora after abdominal operations. Control and prevention include

- good surgical technique
- prophylactic antibiotics immediately before and during operation
  — if postoperative infection likely, e.g. colonic surgery
  — if consequences of infection are devastating, e.g. valve replacement
  — chosen to suit procedure.

ANIMAL AND HUMAN BITE INFECTIONS

Bite injuries vary in degree and in the type of pathogen likely to infect them. They can be classified into:

- human bite injuries
- human fist injuries
- animal bites by different species, possibly transmitting
- specific pathogens.

Causative organisms

The important organisms likely to infect these injuries are:

- *Pasteurella multocida* (p. 51)
- *Eikenella corrodens* (p. 47)
- 'mixed anaerobes' (*Bacteroides, Fusobacteria*, (p. 48), etc.)
- *Capnocytophaga canimorsus* (previously called DF2, rare but kills if untreated (Fig. 1, p. 184)).
- *Staph. aureus* and streptococci ('viridans', group A) (pp. 28–31).

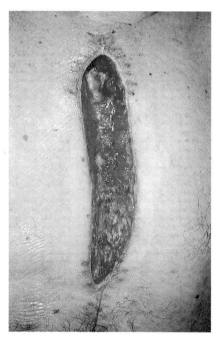

Fig. 2 **Infected sternal wound.**

Clinically, the bite injury is very variable, and deep injury should be sought, especially tendon or joint in fist injuries. Gram stain and culture only help if infection is uncontrolled.

Chemotherapy should be immediate and broad spectrum, e.g. penicillin (for clostridial and *Eikenella* spp.) and cephalothin. Surgery is important to remove dead tissue and repair deep structures, but because of the high risk of infection the wound must not be sutured except for potentially disfiguring facial wounds. Tetanus prophylaxis should be given, and rabies considered.

Seal finger is a curiosity, similar to erysipeloid but following a seal bite. The causative organism is unknown, but tetracycline treatment is effective.

Specific pathogens transmitted by bites include:

- *Clostridium tetani* (p. 35)
- rabies virus
- *Streptobacillus moniliformis,* and *Spirillum minor* (p. 47): **rat bite fever**
- *Rochalimaea henselae*: **cat scratch fever**

Rat bite fever. This is an acute illness with high recurrent fever, chills, rash and, with *S. minor*, lymphangitis and lymphadenopathy, with *S. moniliformis* myalgia, arthralgia and arthritis. Laboratory diagnosis is specialised (p. 47). Penicillin is the treatment of choice.

Cat scratch fever. The local lesion is only a small papule or pustule but impressive regional lymphadenopathy develops and persists 3 months or more. The organism is difficult to visualise in culture, but histology is usually diagnostic. The organism is sensitive to later cephalosporins but no best regimen is proved.

INFECTED BURNS

Infection has been a major cause of morbidity and mortality in burns, as any burn damages or removes the first line of defence (the skin), and extensive burns impair other host defences also.

Causative organisms include *Staph. aureus, Strep. pyogenes* (now fortunately rare), enteric Gram-negative rods (*Escherichia coli, Klebsiella* spp., etc.) and *Pseudomonas aeruginosa.* The sources are the same as for wound infections: the patient, the staff, equipment and the environment.

Clinical features are usually confined to pus, fever, and skin graft loss (Fig. 3) but spread into deeper tissue layers can lead to septicaemia.

Confirmatory diagnosis by Gram stain and culture is important to determine the pathogens and their antibiotic sensitivities, hence the appropriate antibiotic therapy.

The four principles of control by prevention are:

- primary early burn wound therapy to remove necrotic tissue and cover the area by skin grafts or other materials
- cross-infection control
- topical antibacterial use, usually silver sulphadiazine (SSD)
- prophylactic antibiotics used only at times of decreased host resistance and microbial contamination, e.g. the immediate 3 days after the burn injury, or on return to theatre for excision or grafts.

INFECTIONS OF PRE-EXISTING SKIN CONDITIONS

Any eczematous or ulcerating skin disease can become infected, for example:

- acne conglabata
- bullous and vesicular eruptions
- chronic ulcers
- dermatophytosis and intertrigo
- eczematous or exfoliative conditions.

The causative organisms are usually those infecting burns (see above) and pus is the predominant sign. Gram stain and culture should be carried out and appropriate antibiotics given.

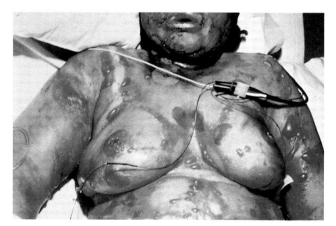

Fig. 3 **Infected burn.**

Wound, bite and burn infections

- Wounds can be traumatic, surgical, bites or rupture of the skin defences by burns or pre-existing disease.
- Infection of wounds comes from skin flora, contaminating foreign bodies or soil, and perforated viscera.
- Management involves drainage of pus, removal of foreign bodies and necrotic tissue, and antibiotic treatment. Antibiotics are chosen empirically until confirmatory tests identify the pathogen(s).

FUNGAL INFECTIONS OF THE SKIN, HAIR OR NAILS

SUPERFICIAL MYCOSES

Superfical mycoses are characterised by infection of the outermost layers of skin and hair, hence there is no host response. Four types are recognised (see also p. 68).

Tinea versicolor (pityriasis versicolor)

Tinea versicolor is caused by *Malassezia furfur (Pityrosporum orbiculare)* and is found worldwide. Clinically, the two major characteristics are hypopigmentation and scaling. Initially perifollicular, the hypopigmentation spreads and coalesces. It is asymptomatic, or mildly itchy. Diagnosis is confirmed by a KOH preparation of the skin scales showing clusters of yeasts with short hyphal fragments, so-called 'spaghetti and meatballs'. Treatment is by 1% selenium sulphide shampoo seven times in 14 days.

Tinea nigra

Tinea nigra is caused by *Exophiala werneckii* and is characterised by dark brown or black macules (Fig. 1). It is particularly common in warm countries. Clinically, the asymptomatic lesions are usually on the palms or soles and enlarge peripherally. The differential diagnosis from malignant melanoma is most important and is made by finding the characteristic dark two-celled oval yeasts and short hyphae in KOH mounts. Treatment is to remove the stratum corneum by scraping, by adhesive tape or by a keratolytic such as Whitfield's ointment.

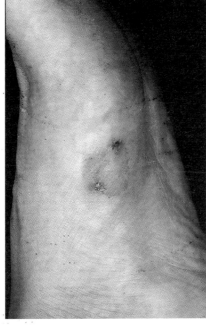

Fig. 1 **Tinea nigra.**

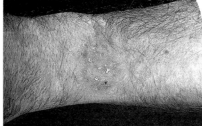

Fig. 3 **Tinea corporis.**

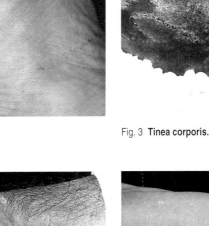

Fig. 2 *M. gypseum* geophilic infection from soil contact in drainworker.

Fig. 4 **Tinea pedis.**

Black piedra

Black piedra is caused by *Piedraia hortai* infection forming hard black nodules along the hair shaft. The diagnosis is easily confirmed by microscopy. Cutting or shaving the hair usually avoids topical anti-fungal agents.

White piedra

White piedra is caused by *Trichosporon beigelii* infection forming soft cream-white sleeves around the hair shafts. Differential diagnosis includes the harder more adherent 'nits' of pediculosis, distinguished by microscopy. Treatment is also by cutting or shaving the hair.

CUTANEOUS MYCOSES (DERMATOPHYTOSES)

Cutaneous mycoses (known as tinea or ringworm) are characterised by infection of the keratinised layer of the skin, hair or nails. The clinical disease depends on the specific infecting fungus and on the host response. The clinical classification depends on the body area involved:

- skin: tinea corporis (of the body), tinea cruris (groins), tinea manuum (hand), tinea pedis (feet)
- hair: tinea capitis (scalp), tinea barbae (beard)
- nails: tinea unguium.

Tinea of the skin

The causative organisms are commonly *T. rubrum, T. mentagrophytes, E. floccosum* and, on the body, *M. canis*, but any of the other species of dermatophytes (pp. 68–9) can infect (Fig. 2). Predisposing factors include warmth and moisture, and hence shoes, tight underclothes and obesity can become causative factors.

The clinical features vary. On the body, groins and hands, the common lesion is a round or irregular ('gyrate') scaly area with a red edge and a healing or scaling centre (Fig. 3). Other forms are vesicular, pustular, granulomatous or, very rarely, mycetoma.

Tinea pedis (Fig. 4) is commonly interdigital (intertriginous), with red scaling and white macerated fissuring but may be hyperkeratotic or vesicopustular.

The differential diagnosis is wide and includes seborrhoeic dermatitis, lichen planus, candidiasis and psoriasis. Confirmatory diagnosis is by KOH mount and culture.

Chemotherapy is usually by a topical anti-fungal, commonly an imidazole such as clotrimazole. Oral terbinafine is replacing griseofulvin for extensive disease. Control and prevention aim at removing predisposing factors, and preventing reinfection from self or others.

Tinea of the hair (capitis, barbae)
Mycoses of the hair can be either non-inflammatory or inflammatory. The causative organisms are from the genera *Trichophyton* or *Microsporom*.

Clinical syndromes are scaling and alopecia, with short broken hairs in ectothrix infections and 'black spots' where a hair has broken off at follicular level in endothrix infections. The inflammatory types begin as pustular folliculitis, which can progress to wide-spread suppuration under a fluctuant scalp (**kerion**), with malaise, fever and local lymphadenopathy. Scarring and alopecia

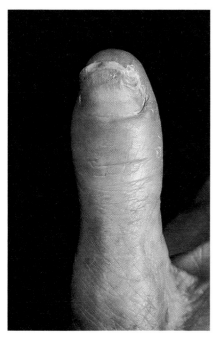

Fig. 5 **Tinea unguium.**

may then be permanent. In the principal differential diagnosis, bacterial folliculitis, alopecia is extremely uncommon.

Confirmatory diagnosis is made by examining the scalp with a Wood's light for fluorescent endothrix-infected hairs, by microscopy of KOH-treated hairs and scales, and by culture.

Chemotherapy with terbinafine or griseofulvin is essential, as topical medications are not curative. Griseofulvin is usually given for a period of 6 to 8 weeks, but a single large (3 g) dose cures 80% of children with ectothrix infections. Control and prevention is by avoiding overcrowding and by good hygiene, including avoiding shared contaminated combs.

Tinea of the nails (unguium)
The causative organisms are usually *Trichophyton* species, most commonly *T. rubrum* or *T. mentagrophytes*. Some other fungi, e.g. *Aspergillus, Candida* (Fig. 6) and *Fusarium* spp., can infect the nails. Clinically, distal white discoloration followed by subungual hyperkeratosis giving a thickened discoloured nail separated from the nail bed is most common (onychomycosis, Fig. 5), though sometimes **proximal** leuconychia (white nail) or **superficial** infection occurs. Co-existent tinea pedis or corporis is common.

Confirmatory diagnosis depends on culture from the infected subungual, proximal or superficial areas.

Chemotherapy, except for superficial disease, is by terbinafine or griseofulvin for 4 months for finger-nails, and 6, 9 or even 12 months for the slower growing toe-nails.

CUTANEOUS CANDIDIASIS

The cutaneous manifestations of *Candida* infection include local infection in the normal host, local infection in immuno-compromised host, and a local manifestation of disseminated disease.

Normal host
Intertrigo (in skin folds), napkin rash, balanitis, folliculitis and vulval, perianal and interdigital *Candida* infections are all similar, with initial vesicles or pustules which break to leave spreading red, moist, itchy areas with irregular white borders. There is often co-existing vaginal or rectal infection. Species other than *C. albicans* are unusual. Diagnosis is confirmed by Gram stain and culture, and local anti-fungal therapy (e.g. with clotrimazole) is usually curative. Prevention and control of recurrences depends on attacking predisposing factors such as moisture, obesity and diabetes, and any vaginal or rectal source. **Paronychia** (see p. 166) (Fig. 6) and **onychomycosis** (see above) are also seen.

Immunocompromised host
Chronic mucocutaneous candidiasis (CMC) is a severe chronic disease in those, usually children, whose T-cells fail to respond to *Candida* antigen in vitro or in vivo. Clinically, oral thrush is usually followed by nail then skin infections of variable severity but unremitting chronicity. In addition, about half have an endocrinopathy, e.g. hypoparathyroidism or Addison's disease, often with autoimmune antibodies.

Confirmatory diagnosis depends on T-cell function tests, Gram stain and culture and appropriate endocrine function tests. Treatment is difficult and disappointing.

Local manifestation of disseminated disease
Macronodular candidiasis is a useful diagnostic sign in some patients with disseminated candidiasis, when scattered pink-red maculopapular lesions appear over the scalp and body. Histopathology stains on punch biopsy are more reliable than culture. Treatment is for disseminated candidiasis (p. 63).

SUBCUTANEOUS MYCOSES
See page 176.

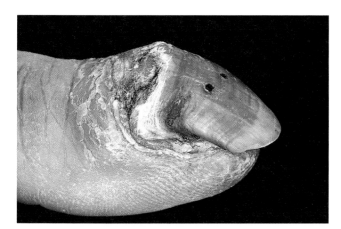

Fig. 6 *Candida* infection of nail, with paronychia.

Fungal infections of the skin, hair or nails

- Superficial mycoses affect the hair (piedra) and outer layer of the skin (tinea versicolor or tinea nigra).
- Cutaneous mycoses affect various areas of the body including hair and nails. The success of treatment depends upon the region infected.
- Candidiasis occurs in the skin and nails of immunocompetent hosts. Chronic mucocutaneous candidiasis and disseminated disease can occur if immune function is compromised.

TROPICAL AND RARE BACTERIAL SKIN AND SOFT TISSUE INFECTIONS

ERYTHRASMA

Erythrasma is a superficial skin infection caused by *Corynebacterium minutissimum*, a Gram-positive filamentous diphtheroid identified by coral-red fluorescence under a Wood's light, and cocci and filaments up to 10 μm long in KOH mounts. Pink or red-brown scales in intertriginous areas or toe-webs are treated with oral erythromycin for 5–21 days.

MYCETOMA (MADURA FOOT)

Mycetomata are chronic infections of the subcutaneous tissues characterised by swelling, sinus formation and suppuration.

They are classified into *Actinomycotic mycetomata* caused by various Actino-mycetes including *Actinomyces*, *Nocardia* and *Actinomadura* spp. (p. 56) and *eumycotic mycetomata*, caused by many fungi including *Madurella* and *Pseudalle-scheria* spp. (p. 69).

Clinically, after implantation of soil organisms, usually into the foot, there is massive swelling and deformity, with pus discharging through numerous sinuses. Pain is not prominent.

Culture and microscopy are essential to distinguish the actinomycete infections, which often respond to chemotherapy, from fungal infections, which rarely do.

Chemotherapy is usually penicillin for *Actinomyces* spp., or sulphonamides and an aminoglycoside for *Nocardia* spp. Total excision or amputation is usually needed for fungal mycetoma.

MYCOBACTERIAL INFECTIONS

Tuberculosis

All five forms of cutaneous tuberculosis caused by *Mycobacterium tuberculosis* are now rare.

- **Primary inoculation TB** in pathologists and embalmers from accidental injury leads to ulceration and lymphadenitis.
- **Tuberculosis verrucosa cutis** with warty plaques occurs also by inoculation, usually in young adults in tropical areas.
- **Lupus vulgaris** (the common wolf) has 'apple-jelly-like' nodules that coalesce to destroy ('eat') tissue.

- **Scrofuloderma** is skin infection around sinuses draining from underlying tuberculous lymph nodes.
- **Tuberculides** are scattered papulo-necrotic nodules denoting tuberculosis elsewhere in the body.

All are diagnosed and treated as tuberculosis elsewhere (p. 204)

Leprosy

The skin manifestations of leprosy range from the single anaesthetic hypo-pigmented tuberculoid plaque through intermediate forms to the numerous, large, deforming and destructive lesions of the lepromatous form (see p. 92).

Bairnsdale (Buruli) ulcer

The Bairnsdale ulcer (Fig. 1) is so named because *M. ulcerans* (p. 55) was first isolated from a patient from Bairnsdale in Australia, and the ulcer is now most commonly seen around Buruli in Africa.

Clinically, a subcutaneous nodule forms at the inoculation site. This breaks down to form a chronic ulcer with under-cut edges, at times slowly healing on one side while continuing to destroy tissue on

Table 1 **Summary of some tropical bacterial skin infections**

Category/classification	Causative organism	Clinical features	Confirmatory diagnosis	Chemotherapy	Control and prophylaxis
Bacterial mycetoma	*Actinomyces* spp. *Nocardia* spp.	Swelling, suppuration, sinuses	Granules, Gram-positive clubs, granulomata	Penicillin or sulphonamides, surgery	Avoid soil inoculation
Mycobacteria					
Tuberculosis	*M. tuberculosis*	Chronicity, ulceration	AFB smear, biopsy, culture	Triple (including isoniazid and rifampicin)	BCG, isonicotinic for acid hydrazide for contacts
Leprosy	*M. leprae*	Anaesthesia, pigmentation, ulceration, deformity	AFB smear, biopsy (snip), clinical	Triple (including dapsone and rifampicin)	BCG?, avoid close contact
Bairnsdale (Buruli) ulcer	*M. ulcerans*	Ulceration, undercut-edge	AFB smear, biopsy	Excise and graft, ABC?	BCG?, avoid soil
Fish-tank granuloma	*M. marinum*	Ulceration, granuloma	AFB smear, biopsy	Cotrimoxazole or combined rifampicin–ethambutol	Avoid fish and trauma
Spirochaetal					
Yaws	*Treponema pallidum* spp. *pertenue*	1. Painless papules 2. Generalised papillomata 3. Gummata, ulceration	Microscopy (dark ground), serology (late), clinical	Penicillin	Treatment (perhaps mass), avoid skin contact
Pinta	*T. carateum*	1. Pruritus, papules, pale areas 2. Pintides 3. Pale areas	As yaws	Penicillin	Treatment, avoid skin contact
Bejel (endemic syphilis)	*T. pallidum* spp. *endemicum*	1. Mucous ulcer 2. Oropharyngeal 3. Gummata	As yaws	Penicillin	Avoid shared utensils
Syphilis	*T. pallidum* spp. *pallidum*	1. Chancre 2. Rash (varies) 3. Gumma,CNS,etc	Microscopy (dark ground), serology	Penicillin	Control copulation
Tropical pyomyositis	*Staph. aureus* (*Strep. pyogenes*)	Pain, tenderness, swelling, fever	Gram stain culture	Drainage flucloxacillin	None known
Tropical ulcer	Anaerobes Spirochetes	Deep ulcer	Microscopy, clinical	Penicillin	Avoid trauma

AFB, acid-fast bacilli; ABC, anti (myco)bacterial chemotherapy the opposite side of the ulcer. There are no systemic symptoms unless secondary infection occurs.

Clinical diagnosis is confirmed by culture (at 33°C) or histopathology, if smears for acid-fast bacilli from the ulcer are negative. Chemotherapy is usually ineffective, though antibiotic 'cover' is usual when definitive excision and grafting are performed.

Fish tank granuloma

M. marinum infections are described on page 170.

SPIROCHAETAL INFECTIONS
Yaws

The causative organism is *Treponema pertenue* (now called *Treponema pallidium*, subspecies *pertenue*) (p. 52), visually and serologically identical with other human treponemes and also not yet cultured in vitro. Yaws is a disease of the tropics. Infection follows contact of exudate with abraded skin. Clinically, like the other treponematoses, there are three stages. The primary lesion is a painless papule that progresses, becomes papillomatous with surface erosions then slowly heals.

The secondary stage has similar but multiple papillomata that heal but may relapse. Lymphadenopathy and osteitis may occur. The tertiary stage has multiple skin lesions of many types (plaques, nodules and especially ulcers) and gummata (chronic destructive ulcers) of bones, especially skull, nose and tibia.

Clinical diagnosis may be confirmed by dark ground microscopy (DGM) of

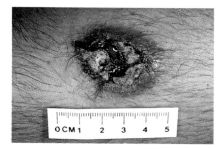

Fig. 1 **Bairnsdale or Buruli ulcer.** Note undermined edge.

Fig. 2 **Tropical pyomyositis.** Note thigh swelling.

exudates, or serology (VDRL or RPR; TPHA or FTA-ABS). Chemotherapy is extraordinarily effective, one injection of benzathine penicillin being curative, and controlling spread.

Pinta

The causative organism is *T. carateum* (p. 52). Transmission is, like yaws, by infectious secretions onto broken skin.

Pinta (Spanish for blemish) is seen in dry rural areas of Central and South America. Clinically, the primary stage shows pruritic papules that merge, persist for months, then heal with hypopigmentation. The secondary lesions are small, multiple, pigmented papules called pintides. These persist or recur for years, until the tertiary stage occurs with multiple disfiguring hypopigmented areas.

Clinical diagnosis is confirmed by DGM or serology. Chemotherapy is by penicillin, and all stages should be treated to control transmission, though early stages respond best, and depigmented (achromic) lesions remain.

Bejel (endemic syphilis)

The causative organism is *T. pallidum, endemicum*. Transmission is person-to-person by direct contact, or via common cooking cauldrons and cutlery through the oral mucosa.

Clinically, the primary lesion is usually unseen, but uncommonly ulcerates. The secondary lesions can be oral mucous patches, perioral papules or perianal condylomata lata. Tertiary lesions are usually obvious gummata of skin or bone. Clinical diagnosis again is confirmed by DGM or serology. Chemotherapy and control again is long-acting penicillin.

Syphilis

The causative organism is *T. pallidum,* spp. *pallidum* (p. 52). Transmission is virtually always venereal, at times transplacental, almost never accidental (by needlestick) as the spirochaete is very susceptible to drying and light, and soon

dies outside the body. The primary stage is the classic chancre (see p. 165). The secondary stage includes a rash of astonishing variability, from macular to maculopapular to papular to pustular (pustular syphilids). It persists for many weeks and typically involves palms and soles, unusual in other diseases. Tertiary involvement of the skin is rare, either by local gumma (chronic destructive ulceration) or from bone disease.

Clinical diagnosis may be confirmed by DGM (rarely), by serology (may be unhelpful), or by histopathology. Chemotherapy is by penicillin, dose and duration depending on the stage. Primary and secondary lesions resolve remarkably. Unlike most other tertiary syphilis, gummata also respond well (see also p. 163).

TROPICAL PYOMYOSITIS

Obviously, this is a pus-forming infection of a patient's muscle in the tropics; the causative organism is usually *Staphylococcus aureus*. Why it occurs is less obvious: predisposing factors may be trauma or parasitic infestation.

Clinically, pain, swelling (Fig. 2) and tenderness develop over 2 or 3 days in a leg or trunk muscle, followed by fever, and local heat and erythema if subcutaneous spread occurs. A minority of patients have multiple abscesses. Very acute onset suggests *Streptococcus pyogenes* as the cause. Diagnosis is by Gram stain and culture of operative pus. Chemotherapy with i.v. penicillinase-resistant penicillin is an adjunct to surgical drainage of all abscesses. Penicillin G is given if *Strep. pyogenes* is present.

TROPICAL ULCER

This is a deep, chronic painful skin ulcer of uncertain aetiology, possibly anaerobes and/or spirochaetes (Fig. 3, p. 189). Known causes should be excluded. Most respond to penicillin, local care and improved nutrition.

Tropical and rare bacterial skin and soft tissue infections

- Mycetoma is easily diagnosed clinically, but culture is essential to distinguish bacterial infections, treatable with chemotherapy, from fungal infections, treated surgically.
- Mycobacterial infections are chronic, characteristic clinical conditions (e.g. leprosy), confirmed by special stains and/or culture, with specific long-term chemotherapy and limited special surgery.
- Spirochaetal infections have primary, secondary and tertiary stages. The geographic area and clinical syndromes combine to give the diagnosis. Confirmation is usually by serology, rarely by microscopy, and never by culture. Chemotherapy is penicillin.
- Tropical pyomyositis is also a clinical diagnosis, confirmed by microscopy and culture of pus at operation. *Staph. aureus* is treated with flucloxacillin, *Strep. pyogenes* with penicillin G.

TROPICAL AND RARE FUNGAL AND PARASITIC SKIN AND SOFT TISSUE INFECTIONS

SUBCUTANEOUS MYCOSES

Sporotrichosis

The causative organism is *Sporothrix schenckii* (p. 69). Clinically, there are three forms:

- **Lymphocutaneous infection** follows neglected implantation of the fungus from soil or vegetation. Slowly a painless papule appears, enlarges and ulcerates, then multiple painless nodules appear along draining lymphatics (not at lymph nodes) and also ulcerate. Systemic symptoms are absent in this chronic disease.
- **Cutaneous infection** occurs as a verrucous or ulcerated plaque.
- **Disseminated disease** to lungs, bones or brain is very rare.

Confirmatory diagnosis by culture of pus or tissue is not hard. Histopathology shows granulomata, micro-abscesses and pseudo-epitheliomatous hyperplasia (excess of normal epithelial cells). Chemotherapy is by oral KI (potassium iodide) solution.

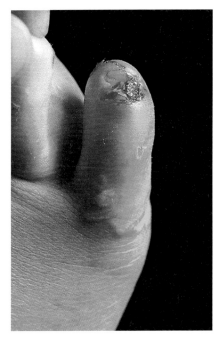

Fig. 1 **Cutaneous larva migrans.** Note elevated tracks of worms at base of toe.

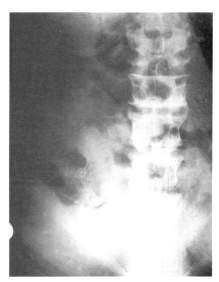

Fig. 2 **Guinea worm: X-ray showing calcification of dead worms above iliac crest.**

Chromoblastomycosis

This mycosis is caused by the implantation of any of a host of melanin-containing 'dematiaceous' fungi, predominantly *Phialophora* spp. (p. 69).

Clinically, early lesions are smooth papules which become verrucous with age. Infection spreads proximally, so an infected leg has proximal smooth lesions and cauliflower-like distal ones. Pain, sinuses and systemic symptoms are absent unless secondary bacterial infection occurs.

Confirmatory diagnosis is by histopathology, showing pigmented fungi ('copper pennies', sclerotic or 'Medlar bodies'), and granulomata, micro-abscesses and pseudo-epitheliomatous hyperplasia. Culture is unreliable as the causative fungi are frequent contaminants.

Chemotherapy with flucytosine is used for extensive disease, while surgical excision is used for localised early disease.

Rare fungal infections

Eumycotic mycetoma. This is described with actinomycotic mycetoma (p. 174).

Lobomycosis. This is a rare, geographically localised disease. The causative fungus, *Loboa loboi*, has not been cultured. Clinically, chronic warty, nodular, subcutaneous lumps form. Microscopy shows budding yeast-like fungi in chains. Treatment is excision.

Phaeohyphomycosis (including phaeomycotic cyst). This is another rare infection. Over 40 different fungi have been described as causal. Clinically, the lesions are single and slow growing. Histopathology shows brown septate hyphal fragments. Chemotherapy is useless: treatment is excision.

Rhinosporidiosis. This is a rare infection with *Rhinosporidium seeberi*. Clinically, there are large vascular friable warts or polyps, often involving the nasal mucosa. Diagnosis is confirmed by histopathology showing endospores within large spherules. The treatment is excision.

PARASITIC INFECTIONS

Cutaneous amoebiasis

Entamoeba histolytica skin infection is rare, but can occur, by extension from rectal or anal infection, from a bowel fistula, or from anal or perhaps vaginal intercourse.

Clinically there is extending ulceration, pain and bleeding, which can mimic a carcinoma. It is, therefore, essential that clinical diagnosis is confirmed by tissue smear or biopsy. Chemotherapy is with metronidazole; surgery can be harmful. Control and prevention depend on hygiene, and early treatment of intestinal and visceral infection.

Cutaneous larva migrans

This is characterised by red, itchy, snake-like tracks (creeping eruption), usually caused by migrating larvae of the dog and cat hookworm *Ancylostoma braziliense* (p. 83), less commonly by larvae of other animal or human hookworms including *Strongyloides stercoralis* (p. 81).

Clinically there is little to find except the multiple tracks, (Fig. 1), but rarely there is pulmonary or systemic involvement. Clinical diagnosis may be confirmed by biopsy, though this rarely identifies the causative parasite.

Chemotherapy is thiabendazole (local or oral) or mebendazole. Control depends on avoiding contact with infected soil and treating infected hosts.

Dracunculiasis (Guinea worm)

This is characterised by a chronic skin ulcer from which the causative worm *Dracunculus medinensis* (Fig. 2) partly

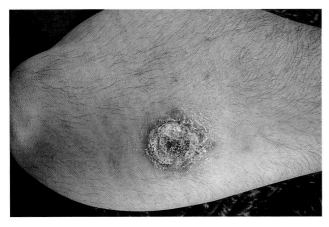

Fig. 3 **Leishmaniasis: ulcerated papule.**

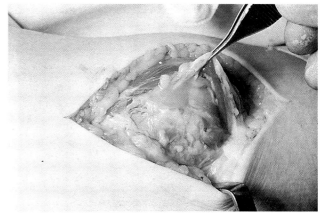

Fig. 4 **Hydatid of the thigh.**

protrudes. Infection results from drinking infected water. (p. 81).

Clinically a painful papule, usually on the leg, ulcerates and the worm becomes visible. Systemic symptoms of itch, nausea, diarrhoea and dyspnoea may occur even before ulceration. The ulcers may be multiple and/or secondarily infected. Clinical diagnosis is easy, but may be confirmed by showing larvae in the ulcer fluid.

Chemotherapy is by thiabendazole or mebendazole, which reduce inflammation and allow the worm to be removed by slowly winding it on to a matchstick over 4–7 days. Control depends on clean drinking water, by preventing contamination of wells and streams with larvae from ulcers.

Filariasis

This is characterised by acute lymphangitis and lymphadenitis, then by chronic lymphatic obstruction. The causative organisms are *Wuchereria bancrofti*, *Brugia malayi* and *B. timori* (p. 80). Clinically, infection may be asymptomatic despite microfilaraemia, or there may be recurrent brief attacks of acute lymphatic inflammation including fever, headache and, for example, epididymitis.

Chronic lymphadenopathy may remain, then chronic lymphatic obstruction develops causing lymphoedema, hydrocoele and eventually elephantiasis (thick fissured elephant-like skin). Clinical diagnosis may be confirmed by finding microfilaria in blood films, taken at midnight except in South Pacific infection. However blood films are often negative even in active infection. Chemotherapy is unsatisfactory, as diethylcarbamazine kills microfilariae but often causes inflammation around adult worms. Control is by avoiding mosquito bites.

Cutaneous and mucocutaneous leishmaniasis

Animal hosts for these zoonoses (p. 76) include rodents and canines, while sandflies are the vectors. A simple classification is:

- **'Old World' disease** (single, multiple or diffuse) cutaneous infection caused by *Leishmania major* or *L. tropica*, occurring in the Middle East (Baghdad boil), India (Delhi boil), Africa and the Mediterranean (Bouton de Crete). Clinically either single or multiple papules ulcerate with a hard base or horn, then slowly heal (Fig. 3). In diffuse disease, many non-ulcerating papules appear.
- **'New World' disease** (single, multiple, diffuse) cutaneous or mucocutaneous infection caused by the *L. mexicana* complex or the *L. braziliensis* complex in Central and South America (Espundia, Chiclero's ear). Single, multiple and diffuse disease again show papules, ulceration and slow healing, though fungating and keloidal forms occur. Any of these may be followed by disfiguring upper lip and nasal mucosal destruction (tapir nose), progressing to buccal, oral and laryngeal involvement, aspiration pneumonia and death.
- **Post kala-azar dermal leishmaniasis** caused by *L. donovani*. In the Indian subcontinent and Africa a few patients show cutaneous depigmentation, sometimes with nodules, weeks, months or years after kala-azar (pp. 76, 130).

Clinical diagnosis of leishmaniasis is often simple but should be confirmed by Giemsa stains on tissue or ulcer scrapings, not surface swabs. The leishmanin test may assist, but serology is not yet reliable. Chemotherapy is specialised, based on pentavalent antimony compounds. Amphotericin B or pentamidine are second-line drugs. Control is difficult, depending on sandfly avoidance.

Onchocerciasis and loiasis

See eye infections on pages 80–1 and 99.

Hydatid cysts

These may be found in soft tissue anywhere in the body (Fig. 4) (p. 83).

Tropical and rare fungal and parasitic skin and soft tissue infections

- Many different fungi and parasites cause skin disease.
- The clinical presentation often suggests the causative parasite but rarely suggests the causative fungus, except in sporotrichosis.
- Confirmatory diagnosis is necessary, often by histopathology with specific tissue stains.
- Chemotherapy is specialised and often unsatisfactory. Surgery is limited to excision of localised fungal disease.
- Control depends on avoiding, protecting or treating the source (soil, water or human), or avoiding the insect vector.

OSTEOMYELITIS

Osteomyelitis is infection of bone and bone marrow. Osteitis means infection of cortical bone only, an uncommon situation. Infection can be direct from a focus of infection (e.g. a compound fracture) or from circulating infection. Osteomyelitis is classified by numerous criteria; classification here is by acuity plus a specific feature:

* acute
  — acute haematogenous osteomyelitis
  — acute contiguous-focus osteomyelitis
  — acute ischaemic-neuropathic osteomyelitis
* subacute
  — fungal osteomyelitis (rare)
  — prosthetic device-related osteomyelitis
  — vertebral body osteomyelitis
* chronic
  — pyogenic ('bacterial') osteomyelitis
  — actinomycosis (rare)
  — brucellosis (rare)
  — tuberculous osteomyelitis

Acute osteomyelitis and chronic pyogenic osteomyelitis are described here. All others are described on pages 180–1.

ACUTE HAEMATOGENOUS OSTEOMYELITIS

Characteristic of this condition is infection in the metaphysis of a long bone, because of slow blood flow in the unusual sinusoidal venous system, poor collateral circulation and few phagocytes. Infection is by bacteraemic spread from a primary source, often in skin but frequently inapparent.

Causative organisms

These are usually *Staph. aureus*, sometimes streptococci or Gram-negative rods; *Haemophilius influenzae* is more common in children.

Clinical features

Disease occurs in five clinical settings: neonates (p. 187) children, adults, sickle cell anaemia, and intravenous drug users. Fever, severe deep 'bone' pain and acute local bony tenderness are hallmarks; limited mobility of the limb is usual, while swelling, except in the fingers or in neonates, is late. Systemic symptoms of bacteraemia – high fever, sweats, rigors – are more common in children and help

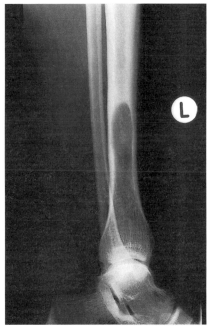

Fig. 1 **Extensive Brodie's abscess in lower third of tibia.**

to distinguish osteomyelitis from a sickle cell crisis. Intravenous drug users often have infected injection sites.

Complications include:

* local destruction: causing dead bone called a sequestrum
* local spread in children: usually laterally because of the avascular epiphyseal plate, causing subperiosteal pus and, later, enveloping new-bone formation called an involucrum
* local spread in adults: usually through the vascular epiphysis into the adjacent joint, causing bacterial 'septic' arthritis
* distant spread: causing bacteraemia, septicaemia or sometimes endocarditis, especially in intravenous drug users
* chronicity: causing a late, localised Brodie's abscess (Fig. 1) or more widespread chronic osteomyelitis, see below.

Confirmatory tests

Blood cultures and bone scan are important, for X-ray examination is usually negative initially. White cell count (WCC), neutrophil count, erythrocyte sedimentation rate (ESR) and C-reactive protein (CRP) are all elevated. Ultrasound detects fluid in adjacent soft tissues or joints. Microbiologic examination of this fluid or aspirated pus from bone is diagnostic.

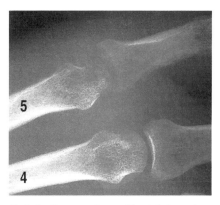

Fig. 2 **Contiguous osteomyelitis of phalangeal head of fifth finger from infective septic arthritis.** Compare with normal joint and bone of fourth finger.

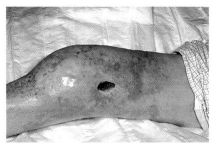

Fig. 3 **Chronic osteomyelitis.** Infection in femur of 68 years duration, showing scarlet swelling, scarring, sinus and stiffness (ankylosed knee).

Open biopsy with microscopy and culture is used when treatment response is poor or diagnosis is uncertain.

Management

Treatment is initially empiric, with flucloxacillin, plus gentamicin if a Gram-negative rod is suspected (e.g. in drug users) or with ceftriaxone in children over 6 months of age. Specific therapy is given after positive culture and sensitivity testing.

As antibiotics penetrate relatively poorly into bone:

* bactericidal drugs are preferred
* initial therapy is high-dose and intravenous for 2 weeks in adults, or 4–7 days in children
* subsequent oral therapy is high-dose, well-supervised, carefully-monitored and continued for 4 weeks in children and 6 weeks in adults for total treatment time.

Pus must be drained, and sequestra removed surgically.

Control and prevention depend on early diagnosis and treatment of primary sources of bacteraemia, and education about safe injection equipment and techniques in intravenous drug users.

ACUTE CONTIGUOUS-FOCUS OSTEOMYELITIS

Characteristic of acute contiguous infection is spread from an adjacent focus. The four major causes are:

- infective arthritis (Fig. 2)
- postoperative infections (open fractures, bone surgery)
- soft tissue infections (skin wounds, pressure ulcers, sinus mucosa in sinusitis)
- puncture wounds (feet, animal bites).

Osteitis (infection of cortical bone only), which is otherwise an uncommon situation, occurs initially because the infection begins *outside* the bone.

Causative organisms

Infections are often mixed and include:

- *S. aureus* and other staphylococci, especially postoperatively
- enteric Gram-negative rods, especially in mandible, pelvis and small bones
- *Pseudomonas aeruginosa*, especially in punctures through smelly 'sneakers' to the calcaneum
- *P. multocida*, especially after animal bites
- anaerobes, especially in facial, pelvic or sacral osteomyelitis, and in human-bite infections.

Clinical features

Symptoms are usually acute, with fever, pain, soft tissue swelling, redness and tenderness. Pus may form in any wound. The features may be subacute in sacral or skull infections, or in deep postoperative infections with less virulent pathogens like coagulase-negative staphylococci.

Confirmatory tests

These include culture of the adjacent focus and blood, bone scans, and X-ray, which is more likely to be positive as the history is longer. Sinus cultures are often misleading because of contaminating surface organisms.

Management

The choice of antibiotic depends on adequate deep cultures; antibiotics according to sensitivity tests are needed for at least 4 weeks.

Pus must be drained, and necrotic tissue or sequestra removed surgically.

Control and prevention depend on early diagnosis and treatment of infections adjacent to bone.

ACUTE ISCHAEMIC-NEUROPATHIC OSTEOMYELITIS

Characteristically, ischaemic-neuropathic osteomyelitis is usually found in diabetics as part of diabetic foot infections (p. 171), resulting from the impaired host response, neuropathic damage and vascular disease. Causative organisms are usually mixed, with staphylococci, streptococci and anaerobes common, and enteric Gram-negative rod colonisation.

Clinical features are predominantly local, with pain, swelling, redness, cellulitis, ulceration and pus. Impaired sensation and poor circulation are common, as are gangrene and offensive odour. Confirmatory tests are appropriate cultures of pus and any available tissue, for surface swabs are often misleading. Bacteraemia is uncommon, and the usual technetium bone scans and X-ray studies are difficult to interpret because of ischaemia, neuropathy, trauma and cellulitis. Indium and gallium scans can be more helpful.

Chemotherapy is directed by sensitivity tests on deep specimens. Surgery should be conservative whenever possible, but pus must be drained, and necrotic tissue or sequestra removed surgically. Control and prevention depend on educating the at-risk patient with diabetes, neuropathy and/or ischaemia, plus early treatment of local infections.

CHRONIC OSTEOMYELITIS (PYOGENIC)

Chronic osteomyelitis is characterized by bone infection lasting months or years. It is caused by the usual pyogenic bacteria such as staphylococci. Tuberculosis and other special chronic or subacute forms are described on pages 180–1.

Chronic pyogenic osteomyelitis sometimes presents as a late, localised Brodie's abscess (Fig. 1) without a history of acute osteomyelitis, or more usually as chronic osteomyelitis following unresolved acute osteomyelitis, of any of the above three types.

Clinical features

Symptoms are usually mild pain with swelling, which may be scarlet, with sinuses, scarring and stiffness if infection spreads to the adjacent joint (Fig. 3).

Confirmatory tests

Tests should include bone biopsy for the definitive cause, as secondary colonising organisms are usual, and misleading, in sinus and surface swabs. X-ray and CT scan define the extent and can show patchy destruction, sequestrum formation, a cloaca (hole in bone leading to a skin sinus), periosteal new bone formation and overall thickening with sclerosis. Blood tests (ESR and CRP) help assess progress during treatment.

Management

Treatment is usually with flucloxacillin, often plus rifampicin, for at least 6 months and, at times, for years. Surgery is needed to drain pus and remove devitalised and dead bone, both at the onset and during chemotherapy. Hyperbaric oxygen therapy and vascularised muscle flaps can improve the modest cure rate. Control and prevention depend on cure of acute osteomyelitis.

Osteomyelitis

- Osteomyelitis is infection of bone and bone marrow; it can be acute, subacute or chronic and caused by local or haematogenous spread.
- Acute haematogenous disease occurs in the metaphysis of a long bone as a result of bacteraemia; it causes fever and severe bone pain.
- Acute contiguous-focus disease is initiated by spread from an adjacent focus; it usually causes pain, fever and soft tissue swelling.
- Acute ischaemic-neuropathic disease is usually found in diabetics, and presents with pain, ulceration and cellulitis.
- Chronic pyogenic osteomyelitis lasts months or years and is usually caused by staphylococcal infection. It presents with mild pain and swelling, and often a sinus.
- Blood and local cultures and bone scans direct treatment, which is long-term antibiotics and may involve surgery for drainage and/or excision.

SPECIAL, TROPICAL OR RARE BONE INFECTIONS

Acute and chronic pyogenic osteomyelitis are discussed on pages 178–9. Here subacute and other chronic forms of the disease are discussed.

SUBACUTE OSTEOMYELITIS

DEVICE-RELATED OSTEOMYELITIS

Bone infection can present weeks or months after insertion of an orthopaedic device. It can be classified into infection around fixateurs, nails, pins, plates, rods, screws and other temporary devices which can later be removed, or around prosthetic joints, intended to be permanent. Infections are usually with *S. aureus* or coagulase-negative staphylococci: either can be methicillin sensitive or resistant. Streptococci, Gram-negative rods or other bacteria are rarely found.

Clinical features

Symptoms are mainly of wound breakdown and purulent discharge (Fig. 1). Fever is uncommon, and pain occurs late.

Confirmatory tests

Early samples of the wound discharge for microscopy and culture are reliable but these become misleading with time owing to secondary colonising bacteria; deep cultures at bone level are then needed. X-ray visualised changes are late; bone scans are unhelpful because of the recent procedure. Erythrocyte sedimentation rate (ESR) is elevated and of some use in following progress.

Management

Chemotherapy depends on the organism isolated. Mixed infections are uncommon. Surgery is needed to drain pus and remove necrotic tissue. In infections around temporary devices, the device is removed when union occurs or the site is stable, and antibiotics then continued for 6 weeks or longer. Infected prosthetic joints are a specialised problem: in brief, infection cannot be cured while the prosthesis is in situ; antibiotics may suppress the infection, but in time the prosthesis must be removed and either replaced or arthroplasty or arthrodesis done.

Control and prevention depend on rigorous operative asepsis and appropriate chemoprophylaxis.

VERTEBRAL BODY OSTEOMYELITIS

Infection can occur in one or more vertebral bodies. The route is usually blood borne but sometimes occurs directly from operation or procedure (see also vertebral disc infections, p. 183). Infection is usually with *S. aureus*, rarely with Gram-negative rods including *Salmonella* and *Pseudomonas* spp. *Candida* spp. can infect in drug addicts. Chronic spinal tuberculosis and brucellosis are described below.

Clinical features

Clinical features are persistent back pain, muscle spasm, referred pain (e.g. sciatica or hip pain) and later spinal cord compression

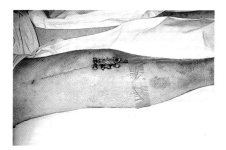

Fig. 1 **Infected total knee prosthesis, with discharging wound (disregard the snail-like suture dressings!).**

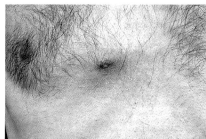

Fig. 2 **Tuberculous sinus from rib.**

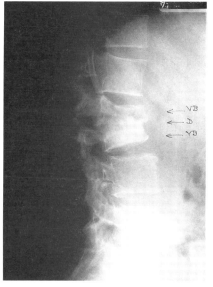

Fig. 3 **TB spine: radiograph showing typical disc destruction and wedge-shaped collapse of adjacent vertebral bodies.**

with paraparesis or paraplegia (weakness or paralysis from the waist down). Fever is common with *S. aureus*, less prominent with other organisms.

Confirmatory tests

Tests are imaging by CT or MRI (see Fig. 4, p. 183) to identify the extent of disease, and microscopy and culture of vertebral body samples taken by needle aspiration or open operation to define the pathogen and exclude alternative diagnoses, including tumour.

Chemotherapy and control

Chemotherapy depends on the pathogen and sensitivity tests (e.g. flucloxacillin for sensitive staphylococci). Control and prevention depend on treatment of any potential primary source, and procedural asepsis.

FUNGAL OSTEOMYELITIS

Characteristics of this rare infection are chronicity, osteolytic ('bone dissolving') lesions and an overlying 'cold' abscess, i.e. it resembles tuberculosis. The route is usually blood borne during disseminated fungal disease; rarely it occurs by direct inoculation in sporotrichosis. Immunosuppression by drugs or disease, and intravenous drug use are predisposing factors. Causative organisms are usually *Candida* spp., but osteomyelitis has been seen in systemic mycoses, aspergillosis, sporotrichosis and zygomycosis.

Clinical features are those of the underlying disease, plus fever, chronic bone pain, bony tenderness and failure to respond to anti-bacterial drugs. A cold abscess overlying superficial bones is a valuable clue. Infection can spread to adjacent joints.

Confirmatory tests are fungal blood cultures, bone scan, X-ray and CT scan with diagnostic aspiration for microscopy, special stains and fungal culture. Serology assists with systemic mycoses.

Management. Chemotherapy is classically with amphotericin B, but the azoles are becoming more prominent as experience grows. Control and prevention depend on early treatment of local and systemic fungal disease.

CHRONIC OSTEOMYELITIS

ACTINOMYCOSIS

Actinomycosis is a very rare chronic bone infection from dental, sinus or lung actinomycosis. The causative organism is usually *A. israelii*, rarely other species (p. 56).

Clinical features are sinus formation, sulphur granules in the pus, scarring and

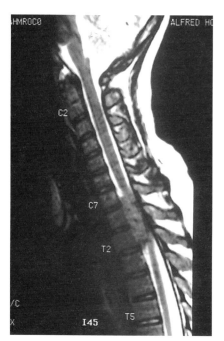

Fig. 4 **TB spine on MRI, showing cord dysfunction at T1–2.**

slow progression across tissue planes. Pain and fever are not prominent.

Confirmatory tests are imaging by X-ray, bone scan or CT as needed; sulphur granules from the pus and bone specimens are taken for microscopy (branching Gram-positive filaments) and anaerobic cultures for at least 10 days. Mixed 'associate' bacteria are not uncommon.

Chemotherapy is high-dose long-term penicillin, initially intravenous for 4–6 weeks, then oral for 6–12 months. Rarely other antibiotics are needed.

Control and prevention depend on treatment of potential primary sources.

BRUCELLOSIS

Brucellosis is a very rare infection in bone, with vertebral osteomyelitis following disc space infection during chronic localised brucellosis. The causative organism is usually *B. melitensis* (p. 50).

Clinical features are back pain, fever, malaise and lethargy. Splenomegaly and hepatomegaly if present are useful clues. Confirmatory tests are bone scan, CT and aspiration for microscopy and special prolonged culture in CO_2. Serology is helpful if positive but can be negative during active bone infection.

Management. Chemotherapy is usually with streptomycin or gentamicin for 2 weeks plus tetracycline for 4 weeks or more. Rifampicin can be used with tetracycline in the second 2 weeks. Cotrimoxazole is an alternative to tetracycline. Surgery is rarely needed. Control and prevention is by animal brucellosis eradication programmes and prompt effective treatment of acute brucellosis.

TUBERCULOSIS

Tuberculosis may affect the spine and joints. Characteristics are chronicity, bone and joint destruction, spread with adjacent

'cold' abscesses (i.e. without palpable heat, and minimal visible inflammation) and TB elsewhere, often pulmonary or renal. The route is usually blood borne, sometimes lymphatic and rarely directly from a caseating node. The causative organism is usually *M. tuberculosis*, occasionally *M. bovis* from unpasteurised milk; rarely an 'atypical' mycobacterium can be involved (pp. 54–5).

Clinical features

Clinical symptoms are relatively late and include pain, deformity or swelling, with little fever or systemic symptoms. A sinus develops from superficial bones (Fig. 2). Spinal TB often starts in the disc and spreads to adjacent vertebral bodies (Fig. 3); it can cause referred pain (e.g. sciatica or hip pain), and later spinal cord compression with paraparesis or paraplegia (weakness or paralysis from the waist down).

Confirmatory tests

- Mantoux, ESR (often useful in following progress) and FBE
- bone, other tissue, sputum and/or urine mycobacterial stains and specific culture and sensitivity tests
- chest X-rays often show the primary focus, which may now be inactive
- bone X-rays can show destruction and collapse (Fig. 3) and adjacent abscesses, while CT shows further detail, and MRI particularly shows disc, bone and CNS involvement (Fig. 4).

Management

Chemotherapy is triple therapy including isoniazid and rifampicin for at least 12 months. Surgery is needed to drain pus, remove necrotic tissue and correct deformity. If expert surgery is unavailable, simpler means can be effective (Fig. 5). Control and prevention depends on BCG, and effective treatment of all active early TB.

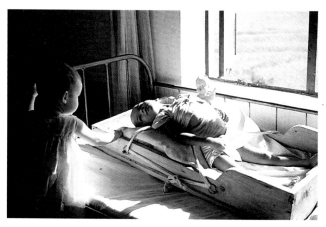

Fig. 5 **TB spine showing effective correction by simple means in Bangladesh.**

Special, tropical or rare bone infections

- Subacute bone infection can occur after insertion of an orthopaedic device and in vertebral bodies. *Staph. aureus* is commonly involved.
- Fungal infection is a rare cause of subacute or chronic osteomyelitis, usually following disseminated disease in the immunocompromised.
- Actinomycosis and brucellosis may both rarely spread to infect bone.
- Tuberculosis of bone can occur secondarily to TB elsewhere; it causes bone destruction and 'cold' abscesses.

JOINT INFECTIONS

INFECTIVE ARTHRITIS

Infection in a joint is termed infective arthritis. Like osteomyelitis, because of the relatively avascular epiphyseal plate, it has different features in neonates and children from those in adults. It is classified clinically into:

- acute, usually from pyogenic bacteria, or viruses
- chronic, usually tuberculous (Fig. 1), spirochaetal or fungal.

The route is usually blood spread from a primary focus: occasionally it can be caused directly by trauma, surgery or animal or human bites.

Predisposing and risk factors are:

- existing joint damage, especially rheumatoid arthritis or gout
- immunosuppression, by disease or drugs
- intravenous drug use
- joint surgery: infected prosthetic joints, where the principal problem is osteomyelitis at the prosthesis–bone interface (p. 180).

Causative organisms

The likely pathogen causing joint infection varies with the age and condition of the patient (Table 1).

Clinical features

A red, hot, swollen, tender, immobile joint is usual except in neonates, the immunosuppressed and in deep joints like the hip. A primary focus may be found in areas such as the skin, heart or genitalia. Skin rash is common in disseminated gonococcal, staphylococcal and streptococcal infections, in early Lyme disease (erythema chronicum migrans) and in viral infections. Adjacent tenosynovitis strongly suggests gonorrhoea in the 15–40 age group, otherwise rheumatoid arthritis, gout or trauma may be responsible.

A postinfectious 'reactive' arthritis occurs after gut infections with some *Campylobacter, Shigella, Salmonella* and *Yersinia* spp. infections in HLA B27-positive patients. This is an immunologically mediated effect and the pathogen is not present in the joint. A similar condition after urethritis, often with uveitis, is Reiter's syndrome.

In rheumatoid arthritis or gout, the differentiation between supervening acute bacterial arthritis or an acute exacerbation of the underlying disease is often difficult and requires careful investigation.

Confirmatory tests

First, *the site of infection* is located by clinical examination and ultrasound. Plain X-ray visualisation initially may be normal (Fig. 2a) or only show soft-tissue swelling and at times a distended joint; it does exclude other possible diagnoses including fractures. Bone scans are unnecessary early unless preceding osteomyelitis is suspected, or clinical examination is equivocal (Fig. 2b).

Secondly, *the pathogen is identified* by blood cultures and joint aspiration (Fig. 3) with microscopy and culture of the joint fluid. If Gram stain shows no pathogen, joint fluid examination

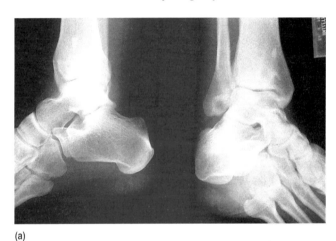

(a)

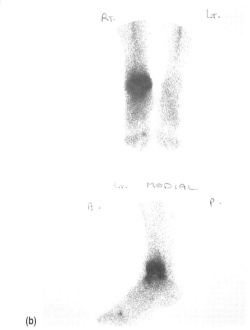

(b)

Fig. 2 **Infective arthritis of the ankle. (a)** Normal X-radiograph; **(b)** abnormal 'hot' bone scan (same patient, same day).

Table 1 **Pathogens involved in joint infections**

	Pathogen
Neonates	
Hospital-acquired	*S. aureus, Candida* spp., Gram-negative rods
Community-acquired	Group B streptococci or gonococci (90% of cases)
Children	*S. aureus, H. influenzae,* streptococci (70% of cases)
Adults (15–40 years)	Gonococci (90% of cases)
Adults (40+ years)	*S. aureus* (70% of cases)
Less common	Mycobacteria, spirochaetes, fungi, viruses

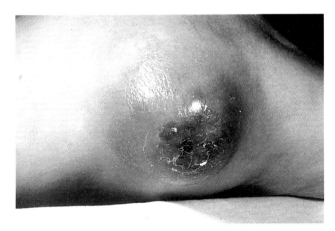

Fig. 1 **Tuberculosis of elbow joint, with overlying 'cold abscess'.**

Table 2 Joint fluid examination

Characteristic	Normal	Bacterial infection	Inflammatory[a]	Non-inflammatory[b]
Clarity	Clear	Opaque	Cloudy	Transparent
Colour	Clear	Purulent	Light yellow	Light yellow
WCC/mm³	<200	>100 000	2000–80 000	200–2000
Neutrophils	<25%	>75%	50–75%	<25%
Gram stain	Negative	Often positive	Negative	Negative
Culture	Negative	Often positive	Negative	Negative
Glucose	Same as in blood	Less than in blood	Much less than in blood	Same as in blood

[a]Inflammatory: rheumatoid arthritis, gout/pseudogout, Reiter's syndrome, rheumatic fever.
[b]Non-inflammatory: osteoarthritis, trauma, neuropathy

Table 3 Gram stain guidance for initial antibiotics in infective arthritis

Gram stain	Likely organisms	First choice	Alternatives
Gram-positive cocci	Staphylococci or streptococci	Flucloxacillin	Cephalothin
Gram-negative cocci	Gonococci	Ceftriaxone	Ciprofloxacin, Spectinomycin
Gram-negative cocco-bacilli	H. influenzae	Ceftriaxone	Amoxycillin–clavulanate
Gram-negative rods	E. coli or other enterics	Gentamicin	Ceftriaxone
No organism:			
Neonate	Various (see Table 1)	Ceftriaxone ± gentamicin	Ceftriaxone + flucloxacillin
Child	S. aureus, H. influenzae	Ceftriaxone	Amoxycillin–clavulanate
Adult to age 40	Gonococci	Ceftriaxone	Ciprofloxacin
Older adult	S. aureus, Gram-negative rods if risk factors	Flucloxacillin (± gentamicin if risk factors)	Ceftriaxone (± tobramycin if risk factors)

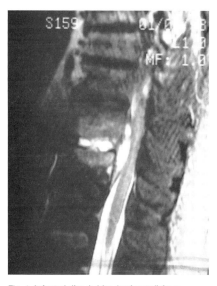

Fig. 4 **Infected disc (white, horizontal) from discography with contiguous vertebral osteomyelitis and intraspinal abscess (white vertical), shown by MRI.**

is still helpful (Table 2). Sometimes open operation is needed, for example in deep joints like the hip.

Management

Therapy has three principles:

- hydration and nutrition of the patient
- surgery, either arthroscopic or open, to remove necrotic tissue, drain pus and lavage the joint
- initial empiric antibiotics, guided by the Gram stain if bacteria are seen (Table 3), then with specific antibiotics directed by culture and sensitivity tests.

Control and prevention depends on early diagnosis and treatment of possible primary foci.

INTERVERTEBRAL DISC INFECTIONS

Infection within the vertebral discs differ from other joint infections in that there is no joint space and there is a poor blood supply in adults. Infection is either blood borne (especially in children) or directly as a result of surgery or discography (especially in adults). Spread is to the vertebral end plates, vertebral body (p. 180) and paraspinal abscesses.

Causative organisms

Pathogens are usually *S. aureus* (blood borne), coagulase-negative staphylococci after procedures, *M. tuberculosis*, Gram-negative rods or, rarely, fungi.

Clinical features

Symptoms include back pain, limited mobility and fever.

Confirmatory tests

Imaging confirms disc involvement and needle aspiration provides samples for microscopy and culture. While bone scans and CT are useful, MRI is most precise in showing spinal canal or cord involvement (Fig. 4). Open operation to provide culture material is necessary if aspiration fails.

Chemotherapy and control

Chemotherapy is guided by vertebral disc culture and is usually flucloxacillin for sensitive staphylococci, vancomycin for methicillin-resistant staphylococci, ceftriaxone or gentamicin for Gram-negative rods or anti-TB therapy. Control and prevention is by treatment of potential primary sources, and procedural asepsis.

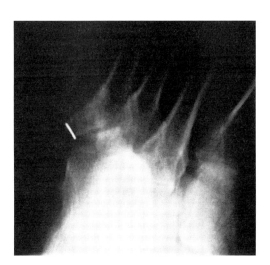

Fig. 3 **Gonococcal arthritis, showing marker for aspiration under X-ray control.**

Joint infections

- Infection of joints occurs usually from blood-borne pathogens but also directly through surgery or trauma.
- Likely pathogens vary with age of patient and the existence of predisposing risk factors.
- Infective arthritis usually produces a hot swollen tender joint; antibiotics are initially empirical then guided by sensitivity tests.
- Intervertebral disc infections cause pain, fever and limited mobility. They result from blood-borne infections or surgical procedures.

ZOONOSES

Zoonoses are infectious diseases transmitted between vertebrate animals and humans in which the human infection is incidental and does not form part of the pathogen's life cycle. Common sources of zoonoses are domestic animals, both farm animals and pets, and sufferers, therefore, are often farmers, veterinarians, abattoir workers and pet-owners; research workers and those visiting or living in areas with infected wild animals may also be at risk (Table 1).

The clinical syndromes often follow logically from the method of infection:

- arthropod bites: often cause multiorgan or systemic disease
- animals bites: local or systemic disease (Fig. 1)
- through wounds or abrasions: local or systemic disease
- ingestion: gut or systemic disease
- inhalation: pulmonary or systemic disease.

Six selected zoonoses are described.

BRUCELLOSIS

Brucellosis is caused by *Brucella* spp. (p. 50), which infect humans as a result of contact with infected animals and their products (particularly milk and cheese).

- *B. abortus*: cattle
- *B. melitensis*: goats and sheep
- *B. suis*: pigs and their products
- *B. canis*: dogs.

Animal infection is often asymptomatic, affecting particularly erythritol-rich tissues like uterus, placenta and breast, which explains the incidence of animal abortion, the risk to abattoir workers of air-borne infection from the placenta, and the high infectivity of milk and cheese. Human infection is usually by ingestion but can be directly through the skin or by inhalation.

Clinical syndromes

Acute brucellosis may be localised and suppurative, or systemic with multi-organ involvement and waves of fever (undulant fever). *B. abortus* and *B. canis* cause mild undulant fever, rarely suppuration; *B. suis* causes severe, suppurative, chronic disease, while the commonest, *B. melitensis*, causes acute, severe, complicated disease. Brucellae are intracellular pathogens of the cells of the reticulo-endothelial system. The inaccessibility of the pathogen explains why the disease

Table 1 **Five modes of transmission**

Transmission	Animal reservoir	Mechanism	Disease
Insect bite	Rodents	Flea bite	Plague
Animal bite	Dog	Direct bite	*Capnocytophaga canimorsus* infection
Directly through skin	Goats, sheep	Touching wool, skin	Anthrax
Ingestion	Cows, goats, etc.	Milk, cheese	Brucellosis
Inhalation	Birds	Droppings	Psittacosis

persists, relapses are common and chemotherapy is often ineffective.

Subacute brucellosis causes intermittent fever with systemic symptoms, myalgia, arthralgia, cough and depression.

Chronic brucellosis causes similar symptoms and in addition, can localise to various tissues, including bone (p. 181).

Confirmatory tests

Special prolonged blood culture and serology are the most important.

Chemotherapy and control

Treatment is usually with streptomycin or gentamicin for 2 weeks plus tetracycline for 4 weeks or more. Rifampicin can be used with tetracycline in the second 2 weeks. Cotrimoxazole is an alternative to tetracycline. Surgery is rarely needed for localised disease. Control and prevention are by animal brucellosis eradication programmes, and prompt effective treatment of acute brucellosis.

LEPTOSPIROSIS

Leptospirosis is characterised by human infection with *Leptospira interrogans* (p. 53) from animals, usually livestock or rodents, with asymptomatic chronically infected kidneys. Infection is by direct animal contact, or indirect contact with animal urine on vegetation (sugar cane) or in soil or water. *L. interrogans* has more than 200 serovars, previously called separate species. Humans are usually infected by serovar *canicola* from dogs, *icterohaemorrhagiae* from rats, *hardjo* from cattle, or *pomona* from pigs.

Clinical syndromes

Leptospirosis has four clinical syndromes:

- asymptomatic infection detected only by serology
- acute flu-like illness with fever and myalgia
- acute 'aseptic' meningitis (p. 87)
- Weil's disease (leptospirosis ictero-haemorrhagica) with fever, jaundice, haemorrhage and hepato-renal failure (Fig. 2).

The first three syndromes are common, but Weil's disease is rare.

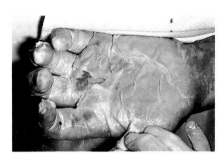

Fig. 1 **Zoonotic infection with *Capnocytophaga canimorsus* through a dog bite, with fatal outcome (p. 171).**

Confirmatory tests

Culture of blood or CSF in the first 10 days, then urine for some weeks is positive only on special media that is unavailable in most laboratories. Serology by the Microscopic Agglutination Test (MAT) is usually diagnostic.

Chemotherapy and control

Penicillin or tetracycline probably shorten the course of the self-limiting flu-like or meningitic forms and improve prognosis in Weil's disease. Control and prevention are by rodent control and prevention of occupational exposure.

PLAGUE

Plague is caused by *Yersinia pestis* (p. 51) which is transmitted animal to animal and animal to human by flea bites; *Y. pestis* exists in two epidemiological forms:

- sylvatic plague: wild mammals including rodents
- urban plague: rats (and humans).

Three clinical forms of plague are recognised:

- **bubonic plague:** infection is via the skin from an infected flea, causing swollen regional lymph nodes called buboes. This form is not readily transmitted person to person.
- **septicaemic plague:** fever, prostration and rapid death
- **pneumonic plague:** caused by spread of the pathogen to the lungs in bubonic plague; this form is highly infectious person to person by inhalation.

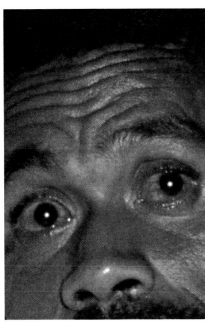

Fig. 2 **Leptospirosis: Weil's disease with jaundice and conjunctivitis.**

Clinical syndromes

Bubonic plague is more common, with fever, rash, swollen lymph nodes (called buboes), prostration from septicaemia, and mortality up to 75% if untreated. Terminal pneumonia from the septicaemia can infect others by inhalation, causing pneumonic plague with overwhelming pneumonia fatal in 90% if untreated.

Confirmatory tests

Gram's or Wayson's stain of bubo aspirate show typical bipolar staining of Gram-negative rods (like safety pins) in over 50%, and blood cultures are positive in two-thirds (bubonic plague) to nearly 100% in pneumonic plague and septicaemia.

Chemotherapy and control

Either streptomycin or tetracycline is effective. Control and prevention are by rodent and flea control, avoidance of wild animals in endemic areas, and by vaccine in specific groups including laboratory workers.

TOXOPLASMOSIS

Toxoplasmosis is characterised by human infections which are severe if congenital, usually asymptomatic in childhood, and systemic or localised in adults. It is caused by *Toxoplasma gondii* (p. 74), which has its sexual life cycle in felines, predominantly domestic cats. Oocysts in cat faeces are the usual source, either directly via contaminated soil, or indirectly through cysts in the raw or undercooked meat (chicken, pork, beef, mutton) of another animal that has ingested the oocytes.

Clinical syndromes

There are five clinical syndromes:

- congenital infection of the fetus, usually from a primary maternal infection during pregnancy (p. 186)
- asymptomatic infection, usually in childhood
- primary systemic infection, giving a 'glandular fever' syndrome
- localised tissue cysts, especially in retina or brain (p. 100 and p. 91)
- re-activation of asymptomatic infection in the immunocompromised, e.g. cerebral toxoplasmosis in AIDS or transplant patients.

Confirmatory tests

Serology is the usual diagnostic method, with CT scans of brain in cerebral infections.

Chemotherapy and control

Pyrimethamine plus sulphadiazine (or clindamycin if intolerant) is the usual treatment for symptomatic disease. Control and prevention depend on avoiding cats and cat faeces in pregnancy, avoiding raw or undercooked meat, and chemoprophylaxis in the sero-positive immunocompromised patient.

TRICHINOSIS

Trichinosis is characterised by the deposition in human striated muscle of larvae of *Trichinella spiralis* (p. 81) from ingested infected meat, usually pork. *T. spiralis* is a nematode parasite of carnivores, including pigs and bears. Humans are a dead-end host, infected by eating raw or undercooked meat containing encysted larvae.

Clinical syndromes

These vary from abdominal distress, fever and 'flu' to severe myalgia and periorbital oedema. Rarely, myocarditis, pneumonitis and encephalitis are fatal.

Confirmatory tests

Clinical findings of eosinophilia, myalgia and periorbital oedema, with an appropriate food history, gives a presumptive diagnosis. Cysts can be found in the infected meat. Definitive diagnosis by human muscle biopsy is seldom necessary.

Chemotherapy and control

Mebendazole or thiabendazole is used, with steroids to control marked allergic symptoms.

Control and prevention depend on freezing meat at −40°C, and proper cooking (not by microwave). Pigs or bears should not eat garbage, which may contain infected meat.

TULARAEMIA

Tularaemia is caused by *Francisella tularensis* (p. 51); infection is transmitted at least four ways to humans: directly through the skin or conjunctiva when handling infected mammals, by tick bite, by ingestion of infected meat, or by inhalation, especially in laboratories. *F. tularensis* is found in many reservoir hosts, including mammals (rabbits, rodents) and ticks, and in water. Hunters, trappers and campers in tick areas are, therefore, at risk.

Clinical syndromes

Clinical features depend on the portal of entry:

- ulceroglandular, with local ulcer at the tick-bite site, and regional lymphadenopathy (inguinal, axillary), in 80%
- typhoidal with systemic disease and pneumonia, from ingestion or inhalation, in 10%
- oculoglandular, oropharyngeal and glandular (without ulceration) are all rare.

Confirmatory tests

Microscopy is difficult and culture is dangerous, so serology is the method of choice.

Chemotherapy and control

Streptomycin is the usual chemotherapy, or gentamicin. Control and prevention depend on avoidance of ticks and infected or sick animals, especially rabbits. Gloves minimise skin contact.

Zoonoses

- Infections of humans by animal pathogens are called zoonoses.
- Infection occurs through animal bites, insect bites, ingestion, inhalation and contamination of skin injuries.
- Brucellosis is caught from infected animals and their products (milk, cheese).
- Brucellae are intracellular pathogens which cause acute and chronic disease.
- Plague is transmitted from rats by flea bites; the pneumonic form is infectious by inhalation.
- Toxoplasmosis is a common latent infection from cats; congenital disease is often fatal.

INFECTIONS OF MOTHER, FETUS AND NEWBORN CHILD

Infections of the mother and baby can be classified as occurring

- before birth: prenatal/antenatal in mother, prenatal/congenital in fetus
- during birth: puerperal or perinatal
- after birth: postnatal in mother, neonatal in newborn baby.

MATERNAL INFECTIONS

Prenatal infections

The hormonal, immunological and anatomical changes of pregnancy make maternal infections during pregnancy:

- more common, e.g. urinary tract infections (p. 148)
- more obvious, when latent viral infections re-activate, e.g. herpes simplex, CMV
- more severe, e.g. candidiasis, UTI, listeriosis, malaria, viral hepatitis, polio and influenza
- more serious, because of the effect on the fetus, see below.

Three new tissues appear which can be infected: the fetus, the placenta with membranes, and the lactating breast.

Perinatal infections

Puerperal sepsis is characterized by bacterial endometritis acquired in childbirth. Causative organisms now are usually mixed aerobic and anaerobic bowel flora, rather than the fearsome *Streptococcus pyogenes* or *C. perfringens* of former years. The route is ascending, and predisposing factors are premature rupture of the membranes, prolonged labour, instrumentation and retained placenta or membrane fragments.

Clinical features. Clinically there is fever, offensive lochia (the vaginal blood and fluids postdelivery) and systemic upset, even septicaemia. Lower abdominal pain and a tender uterus develop.

Confirmatory tests are Gram's stain and cultures of high vaginal swabs and of blood.

Management. Broad-spectrum antibiotics are required, such as ampicillin or ticarcillin with clavulanate, a later cephalosporin, or clindamycin with gentamicin. Penicillin is given for *S. pyogenes*. Surgery is necessary for myonecrosis or abscesses. Control and prevention depend on chemoprophylaxis during prolonged rupture of the membranes, on hand-washing and asepsis.

Postnatal breast infections

Breast infections are characterised by local inflammation and fever. They may occur soon after delivery from epidemic or endemic hospital-acquired staphylococci, or weeks later as sporadic mastitis from a combination of milk stasis and infection through cracked nipples.

Causative organisms are usually staphylococci, rarely beta-haemolytic streptococci or bowel flora.

Clinical features are local pain and tenderness, segmental redness and fever. Initial milk stasis progresses to cellulitis then abscess, with extreme localised tenderness and high fever.

Confirmatory tests are microscopy and culture of pus from the nipple, from aspiration or from operation. Ultrasound is very helpful in abscess identification (Fig. 1).

Management. Chemotherapy is usually with (flu)cloxacillin or a first-generation cephalosporin. Drainage is essential for an abscess. Control and prevention depend on nipple hygiene and massage to prevent milk stasis.

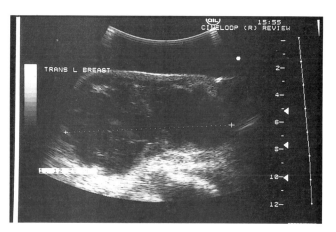

Fig. 1 **Breast abscess shown on ultrasound.**

FETAL AND NEONATAL INFECTIONS

Prenatal (congenital)

Any severe maternal infection in pregnancy, especially with septicaemia can infect the fetus, usually killing it. Numerous viruses also cause fetal infections (chiefly CMV, hepatitis B, HIV, HSV, rubella and VZV). Congenital infection (usually not fatal to the fetus) can be caused by four unusual intracellular organisms: a bacterium (congenital listeriosis), a mycobacterium (leprosy), a spirochaete (syphilis) and a parasite (toxoplasmosis).

Congenital listeriosis. *Listeria monocytogenes*, a beta-haemolytic Gram-positive rod (p. 32) causes a zoonosis from many animals including cattle, and infection is often from unpasteurised milk or cheese. Clinically, maternal infection is often asymptomatic or like mild influenza, but it is bacteraemic, and fetal infection is severe, causing death or premature birth, often with neonatal septicaemia, pneumonia and abscesses. Confirmatory tests are blood, CSF, skin or abscess cultures. Chemotherapy is by penicillin or ampicillin. Control and prevention depend on animal and food hygiene and avoiding products made from unpasteurised milk.

Congenital leprosy. *Mycobacterium leprae* can spread to the fetus during bacteraemia in maternal lepromatous leprosy. Clinically it is rare, but the child is born with leprosy (p. 92) and needs specialist treatment. Control depends on treatment of adults.

Congenital syphilis. *Treponema pallidum* infection of the fetus occurs in areas where routine antenatal serological screening for syphilis is not done. Clinically an infant with congenital syphilis characteristically has rhinitis and a depressed nasal bridge, with other cartilage and bone defects, skin rash and mucosal lesions, lymphadenopathy and hepatosplenomegaly. Subsequent dentition shows notched incisors called Hutchinson's teeth. Confirmatory tests are positive serology in the mother and *T. pallidum*-specific IgM in the baby. Chemotherapy is with penicillin for mother and child. Control and prevention depend on antenatal screening and therapy before the fourth month.

Congenital toxoplasmosis. *Toxoplasma gondii* (Fig. 2) causes a zoonosis principally from cats (pp. 72, 185). Clinically, maternal infection is usually primary and asymptomatic (like listeriosis), but the fetus may die or become severely affected (often months after birth), with convulsions, chorioretinitis, cerebral atrophy and hepatosplenomegaly, with consequent microcephaly,

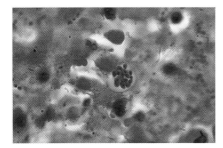

Fig. 2 **Toxoplasmosis showing tissue cysts in the brain.**

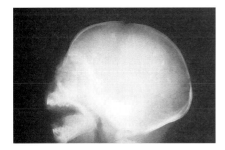

Fig. 4 **Congenital toxoplasmosis showing microcephaly and scattered calcification on skull radiograph.**

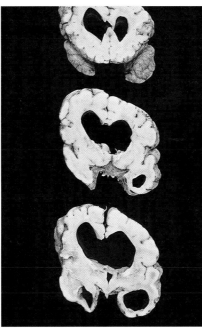

Fig. 3 **Congenital toxoplasmosis showing massive ventricles in hydrocephalus.**

massive ventricles (Fig. 3) and mental retardation. Confirmatory tests are toxoplasma-specific IgM in cord blood, and positive maternal serology. Skull X-ray may show microcephaly and calcification (Fig. 4). Chemotherapy is with sulphonamide and pyrimethamine. Control and prevention are by avoiding cats and undercooked meat (with tissue cysts) during pregnancy.

Perinatal infections

Perinatal infections are acquired by the neonate during labour (mainly during prolonged rupture of the membranes) or during travel down an infected birth canal.

Causative organisms are:

- from maternal bacteraemia or viraemia (hepatitis B, HIV)
- bowel flora from the rectum
- specific pathogens from cervix or vagina, including group B streptococci, *N. gonorrhoeae*, *C. trachomatis*, *C. albicans* or viruses (including HSV and papilloma virus).

Clinical features vary with the pathogens:

- maternal bacteraemia is usually fatal to the baby
- inhaled bowel flora or group B streptococci cause severe pneumonia, septicaemia, meningitis and death
- gonococcal ophthalmia neonatorum ('eye infection of the newborn') is an acute purulent conjunctivitis and blepharitis, serious if untreated

- chlamydial ophthalmia neonatorum is less severe than gonorrhoeal; inhaled chlamydia cause a severe neonatal pneumonia
- neonatal candidial infection is mainly oral or perianal thrush

Confirmatory tests include local eye, skin, blood, CSF and endotracheal microscopy and cultures as appropriate, with chest X-ray examination for pneumonia.

Chemotherapy depends on the cause and clinical features: penicillin or ceftriaxone for gonorrhoea; empiric broad-spectrum cover such as ceftriaxone and gentamicin for pneumonia, septicaemia and/or meningitis; nystatin or an azole for candidial infection.

Control and prevention depend on prophylactic antibiotics for labours involving prolonged rupture of the membranes and at-risk babies, with aseptic technique for delivery and neonatal care.

Postnatal (neonatal) infections

Pneumonia, septicaemia, meningitis. Characteristically, these most serious neonatal infections commence at or before birth (see above) but may be postnatal. They are caused by inhaled bowel or vaginal flora or other pathogens; neonatal pneumonia (p. 112) and septicaemia can disseminate causing meningitis, infective arthritis (p. 182) or osteomyelitis (p. 178). Clinically, early signs are often non-specific with listlessness, pallor, poor feeding and diarrhoea, while fever may be absent, and specific signs too late. Confirmatory tests include local eye, skin, blood, CSF and endotracheal microscopy and cultures as appropriate, with chest X-ray for pneumonia. Chemotherapy is initial empiric broad-spectrum cover such as ceftriaxone and gentamicin, then as directed by sensitivity tests. Control and prevention depend on prophylactic antibiotics for prolonged rupture of the membranes and at-risk babies, with aseptic technique for delivery and neonatal care.

Umbilical stump infections. These are caused by unsterile dressings, including soil and manure; neonatal tetanus is usually fatal. Prevention is by maternal immunisation during pregnancy and the use of sterile dressings, which also prevent staphylococcal and other umbilical infections.

Cross-infection from hospital staff or other babies. Cross-infection can involve staphylococci, streptococci, bowel flora and viruses, including herpes simplex. Skin, eye and gut infections result. Control is by good infection control, particularly hand disinfection.

Infection from mother. Sometimes mothers infect their babies postnatally with viruses via maternal milk, blood or saliva

Oral pathogens. Babies can ingest pathogens, usually in unsterile feeds; this occurs chiefly in developing areas when babies are not breast fed, and causes diarrhoeal disease (pp. 136–141).

Infections of mother, fetus and newborn child

- Infections can be more severe or re-activate during pregnancy.
- A few non-viral infections can pass to the fetus causing congenital disease: listeriosis, leprosy, syphilis, toxoplasmosis. The fetus may die or may suffer serious postnatal illness/malformation.
- Puerperal sepsis is bacterial endometritis acquired in childbirth, which is largely prevented by asepsis and chemoprophylaxis.
- Neonates can be infected during labour, particularly if the birth canal is infected (e.g in gonorrhoea), or may become infected after delivery (from the mother, hospital staff, equipment or food).

TRAVELLERS AND RECENT IMMIGRANTS

Travel often tilts the balance against host defences and in favour of microbial attack when unprepared host defences face new attacks from:

- unfamiliar microbes, e.g. parasites, fungi
- unfamiliar vectors, e.g. mosquitoes, ticks, bed-bugs
- unfamiliar vehicles, e.g. foods, drinks
- unfamiliar activities, e.g. casual sex, water sports.

The common result, therefore, is unfamiliar illness in travellers and recently returned travellers; illness unfamiliar in industrial countries occurs in recent immigrants, particularly from developing countries. Some of these infections can be avoided by preparing host defences (e.g. immunisation). Once infection has occurred, diagnosis and treatment must follow.

PREPARING HOST DEFENCES

Malaria prophylaxis
Travellers to areas with endemic malaria should institute two measures, neither of which is 100% effective. First, the traveller should avoid mosquito bites. This involves screens to windows and doors, repellents containing DEET (diethyl toluamide), protective clothing and minimising outdoor after-dark activities, when most malarious mosquitoes bite.

Chemoprophylaxis is the second measure. Chloroquine weekly is used where chloroquine resistant *P. falciparum* is unknown, while mefloquine weekly is currently recommended for areas with chloroquine resistance. If mefloquine is contraindicated, doxycycline daily, or daily proguanil plus weekly chloroquine may be used. As resistance is spreading, up-to-date advice must be sought before travel to a particular area.

Food and drink precautions
'Peel it, boil it, or forget it' are the general principles to minimise gut infection from food or drink contaminated with human faecal pathogens, either directly from the unwashed hands of infected food-handlers or from contaminated soil. Specifically, the traveller should avoid uncooked meat, salads, vegetables, fruit, dairy products (including ice-cream), unboiled or bottled water (even for teeth-brushing), uncarbonated soft drink, ice, iced tea or coffee. The following are usually safe: cooked food while hot, boiled water, carbonated drinks, alcohol, hot tea or coffee.

Sexually transmitted disease precautions
There are four principles:

- abstinence is absolutely safe
- condoms are compulsory in casual sex.
- prostitutes are perilous (over 50% are infected with HIV in some areas)
- barrier contraceptives in women are *no* barrier to pathogens.

Skin, respiratory, ear and eye precautions
Bathing, swimming, diving and boating all give water exposure and the risk of skin injury. Hazards include water-borne skin infections (aeromonads, group A streptococci, *M. marinum*, schistosomiasis, vibrios), ear infections (especially pseudomonal), eye infections (including conjunctivitis) and gut infections (pp. 136–41). Respiratory infections especially viral are common in travellers, but few precautions are useful, except avoidance of the obviously infected.

Immunisation
There are three categories of immunisation:

- routine for all travellers: diphtheria, tetanus and measles vaccination should be updated, plus polio and BCG for travel or residence in rural developing areas, and rubella for women of child-bearing age
- required for entry to some countries: cholera or yellow fever
- special depending on destination, duration and details of visit: cholera, hepatitis A and B, Japanese B encephalitis, meningococcal, plague, rabies, typhoid and yellow fever vaccines.

Check recent recommendations well before travel to allow multiple doses.

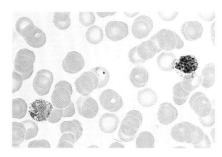

Fig. 1 **Blood film of *P. vivax* showing two schizonts and (centrally) one ring form.**

Special precautions
Infants and young children, the pregnant traveller, the immunocompromised traveller and those with chronic disease all need specialised advice.

Personal medical kit
The need and contents will depend on the destinations and duration of travel and may include insect repellent and anti-malarials, water purifier, rehydration salts, anti-diarrhoeal and anti-emetic tablets, condoms, anti-infectives, both systemic and local for skin, eye and ears, and sunscreens and drugs for non-infectious hazards.

DIAGNOSING AND TREATING INFECTION

Illness in returned travellers and recent immigrants should always be considered in the light of the involved geographic areas. Fever characteristics can also be indicative (Table 1).

Malaria and other fevers
Malaria must be diagnosed or excluded immediately, as delay may be fatal. Enteric fever (typhoid and paratyphoid) and viral haemorrhagic fevers are the other major infections needing rapid diagnosis and infection control. Causes are many; the commonest are shown in Table 1 (see also pp. 130–5). Clinically,

Table 1 **Fever in travellers**

	Acute fevers (within 3 weeks)	Chronic fevers (after 3 weeks)
Major causes	Malaria, typhoid and paratyphoid fevers Viral infections including dengue, hepatitis A, HIV, yellow fever, viral haemorrhagic fevers	Malaria, typhoid and paratyphoid fevers, amoebic liver abscess, filariasis, kala azar, tuberculosis, viral hepatitis, HIV
Less common causes	Typhus, urinary tract infections, prostatitis, rickettsial infections, African trypanosomiasis, brucellosis	Sexually transmitted disease (syphilis), endocarditis, brucellosis, non-infectious causes (e.g. collagen diseases, drug fever, lymphoma)

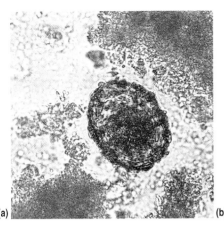 (a)
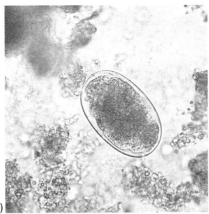 (b)

Fig. 2 **Stool microscopy showing (a)** *Ascaris lumbricoides* **egg and (b) hookworm egg.**

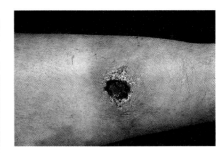

Fig. 3 **Tropical ulcer.**

the pattern of fever may give some clue to the cause:

- tertian: every third day (i.e. 48 hourly) in falciparum or vivax malaria is classical, but fever is often daily early in infection
- persistent, always above normal: in typhoid and other bacteraemias
- intermittent, with afebrile hours: in kala azar, TB and abscesses
- relapsing, with afebrile days: in borrelioses, brucellosis, malaria, trypanosomiasis and viral infections like dengue.

A full clinical examination can show other helpful features including:

- rash: rare in typhoid, common in typhus and viral infections
- tender or enlarged liver: amoebic abscess, hepatitis, kala azar
- tender or enlarged spleen: malaria, typhoid, kala azar, brucellosis
- jaundice: hepatitis, yellow fever
- bleeding: haemorrhagic fevers
- chest signs: typhoid, TB.

Confirmatory tests must include thick and thin blood films for malaria (Fig. 1 and p. 131); if negative, do full blood analysis with differential white blood cell count and film, blood, urine and stool cultures, liver function tests, chest X-ray and imaging of any mass or unexplained tenderness. Serology is more useful in chronic fevers as two specimens two or more weeks apart are usually needed.

Chemotherapy obviously depends on the cause.

Diarrhoeal disease

There are many causes of diarrhoeal disease (bacterial and amoebic diarrhoea; Table 1, p. 138) including intestinal parasitic infestation (Table 1, p. 140). Traveller's diarrhoea, especially from enterotoxigenic *E. coli* (ETEC), is

commonest during travel, while *Campylobacter*, *Salmonella*, *Shigella* and amoebic infection are more commonly found in returnees and immigrants. The characteristics of the symptoms are also indicative.

- preformed toxin: short incubation, vomiting prominent
- endotoxin made in gut: medium incubation, marked diarrhoea, no fever
- invasive pathogen: longer incubation, fever, pain, blood in faeces.

Other indicators are

- possible exposure: developing country and particular area, shellfish, homosexual contact, immunocompromised patient (especially HIV infected)
- symptoms and signs: fever, abdominal pain, blood or mucus in the stools, and the character of the diarrhoea (acute, chronic, watery, foul).

Confirmatory tests are initially stool microscopy for cysts, ova and parasites (Fig. 2) plus stool culture. If negative, special stains, serology and other tests may be needed.

Chemotherapy depends on the cause. If none is found, symptomatic treatment

may suffice, or empiric therapy with, for example, norfloxacin or metronidazole.

Sexually transmitted diseases

These usually present as genital discharge (pp. 154–7) or genital ulceration (pp. 162–3) and are investigated and treated in the usual ways. HIV infection is outside the scope of this book.

Skin infections

There are a number of skin infections, particularly fungal and parasitic infections as well as bacterial (pp. 174–5) that are tropical in origin (p. 176). These may cause local disease or disseminated systemic symptoms. 'Tropical ulcer' is a progressive, offensive, destructive ulcer (Fig. 3), at times following skin trauma, in which anaerobes play an important role. It is usually responsive to penicillin and/or metronidazole

Eye and ear infections

These are described on pages 96, 100–101.

Respiratory infections

Tuberculosis and tropical respiratory infections are described on pages 118–21.

Immigrants

Recent immigrants may import many infections, which may be acute or chronic, as above, or asymptomatic. Specific screening programmes are usually aimed at the major infections in their previous homeland.

Travellers and recent immigrants

- Travel precautions against microbial attack begin before travel, including immunisations, prophylactic drugs and education.
- Travel precautions continue during travel, including prophylactic drugs and behaviour modification towards food, drink, leisure and sex.
- Travel-acquired infections may present after return, chiefly with fever (malaria, enteric fevers and haemorrhagic fevers), diarrhoea, sexually transmitted disease, or skin, eye, ear or respiratory infection.
- All need skilled diagnosis and treatment.

HOSPITAL-ACQUIRED INFECTIONS

Hospital-acquired infection, also called nosocomial infection, is characterised as any infection acquired in hospital. Community-acquired infections incubating on admission to hospital are excluded, though they may subsequently cause hospital infection in another patient. The principles are applicable to all health-care institutions including private medical practices. Recently, greater attention has been paid to the risks of hospital-acquired infections in staff as well as in patients.

Host–microbial relationships and the establishment of disease are covered on pages 14–19; host defences are covered on pages 20–25.

CLASSIFICATION

Nosocomial infections can be classified by four sequential steps:

- reservoirs and sources of infection
- routes of transmission
- rupture of our host defences
- resultant major infections.

Reservoirs and sources of infection

Infectious sources are either endogenous (from the patient's own normal flora, pp. 16–17) or exogenous. The latter derive from people (hospital staff or another patient), usually directly, or from the environment in hospital, usually by indirect spread from sources such as air, dust, linen, food, water and other fluids (including disinfectants (!) and intravenous fluids) or equipment such as endoscopes and ventilators. Some of these can be a *reservoir* (e.g., infected fluid) which contaminates sources (bottles or bowls), so eradication of the source is insufficient and the reservoir must also be found.

Route of transmission

Spread of infection can occur by:

- direct contact: person-to-person *cross-infection*, usually via the hands (sexually transmitted infection is rare, although the author has seen one youth with gonorrhoea after 7 weeks in hospital, in traction!)
- indirect contact: via some object, as above
- common vehicle: a special form of indirect contact in which one infected vehicle, commonly food or fluid, infects many
- airborne spread: either by *droplets* travelling 1–2 metres, or smaller *droplet nuclei* travelling a kilometre or more
- vector borne spread: unusual but not unknown in hospitals.

Rupture of host defences

The portals of entry are the urogenital tract, skin, respiratory tract and the mouth and the gastrointestinal tract. High-risk patients have impaired host defences from:

- age: the very young and the elderly
- immune defects from disease: e.g. HIV, diabetes, hepatic or renal impairment, cancer or lymphoma, especially with neutropenia
- immune defects from drugs: e.g. cytotoxics, steroids or immunosuppressives (in transplant patients)
- immune antibody defects: lack of vaccination or previous exposure (chicken-pox, CMV, hepatitis B)
- organ defects: e.g. pre-existing urinary, lung or skin disease
- skin and tissue defects: trauma, surgery and treatment; multitrauma and/or multiaccess results in multirisk (Fig. 1).

Resultant major infections

The major four types of infection are:

- urinary tract infections (about 40%)
- surgical wound infections (about 25%) (Fig. 2)
- lower respiratory infections (about 10%)
- bacteraemias (about 5%).

Skin ulcer, pressure sore (Fig. 1, p. 168), gut and other infections make up the remaining 20%.

CAUSATIVE ORGANISMS

The major pathogens are *S. aureus*, *E. coli* and other enteric Gram-negative rods and *P. aeruginosa*, though their relative importance differs in the major infections (Table 1). Less important pathogens are enterococci, coagulase-negative staphylococci, anaerobes and *Candida* spp. Multiresistance is common now and important.

CLINICAL FEATURES

Urinary tract infections. Predisposing factors are intermittent or indwelling catheters, operative procedures or partial obstruction. Route of infection is usually ascending, with the usual uropathogens (p. 148). Symptoms and signs are often only cloudy offensive urine: fever is often absent in the older patient and dysuria is noted only when there is no catheter.

Table 1 **Major pathogens in hospital-acquired infections**

Category/classification	Causative organism (%)	Chemotherapy
Urinary	*E. coli* (40)	Gentamicin or directed
	Enteric GNRs (25)	by sensitivity tests
	Enterococci (15)	Ampicillin
	P. aeruginosa (10)	Tobramycin
Surgical wound	*S. aureus* (20)	Flucloxacillin or directed
	Enteric GNRs (15)	by sensitivity tests
	E. coli (12)	Gentamicin
	Enterococci (12)	Ampicillin
Pneumonia/lower respiratory	Enteric GNRs (35)	Timentin
	S. aureus (15)	Flucloxacillin
	P. aeruginosa (15)	Tobramycin
	E. coli (10)	Gentamicin
Bacteraemia	Enteric GNRs (20)	Cephalosporin
	S. aureus (12)	Flucloxacillin
	E. coli (12)	Gentamicin
	Coagulase-negative staphylococci (10)	Vancomycin

GNR, Gram-negative rods.

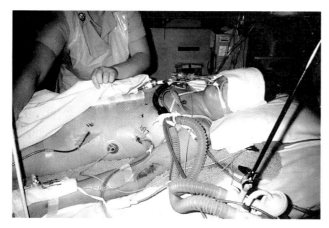

Fig. 1 **Multitrauma and multiaccess gives rise to a multirisk situation.**

Surgical wound infections. Redness, swelling and local pain and heat occur around the wound. This progresses to a purulent discharge from the wound or suture site (p. 170).

Lower respiratory infections (p. 116). Predisposing factors are recumbency, anaesthesia, other pulmonary disease or trauma, and mechanical ventilation. The route of infection is by inhalation. Symptoms and signs vary. In non-ventilated patients, fever and dyspnoea are common. Cough and sputum are less common in ventilator-associated pneumonia (VAP), where increasing difficulty in oxygenation with opacities on chest radiographs may overshadow fever and increasingly purulent tracheal aspirate.

Bacteraemias (p. 128). Predisposing factors are intravascular devices and infection elsewhere, causing bacteraemia or fungaemia. Route of infection is direct contact from imperfect aseptic technique or blood borne from infection elsewhere. Symptoms and signs are fever, then septicaemia. Look for a source elsewhere and signs of infection at the device entry site.

Infected burns. Infected burns may be considered with surgical wounds in that the skin barrier is broken. In burns there are also abnormalities of immune response and loss of fluid. Although initially sterile, burns are quickly colonised by mixed bacterial flora (p. 171).

CONFIRMATORY TESTS

Tests vary with the site, but obviously include local microscopy and culture (urine, wound, sputum, burn, etc.) and blood cultures. Chest X-ray is needed in suspected pneumonia. Sources in, or spread to, other organs need specific imaging.

CHEMOTHERAPY

Antibiotic treatment varies with the site; initial empiric therapy is outlined in Table 1 and is replaced by directed therapy when the infecting organism and sensitivities are known.

THE PRINCIPLES OF INFECTION CONTROL

Sources

Sources of infection are diminished by providing sterile equipment, dressings, fluids and drugs, and clean food, drink, linen and environment. Hospital staff with communicable infections should not have patient contact, while staff with HIV infection, hepatitis B or C need specific review, counselling and exclusion from procedures. Specific staff need hepatitis B, BCG and other immunisation.

Spread

Spread is diminished by:

- building design, air-conditioning and isolation of selected patients to limit airborne spread
- handwashing, 'no-touch' and aseptic techniques, 'blood and body substance precautions' ('universal precautions') and careful placement of infected and susceptible patients in wards to limit contact spread
- educated staff behaviour to prevent contamination of equipment (Fig. 3) to limit indirect spread
- screening in tropical areas to limit spread by vectors.

Handwashing or disinfection before touching an uninfected patient, and after touching an infected patient, are crucial in infection control.

'Blood and body substance precautions' ('universal precautions') mean that all blood and body substances of all patients are considered potentially infectious at all times; this has become necessary because it is not feasible to detect all infectious patients, especially with HIV, hepatitis B and C. Previous 'source isolation' is now needed less for infectious patients, for example respiratory isolation to control airborne spread.

Susceptible patients

Susceptible patients need:

- portal protection
- proper procedural techniques

- prophylactic antibiotics when indicated
- protective isolation for immunocompromised patients.

For all patients, the major infection risks can be avoided by careful management.

Urinary infections. Aseptic technique is required for operations, procedures and catheterisation, closed catheter drainage and aseptic catheter care. Chemoprophylaxis usually fails and is seldom indicated.

Surgical wound infection. Procedural techniques and, when indicated, prophylactic antibiotics reduce infection rates (p. 170).

Respiratory infections. Preanaesthetic assessment, postoperative physiotherapy, and tracheostomy and ventilator care including 'no-touch' during endotracheal aspiration all contribute to avoidance of infection.

Bacteraemia. Aseptic preparation of all fluids and equipment is essential as is care in insertion, additions and changes of intravascular devices and fluids (especially total parenteral nutrition). Potential sources elsewhere must be treated.

Staff

Staff need protection by education in blood and body substance precautions, in care with sharps and in needlestick prevention, and by immunisation when appropriate.

Surveillance

Surveillance helps control by monitoring the incidence of the major four infection types, monitoring the observation of infection control measures, devising protocols for new equipment and procedures and, when any of these fail, investigating outbreaks.

> **Hospital-acquired infections**
>
> - Sources are people (hospital staff or patients) or things (equipment, food, fluids, drugs or equipment).
> - Spread is by direct or indirect contact, common vehicle, airborne spread or vectors.
> - Susceptible patients include the very young and old and those with defects in the immune defences or in organs.
> - Subsequent infections are chiefly urinary, surgical wound, respiratory and bacteraemia.
> - Surveillance and infection control measures limit infection.

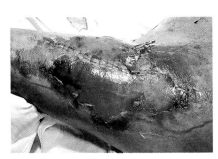

Fig. 2 **Necrotising fasciitis after hip surgery.**

Fig. 3 *P. aeruginosa* growing in an anaesthetic mask.

INFECTIONS IN COMPROMISED PATIENTS

BASIC PRINCIPLES

Immune defences can be divided into two groups: non-specific and specific (pp. 20–3). Non-specific defences (innate, constitutive, congenital) are:

- 1st line: skin and mucous membranes (especially against surface organisms)
- 2nd line: phagocytosis by neutrophils and macrophages, and complement and other chemicals (especially against pyogenic bacteria).

Specific immune defences (inducible, adaptive) are:

- 3rd line: antibodies from B-cells (especially against encapsulated bacteria)
- 4th line: cell-mediated immunity (CMI) from T-cells (especially against intracellular organisms).

Any one (or more) of the four lines can be compromised, either by rare congenital defects (primary immuno-deficiency) or, commonly, by acquired (secondary) defects from disease, drugs, procedures or surgical treatment. Certain organisms are common pathogens in each compromised state. All need swift and specialised investigation and treatment; correction of the underlying defects should be attempted when possible.

Control and prevention (p. 191) depend on asepsis, hygiene, removal of devices, decrease or cessation of causative drugs or radiation, and oral anti-microbials at times of greatest risk. Immunisation is only useful when the 3rd and/or 4th lines (B- and T-cell function) are intact.

The loss of immune defences can occur for a variety of reasons and leaves the individual open to infection by organisms that would not normally be hazardous. Minor infections can become life threatening, and the usual symptoms of infection may be absent, making prompt diagnosis and treatment of infection particularly important.

SPECIFIC DEFECTS

Congenital skin disease

Several rare diseases such as epidermolysis bullosa atrophica cause excessively fragile skin, easily invaded by pyogenic bacteria, including staphylococci and streptococci, which cause severe infections, often fatal even with expert care.

Acquired breaches in the skin or mucous membranes

Acquired breaches are very common and have many causes. All enable surface organisms including staphylococci, streptococci (Fig. 1), *Pseudomonas aeruginosa*, enteric gram-negative rods and fungi such as *Candida* to reach and infect deeper tissues. Resultant infections include:

- burns, surgical and other wound infections (p. 170)
- bacteraemia from intravascular devices (p. 128)
- prosthetic infections (p. 180)
- skin infections (pp. 164–71)
- failure of clearance: urinary tract infections (p. 148), cystic fibrosis (p. 111).

Congenital phagocyte defects

The commonest (but still very rare) phagocyte deficiency is in chronic granulomatous disease, actually a group of disorders characterised by failure of the oxidative burst in neutrophil phagocytosis (p. 20). This results in recurrent pyogenic infections including dermatitis, adenitis, pneumonia, enteritis and colitis, all with characteristic abscesses. *S. aureus* and enteric Gram-negative rods, including salmonellae and *Serratia* spp. are common. Death is usual by age 10.

Acquired phagocyte defects

Acquired phagocyte defects are common, chiefly neutropenia from leukaemia (Fig. 2) or its treatment with cytotoxics, steroids or radiation before bone marrow transplantation. Pyrexia of unknown origin (p. 134), often with septicaemia, is common. Intravascular devices are a common source. Staphylococci and enteric Gram-negative rods are usual early in neutropenia, and fungi infect later. Tests and therapy must be swift and sure, or mortality is high.

Congenital complement defects

Deficiency of any component from C1 to C9 can occur. Because of the alternate pathway, defects in C1, C2 and C4 have little effect on infections, while defects in C3 or C5–8 cause repeated serious infections, usually with encapsulated bacteria such as pneumococci, meningococci, gonococci and *H. influenzae.*

Acquired complement defects

Acquired defects in the complement system are uncommon alone but do occur in combination with other immune defects in chronic liver disease, systemic lupus erythematosis (SLE), some types of nephritis, sickle cell disease and malnutrition (see below). They can cause infections, chiefly with encapsulated bacteria.

Congenital antibody defects

Many rare disorders are known in which either individual classes (IgG, IgA, IgM) of antibodies or all gammaglobulins are defective. They include Bruton's agammaglobulinaemia and 'common variable immunodeficiency'. Patients are prone to infection, often with encapsulated bacteria, hence they often have ear, sinus or chest infections (Fig. 3). Autoimmune disease often co-exists.

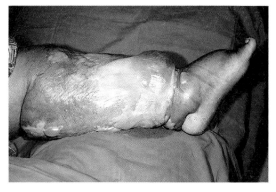

Fig. 1 **Acquired breach in skin during surgery, with resultant group A streptococcal cellulitis.**

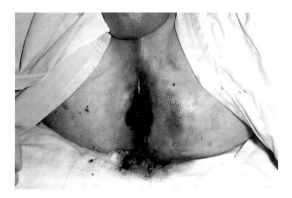

Fig. 2 **Acquired phagocyte defect: neutropenia caused by leukaemia, with resultant perianal spreading suppuration.**

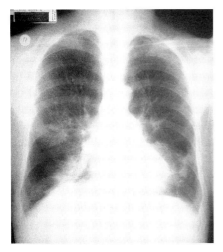
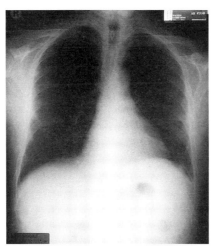

Fig. 3 **Antibodies are absent in agammaglobulinaemia, with resultant chest infection (a), which needed lung transplantation (b).**

Acquired antibody defects

Antibody defects are commonly acquired in lymphoid malignancies such as lymphoma, chronic lymphocytic leukaemia and multiple myeloma, with diminished or abnormal B-cell function and antibody synthesis. Other causes are splenectomy, which removes many B-cells (and diminishes clearance of intravascular organisms); burns, in which immunoglobulins are catabolised; and protein losing enteropathy and nephrotic syndrome, in which immunoglobulins are lost. Infection with encapsulated bacteria or Gram-negative rods is common and can be catastrophic.

Congenital CMI defects

The DiGeorge syndrome of thymic hypoplasia and absent parathyroid glands is one of at least ten rare syndromes with defects in CMI, and consequent life-threatening infections from a wide range of intracellular bacteria (mycobacteria, *Listeria*), fungi, parasites and viruses. In contrast, chronic mucocutaneous candidiasis shows selective failure of T-cell CMI against *C. albicans*, with recurrent and progressive oral, nail and skin infections. Endocrinopathy and auto-antibodies commonly co-exist. Life is unpleasant and short.

Acquired CMI defects

There are four major common causes of acquired CMI defects:

- drugs and radiation
- infections, especially HIV
- malignancies
- malnutrition.

All have other effects on host defences so are discussed below.

Congenital combined defects

Severe combined immunodeficiency (SCID) has both B- and T-cell dysfunction, with consequent serious infections by encapsulated and intracellular bacteria, and fungi, parasites and viruses. Bone marrow transplantation offers hope of survival.

Acquired combined defects

Acquired combined defects are found in six common situations.

Drugs and radiation:

- steroids decrease numbers and function of lymphocytes, monocytes and eosinophils and decrease neutrophil accumulation in inflammation
- cytotoxics cause leucopenia, and impair both T- and B-cell function
- cyclosporin specifically suppresses T-cell function
- irradiation decreases lymphoid cell proliferation.

HIV infection causes major T-cell and CMI deficiency but also impairs B-cell function; neutropenia is common in AIDS

Malignancies cause defects in neutrophil numbers and function, in antibody production and (particularly in Hodgkin's disease and other lymphoid tumours) in CMI. In addition, drugs and irradiation (as above) further impair host defences, and skin and mucosal surfaces are often damaged by disease or drugs.

Malnutrition (p. 27) can affect all host defences: in vitamin deficiencies and protein–energy malnutrition (PEM), skin and mucosal damage occurs, complement synthesis is severely decreased, neutrophil chemotaxis is slow, antibody formation and function are widely impaired, and T-cells are grossly decreased.

Systemic infections (pp. 130–35) particularly TB, leprosy, brucellosis and viruses (including CMV, EBV and HIV) can impair host defences, especially CMI.

Transplantation involves major immunosuppressive drug use (see above) and causes widespread compromise of host defences, and a wide range of resultant infections (Fig. 4).

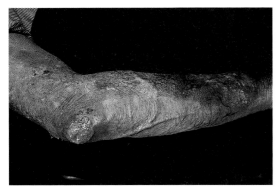

Fig. 4 **All defences breached after heart transplantation, with resultant cryptococcal skin infection.**

> ## Infections in compromised patients
>
> - Defects in any of the four lines of host defences can be congenital or acquired; defects of several lines may be combined.
> - 1st line skin and mucous membrane defects result in infections by surface microbes: pyogenic bacteria, pseudomonads or common fungi.
> - 2nd line neutrophil ('pus cell') defects result in infections with the pyogenic organisms, or Gram-negative rods (GNR).
> - 2nd line complement defects result in infections with encapsulated bacteria, or *S. aureus.*
> - 3rd line antibody defects also result in infections with encapsulated organisms, GNR, or some parasites.
> - 4th line CMI defects result in infections with intracellular bacteria, fungi, parasites and viruses.
> - All need swift and specialised tests and treatment, with correction of the underlying defects(s) when possible.

GENERAL PRACTICE PATIENTS

Most infections in general practice are common, relatively easily diagnosed and treated, and seldom severe: these include otitis media, sinusitis, pharyngitis, gastroenteritis, urinary tract infections and tinea. Conversely in each organ system there is at least one important infection which is superficially similar but serious, which is rare but rapidly progressive, which demands rapid recognition, timely treatment and swift specialist supervision (Table 1). These infections include meningitis with meningococcaemia, where the early rash (Fig. 1) can appear like a common viral infection until rapid progress with gangrene occurs. Similarly, common acute conjunctivitis (Fig. 2) is superficially like iritis but note the localised limbal injection and irregular iris (Fig. 3).

Causative organisms

Most infections in general practice will be caused by staphylococci, streptococci, *H. influenzae*, enteric Gram-negative rods (especially *E. coli*) and fungi, usually dermatophytes or *Candida albicans* (Table 2). Local sensitivity patterns should be obtained from the local laboratory. Penicillin resistance in pneumococci is an increasingly important threat.

Confirmatory tests

These are often unnecessary initially but are needed for recurrent or persistent infection (Table 2).

Chemotherapy

Each GP will be familiar with a small range of antibiotics, including penicillin, flucloxacillin, erythromycin or roxithromycin, amoxycillin (+/– clavulanate), a first-generation cephalosporin, trimethoprim (+/– sulphamethoxazole), and an azole; rarely others are needed.

Control and prevention

The GP can advise the patient, household and contacts on immunisation, hygiene, and early diagnosis and treatment to limit the spread of infections.

Table 1 **Common infections, and superficially similar serious infections**

Organ system	Common infection	Superficially similar, serious infection
CNS	Frontal sinusitis	Bacterial meningitis
Ear	Otitis media	Mastoiditis
Eye	Conjunctivitis	Keratitis, iritis
Respiratory	Sinusitis Pharyngitis, Viral URTI/LRTI Exacerbation COAD	Parameningeal abscess Epiglottitis Pneumococcal pneumonia, TB TB
Systemic	'Viral infection'	Meningococcaemia, endocarditis
Gut	Gastroenteritis	Bacillary dysentery, appendicitis
Urinary	Uncomplicated UTI	Pyonephrosis, abscess
Genital	Vulvo-vaginitis	Pelvic peritonitis
Skin	Tinea Bacterial skin/wound	Nocardiosis Streptococcal cellulitis
Skeletal	Trauma, sprain, gout	Osteomyelitis, infective arthritis

Fig. 1 **Early meningococcal rash.**

Fig. 2 **Acute staphylococcal conjunctivitis is common.**

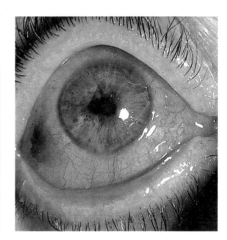

Fig. 3 **Iritis has limited limbal injection, and irregular pupil.**

Table 2 **Common infections in general practice**

Category/ classification	Causative organism	Clinical features	Confirmatory laboratory tests	Chemotherapy
Skin	*S. aureus* *S. pyogenes* Dermatophytes	Boils Impetigo Tinea	None or skin swab M&C	Flucloxacillin Penicillin Azole
Respiratory	*S. pyogenes* *S. pneumoniae* *H. influenzae* (viruses)	Pharyngitis Otitis, sinusitis, bronchitis	Throat swab Nasal M&C? Sputum M&C	Penicillin Amoxycillin, cefaclor, or augmentin
Gut	*Campylobacter, E. coli,* *Shigella* spp. (viruses)	Gastroenteritis	None (stool M&C if persists or traveller)	None initially
Urinary	*E. coli*, other GNRs, *S. saprophyticus*	Cystitis or pyelonephritis	None initially, urine M&C if recurrent	Augmentin or cephalexin or trimethoprim
Genital	*C. albicans, Gardnerella*	Vulvo-vaginitis	None initially, M&C if recurrent	Azole
	C. trachomatis, *N. gonorrhoeae* }	Urethritis, cervicitis	M&C, DIF	Ceftriaxone + doxycycline

M&C, microscopy and culture; DIF, direct immunofluorescence.

> **General practice patients**
>
> - Common infections in general practice are relatively easily diagnosed and treated and seldom severe, but the few important infections which are superficially similar but serious require rapid recognition.
> - Confirmatory tests are often unnecessary initially but are needed for recurrent or persistent infection.

INFECTIONS IN ELDERLY PATIENTS

Infections in elderly patients are more common, more serious and more difficult to treat. This is because of numerous predisposing and perpetuating factors, and because of pathophysiological defects.

Predisposing and perpetuating factors

Impaired defences include:

- skin and mucous membrane defects: collagen aging
- body fluids less bactericidal: gastric, prostatic vulnerability
- antibodies and CMI wane
- clearance mechanisms wane (urinary and respiratory tracts vulnerable).

Anatomic changes include:

- prostatomegaly causes urine stasis in men (Fig. 1)
- cystocoele causes urine stasis in women.

Co-existent disease. Diabetes mellitus, chronic lung disease, heart failure, strokes, mental impairment and fractures all impair structure and function.

Medication. Sedatives, hypnotics, etc. also impair function.

Pathophysiological defects

- fever is often absent or slight in infection
- immunisation, while recommended, gives lower and less prolonged antibody response
- co-existent disease of heart, lungs or kidneys makes complications of infection more likely and more serious
- renal impairment, often silent, alters drug excretion, especially of aminoglycosides, which require dosage modification.

MAJOR INFECTIONS

Urinary tract infections

UTIs (pp. 148–51) are common in the elderly; they are often asymptomatic and usually associated with prostatomegaly and consequent obstruction in men (Fig. 1), or residual bladder urine and stasis in women. Causative organisms are the usual uropathogens, *E. coli* and other enteric bacteria. Clinical features may be absent, or infection may present simply with cloudy or offensive urine. Fever and dysuria are rarer. Confirmatory tests are urine microscopy and culture plus ultrasound to show prostate size and residual urine in recurrent infections. About 10% of renal function is lost each decade from age 30, and over 60% is lost before serum

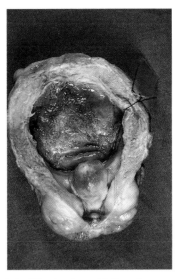

Fig. 1 **Cystitis, showing red, inflamed mucosa and greatly thickened wall, from prostatic obstruction.**

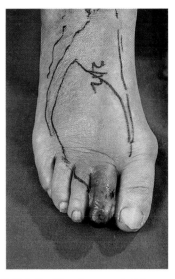

Fig. 2 **Diabetic with early gangrene and progressive cellulitis: inner line is 2nd December, outer line is 3rd December.**

creatinine rises. Chemotherapy is guided by sensitivity testing, usually amoxycillin +/– clavulanate, or cephalexin. If possible causative conditions should be removed; chemoprophylaxis is only used for very frequent recurrences.

Lower respiratory infections

Characteristics of lower respiratory infections (pp. 112–15) include gradual or terminal illness with bronchopneumonia ('the old man's friend') or aspiration pneumonia; less often, acute illness may occur with pneumococcal pneumonia. Causative organisms are the usual respiratory pathogens, chiefly *S. pneumoniae, H. influenzae, C. pneumoniae, M. pneumoniae* and viruses. Enteric Gram-negative rods are commonly found but are less commonly pathogenic, and *M. tuberculosis* must be remembered.

Clinical features vary from mild dyspnoea, cough and mucopurulent sputum with little or no fever to (rarely) high fever, marked cough and haemopurulent sputum. Confirmatory tests are sputum microscopy and culture with chest X-ray, noting that pulmonary opacities may result from collapse, infarction or heart failure rather than infection.

Chemotherapy is usually initially with penicillin or ceftriaxone, but metronidazole for aspiration, doxycycline for 'atypical' organisms, gentamicin for Gram-negative rods, or anti-tuberculous therapy is needed when relevant. Control and prevention depend on removal or treatment of predisposing factors when possible.

Skin and soft tissue infections

Skin and soft tissue infections (pp. 164–73) tend to be progressive because of ischaemia, oedema and/or diabetes. Causative organisms are streptococci, staphylococci, anaerobes including clostridia, and enteric Gram-negative rods. Clinical features are pus, putrid odour and progression (Fig. 2). Confirmatory tests are aerobic and anaerobic examinations of pus. Chemotherapy is with amoxycillin plus clavulanate, with metronidazole or, for example gentamicin, as clinical response and cultures indicate. Control and prevention depend on minimising predisposing factors. Diabetics should have prophylactic antibiotics including penicillin before lower limb surgery.

Infections in elderly patients

- Infections in elderly patients are more common, more serious and more difficult to treat. Complications are common.
- Precipitating and perpetuating factors include impaired defences, anatomic changes, co-existent disease and medications.
- The three commonest infections are in urinary tract, lower respiratory tract and skin/soft tissues.
- Antibiotic dosage often needs modification, especially aminoglycosides, as silent severe renal impairment is common.

STERILISATION AND DISINFECTION

Health requires the absence of infection and hence methods of killing or at least reducing the numbers of pathogenic microorganisms are important factors in reducing the incidence of disease. Like much essential knowledge, it is boring but beneficial.

Sterilisation is the process of killing *all* viable (able to multiply) microbes, including viruses, fungi, bacteria and, in particular, bacterial spores.

Disinfection is killing or removing *most but not all* viable microbes. It is divided into high-level, intermediate and low-level disinfection (Fig. 1). A disinfectant is a substance or process which disinfects.

Antisepsis is disinfection of the skin and uses an antiseptic. Note that skin cannot be sterilised and remain alive.

Asepsis is the absence of microorganisms and infection; hence the aseptic methods in surgery and medical procedures aimed to produce this aseptic state (pp. 170, 191).

Pasteurisation is a gentle disinfection by low-temperature heating.

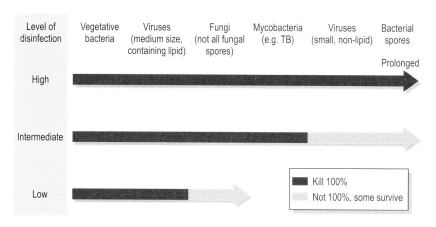

Fig. 1 **Levels of disinfection and microbial resistance.**

Principles

1. 'Clean first, disinfect/sterilise next', i.e. reduce the bio-burden of organisms and organic material first, so the process is easier, shorter and reliable.
2. 'Strong enough for long enough', i.e. rate of kill is proportional to *time* multiplied by *concentration*.
3. 'Horses for courses', i.e. choose the appropriate process, including the appropriate disinfectant, depending on the purpose and degree of cleanliness needed, the materials and risk of damage, the time, the cost and the resources available.
4. 'Control the process, not the product': adhere to a known, proven and tested process rather than look for unsterile or dangerous equipment afterwards.

STERILISATION METHODS

There are four methods of sterilisation:

- irradiation
- filtration
- chemicals
- heat: the most certain and widely used.

Irradiation

Irradiation may be by *ultraviolet light* (less than 330 nm wavelength); this is used in microbiological safety cabinets in laboratories, but eyes must be shielded to prevent damage.

Irradiation is also used as *ionising radiation* by electrons from cobalt-60 or a linear accelerator to sterilise heat-sensitive prepacked single-use plastic items, including syringes and catheters.

Fig. 2 **Filtration of water for microbiologic media.**

Filtration

Filtration was performed with different types of filter including asbestos, earthenware and sintered glass but is now done most often using *nitrocellulose membrane filters* (Fig. 2). It is used for sterilising liquids that heat would damage, including serum and antibiotics. However many filters labelled 'sterilising' only filter bacteria, so the filtered fluid may contain mycoplasmas or viruses.

Chemicals

Formaldehyde (gas), formalin (liquid), glutaraldehyde and ethylene oxide are sporicidal and virucidal and hence can sterilise. They are toxic and irritant so their use is limited.

Formaldehyde was occasionally used to disinfect empty rooms after highly infectious patients.

Formalin is used to fix tissues for safety in laboratories, but not in microbiology laboratories because it prevents growth in cultures.

Glutaraldehyde is used for invasive instruments which cannot otherwise be sterilised, for example bronchoscopes, cystoscopes, anaesthetic equipment and some plastics. It can be used in specially ventilated procedure rooms and operating theatres.

Ethylene oxide needs a special chamber, often called a 'bomb' as the gas is explosive in air! It diffuses well into materials and is used for heat-sensitive articles such as plastic, rubber and complex equipment.

Heat

This can be dry or moist heat. Dry heat is used as:

- red heat: used for metal loops during bacterial cultures
- flaming: for example, lighting microscope slides wetted by methylated spirit
- hot air ovens: 160°C for 60 minutes for scalpels, scissors and substances (oil, grease, wax) impermeable to moist heat
- infrared radiation: seldom used now.

Table 1 **Disinfectants**

Class	Chemical	Advantages	Disadvantages	Uses
Alcohols	Ethyl alcohol, isopropyl alcohol	Quick action, broad spectrum; can combine with others, e.g. chlorhexidine	Damage mucous membranes	Skin and surfaces
Aldehydes	Formalin (liquid, gas[a])	Excellent spectrum	Very irritant	Preserve dead tissues (liquid), room fumigation (gas)
	Glutaraldehyde [Cidex, Aidal]	Excellent spectrum, less irritant	Expensive, needs exhausts	Endoscopes
Diguanides	Chlorhexidine	Broad activity, combines with alcohol etc.	Easily inactivated	Surfaces, mucous membranes, hands
Halogens	Hypochlorites [Milton]	Spectrum includes organic material, inactivates viruses, cheap	Corrodes metal	Blood spills and soiling
	Iodine and iodophores	Sporocidal	Coloured	Skin disinfection
Heavy metals	Mercurics[a] ([Mercurochrome]	Broad spectrum	Tissue and other toxicity	Wound antiseptics (outmoded)
Hexachlorophane and replacements	Hexachlorophane[a], irgasan, triclosan [pHisohex]	Combine with soap, anti-staphylococcal	? Neonatal toxicity, poor killing of GNRs	Skin disinfection
Phenolics	Numerous soluble	Wide activity	Not sporocidal	General purpose on surfaces
	Chloroxylenols[a] [Dettol]	Moderate activity	Poor killing GNRs, easily inactivated	Mild antiseptic
Quaternary ammonium compounds	Benzalkonium[a] [zephiran]	Moderate spectrum, detergent activity	Poor killing GNRs, easily inactivated	Environmental cleaning
	Cetrimide [cetavlon]	Combine with chlorhexidine, detergent activity	Poor killing GNRs, easily activated	Wound cleansing, antiseptics

Trade names in brackets; GNR, Gram-negative rod.
[a] Superseded if better alternatives are available.

Moist heat is achieved by boiling water or autoclaving. Boiling in water for 5, 10 or 20 minutes kills non-sporing microbes but not necessarily all spores. Hence, boiling is unreliable for sterilisation. The varying times quoted are further evidence of its uncertainty.

Autoclaving is the most reliable and efficient sterilising process. It depends on steam under pressure, hence it is hotter than 100°C (e.g. 121°C for 15 minutes or 132°C for 3 minutes). These time–temperature combinations have been chosen because experiments show they kill clostridial and other pathogenic spores.

Special autoclaves are used, often in central sterile supply departments (CSSU), operating theatre suites or laboratories (Fig. 3). The load (instruments, drapes, dressings, etc.) is first cleaned to reduce the bio-burden of organisms, increasing the ease and certainty of sterilisation. The load is then carefully packed to ensure steam can penetrate throughout it, locked in the autoclave, air is exhausted for greater efficiency, then steam under pressure admitted for the correct ('holding') time. Modern autoclaves then dry the load; the pressure must reach atmospheric before the door is opened.

When all steps in each autoclaving cycle are controlled and the correct temperature is held for the correct time, sterility should be assured. In addition, the process is checked, usually by culturing special spore strips of known heat resistance; autoclave tape, which changes colour, is less exact.

DISINFECTION METHODS

Washing removes some microbes from surfaces and so has some disinfectant action by itself. It is made more effective by detergents including soap. In addition, this microbial ('bio-burden') removal makes subsequent disinfection by heat or chemicals more effective, and sterility easier to obtain.

Heat by boiling, as discussed above, disinfects but does not reliably sterilise.

Pasteurisation is a gentle disinfection by low-temperature heating (63°C for 30 minutes or 72°C for 20 seconds) of milk or other liquid foods which would be unpalatable after more vigorous disinfection such as boiling.

Chemicals are the usual disinfectants (Table 1). They are used specially on skin, surfaces and some instruments and supplies which cannot be sterilised.

Disinfectants
The success of a disinfectant depends on:

- the microbial type, state (especially spores) and number
- the surface disinfected (skin, metal, porous materials)
- the environmental pH, temperature and moisture
- the concentration of disinfectant
- inactivation by minerals (hard water) or microbes.

Control of all these variables is difficult: the process recommended by the manufacturer and relevant authorities must be observed. 'In-use' tests by adding known organisms then culturing samples after disinfectant dilution or neutralisation can be done in hospitals, especially in infection control investigations.

Fig. 3 **Autoclave.**

Sterilisation and disinfection

- Sterilisation is the process of killing *all* viable microbes, including their spores, while disinfection is killing or removing *most but not all* viable microbes.
- Four principles are: clean first, disinfect/sterilise next; the method chosen should be the right process for the purpose; it should be controlled; it should be strong enough for long enough.
- The four methods are irradiation or filtration (used for special purposes), chemicals (chiefly for disinfection) and heat (the most certain and widely used for sterilisation).

ANTI-MICROBIALS: GENERAL PROPERTIES

Throughout this book, the importance of control of sources of infection has been stressed; in some instances specific measures are effective, in others general environmental hygiene is required. Once a pathogen has invaded and is causing disease then anti-microbial treatment may be needed to supplement the body's natural defences. Empiric non-scientific medicine produced four important systemic anti-microbial agents before microbes were known to cause infections:

- emetine for amoebic dysentery in China from 500 BC
- quinine in cinchona bark for malaria in Peru before the 16th century
- mercury for syphilis by Paracelsus in Europe in 1530
- potassium iodide for syphilis in 1836.

The scientific development of synthetic anti-microbials began with P. Ehrlich in Germany with methylene blue for malaria in 1891, the organic arsenicals 'atoxyl' (1902), and 'trypan red' (Ehrlich and Shiga, 1904) for trypanosomiasis, and 'salvarsan 606' for syphilis (Ehrlich and Hata, 1909). Atebrin for malarial prophylaxis was made in 1932, and the first effective sulphonamide, prontosil red, was made by G. Domagk in 1935.

The third great phase, antibiotics from living organisms, began with A. Fleming's observation in 1928 of the anti-staphylococcal activity of what he called penicillin, made by an air-borne contaminating fungus, *Penicillium notatum*. Penicillin was developed to clinical use by H. Florey and E. Chain and gave rise to a large group of antibiotics. A second great group of antibiotics, aminoglycosides, was discovered in 1944 by S. Waksman, who identified streptomycin from *Streptomyces griseus*. A third group, cephalosporins, was developed initially from *Cephalosporium acremonium*, found by G. Brotzu in 1945.

CLASSIFICATION

Anti-microbials are classified by the type of pathogen targeted, e.g. anti-bacterials or anti-fungals. This grouping may be further subdivided, e.g. anti-bacterials include urinary antiseptics and anti-mycobacterial (including anti-tuberculous) drugs.

Anti-microbials, especially anti-bacterials, are strictly classified into chemotherapeutic agents, which are synthetic chemicals, and antibiotics, which are produced from living ('bios' = life) organisms, usually fungi. However, 'antibiotic' is often used loosely to mean all anti-bacterials.

Anti-bacterials can be further described by their:

- chemical structure (penicillins, cephalosporins, etc.)
- effect on bacterial growth (bacteriostatic or bactericidal)
- target site (see below).

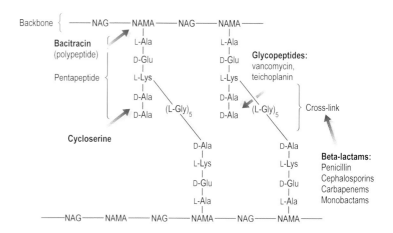

Fig. 1 **The structure of peptidoglycan and the site of action of antibiotics.**
Structure varies slightly among species; this is the structure of *Staph. aureus*.
NAG, *N*-acetylglucosamine; NAMA, *N*-acetylmuramic acid.

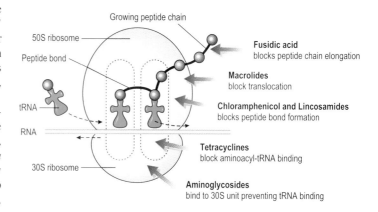

Fig. 2 **Specific site of action of anti-bacterials inhibiting protein synthesis.**

TARGET SITE CLASSES

Cell wall synthesis inhibitors

The cell wall synthesis inhibitors are bactericidal because they block the synthesis of the rigid peptidoglycan component (Fig. 1) of the wall; as a result growing cells lyse and die. They do not affect eucaryotic cells, nor microbes that do not have peptidoglycan, or can prevent the antibiotic reaching their peptidoglycan. These antibiotics are:

- beta-lactams: penicillins, cephalosporins, carbapenems, monobactams
- glycopeptides: vancomycin, teichoplanin
- polypeptide: bactrican
- cycloserine.

Protein synthesis inhibitors

Protein synthesis inhibitors are selective for varying stages of protein synthesis; (Fig. 2); if these are unique to bacteria (e.g. affect the bacterial 70S ribosome rather than the eucaryotic 80S ribosome)

they will be selectively toxic. However, eucaryotic mitochondrial protein synthesis occurs on 70S ribosomes and will be affected. This group includes:

- aminoglycosides: streptomycin, gentamicin, tobramycin, netilmicin, amikacin, spectinomycin, neomycin
- tetracyclines: tetracycline, doxycycline, minocycline
- macrolides: erythromycin, roxithromycin, azithromycin, clarithromycin, spiramycin
- chloramphenicol
- lincosamides: clindamycin, lincomycin
- fusidic acid.

Aminoglycosides cause faulty proteins to form and so are bactericidal. All the others in this group have a reversible action and so protein synthesis begins again when antibiotic levels decrease (bacteriostatic action).

Nucleic acid synthesis inhibitors

Nucleic acids are made by all cells so the possibility of selective agents toxic only for microbes is limited. Some pathways have distinct features that can be targeted or some enzymes are sufficiently different for a selective effect to occur.

- Folic acid synthesis: folic acid is a precursor of purines and pyrimidines which is only synthesised in microbes; humans obtain folic acid as a vitamin in food. Sulphonamides and trimethoprim interfere with the synthesis of folic acid
- RNA polymerase: inhibited by rifamycins (rifampicin, rifabutin)
- DNA structure: disrupted by nitro-imidazoles (e.g. metronidazole).
- topo-isomerase: blocked by quinolones (norfloxacin, ciprofloxacin)

Cell membrane function inhibitors
Drugs that destroy the selective permeability of membranes will kill both microbial and human cells. As a result, they will be relatively toxic if given systemically. Colistin acts like a detergent, disrupting the cell membrane phospholipid. The polyene anti-fungal drugs (e.g. amphotericin B and nystatin) act by damaging sterols in eucaryotic membranes; they are particularly toxic to fungi through their action on ergosterol but also affect human cells.

Uncertain target
The target of some anti-mycobacterial drugs is uncertain: isoniazid may act on mycolic acid synthesis, which would explain its specific activity, while ethambutol may inhibit RNA synthesis.

CHARACTERISTICS
Physicochemical properties
These are important in relation to the effectiveness and mode of administration of a drug, particularly whether they are stable to gastric acid and are absorbed from the gut and hence can be given orally; if unstable or not absorbed, they need injection. Other important factors are whether the drug will cross barriers within the body: into cells, into the brain across the blood–brain barrier, into other protected tissues like prostate, or into cysts.

Spectrum of activity
The spectrum of activity is the range of organisms against which an anti-microbial is usually active. The minimal inhibitory concentration (MIC) is the smallest concentration of antimicrobial which is bacteriostatic, inhibiting bacterial growth; this is reversible, so regrowth occurs if the antimicrobial is removed by excretion or inactivation. By contrast, the minimal bactericidal (or fungicidal) concentration

(MBC, MFC) is the smallest concentration irreversibly killing the microbes, so they do not regrow if the anti-microbial is removed. To be effective an anti-microbial must be able to access the site of the bacteria and be one to which the bacteria are sensitive.

Mechanisms of resistance
Some bacteria are *innately* resistant to certain antibiotics because they lack a target site or are impermeable to the antibiotic; other bacteria *acquire* resistance, by one of three mechanisms.

1. Altered target site. Changes in the target site may result in lower affinity for the antibiotic, or additional target enzymes may emerge that are unaffected by the drug.
2. Altered uptake. Effective drug concentration in the bacterial cell can be decreased either by decreasing permeability or by actively pumping the drug out of the bacterial cell.
3. Antibiotic-inactivating enzymes. These occur particularly against penicillins, cephalosporins and aminoglycosides.

Resistance spreads between bacteria in three genetic ways:

1. Chromosomal mutation, usually random, causes an altered protein, e.g. a ribosomal protein (streptomycin resistance), or altered enzyme (sulphonamides). Selection by the antibiotic after each cell division will result in a resistant population.
2. Transmissible plasmids are small circular DNA units replicating independently of the chromosome and transferred between cells. They have four advantages over chromosomal mutation: transfer between bacteria is more rapid than cell division, resistance to up to six antibiotics can

be carried at once, resistance to several classes of antibiotic can be carried at once, and one class of plasmid can enter numerous genera, e.g. TEM-1 beta-lactamase in enteric Gram-negative rods, in *N. gonorrhoeae* and in *H. influenzae*.
3. Transposons, called 'jumping genes', move from the security of the chromosome to the mobility of a plasmid, and from one plasmid to another.
4. Transformation (the direct transfer of DNA) and transduction (by phages) are less important.

Pharmacokinetics
The pharmacokinetics of a drug describe its behaviour in the body: absorption, distribution, protein binding, serum and tissue concentrations, serum half-life, metabolism and excretion. Important factors that will alter the effective 'life' of the drug (and its toxicity) include the age of the patient, concurrent diseases, particularly of organs which metabolise or excrete the drug (usually liver and kidney), genetic factors (slow and fast drug metabolism) and interaction with other drugs.

Side effects, toxicity
Even safe effective antibiotics like penicillins fall short of Ehrlich's ideals of a 'magic bullet' which would not affect humans yet would eradicate germs by a single 'dosa sterilisa magna' (great sterilising dose). Side effects and toxicity are often similar within a group of antibiotics, e.g. all aminoglycosides are ototoxic and nephrotoxic but vary in degree.

Clinical use
The indications, dosage and routes used for each antibiotic follow from *all* the above properties (see pp. 200–5).

Anti-microbials: general properties
- Anti-microbials are classified by the type of pathogen targeted, the disease for which they are effective and by the mechanism of action.
- True antibiotics are the products of living organisms, although the term is used loosely to include synthetic and semi-synthetic chemicals.
- Anti-bacterials can act on cell wall, protein and nucleic acid synthesis, on the cell membrane or as anti-metabolites blocking metabolic pathways.
- Antibiotic resistance may be innate because bacteria lack the target site or are impermeable to the antibiotic, or may be acquired. Bacteria acquire resistance by one of three mechanisms: an altered target site, altered uptake or by antibiotic-inactivating enzymes.
- Resistance spreads between bacteria in three genetic ways: chromosomal mutation, transmissible plasmids or transposons.
- Important characteristics of an anti-bacterial include its physicochemical properties, mode of action, spectrum of activity and resistance, pharmacokinetic properties, clinical indications, dose and route, and side effects and toxicity.

ANTI-MICROBIALS: SPECIFIC ANTI-BACTERIALS I

The detailed mechanism of action of anti-microbials is covered on pages 198–199. Here the specific characteristics of the drugs are described.

CELL WALL SYNTHESIS INHIBITORS

Action is bactericidal by blocking cross-linking of peptidoglycan chains in the dividing bacterial cell wall.

Penicillins

The penicillins all have a similar structure, differing side chains giving different drugs (Fig. 1). The side chains of the natural product can be modified chemically to give a wider spectrum of activity. A common bacterial drug resistance is via beta-lactamases (Fig. 2).

Pharmacokinetics. Penicillins can be stable to gastric acid and absorbable (penicillin V, ampi/amoxycillin, (flu) cloxacillin) and so can be given orally. Others must be given by injection: penicillin G, ticarcillin and piperacillin. All penetrate widely, though poorly to CSF and brain; all have relatively short half-lives, are little metabolised and have renal excretion.

Spectrum/use. This varies and is shown in Table 1 and Table 2.

Toxicity. This is minimal, almost limited to hypersensitivity, chiefly rash (Fig. 3) or fever, rarely anaphylaxis.

Cost. This is small except for (flu) cloxacillin, ticarcillin and piperacillin.

Cephalosporins

Structure of cephalosporins differs from the penicillins in having a *six*-member ring attached to the beta-lactam ring (Fig. 1): different side chains give the different cephalosporins, which are often classified into first, second, third and fourth 'generations' (Table 1) by their date of introduction; more logical classifications exist but are less used.

Pharmacokinetics. Cephalosporins behave like penicillins, although the later drugs have longer half-lives, especially ceftriaxone.

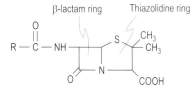

β-lactam ring　**Thiazolidine ring**

Penicillins

Side chain varies, e.g.

R = [phenyl]—CH₂—

penicillin G

Cephalosporins

β-lactam ring attached to a six-membered ring
Side chain varies

Tetracycline

p-Aminobenzoic acid　　　Sulphanilamide

Fig. 1 **The structures of some antimicrobials.**

Fig. 2 *Penicillium notatum*, a source of penicillin, killing the (yellow-white) colonies of *Staphylococcus aureus*, except where protected by the beta-lactamase from a *Bacillus* (dark colony, lower quarter).

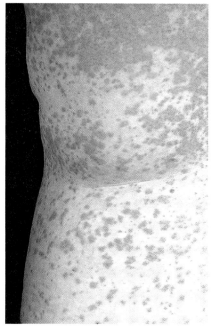

Fig. 3 **Penicillin rash.**

Table 1 **The ABC of antibiotics**

Antibiotic	Bacterial spectrum							Clinical uses
	AnO	Sta	Str	Enc	GNC	GNR	Ps	
Penicillins								
(Flu)cloxacillin	++	3+	++	+	–	–	–	Staphylococcal infections (not MRSA)
Penicillin V, G	++	+	3+	++	+	+	–	Erysipelas, gas gangrene, endocarditis, meningitis, pharyngitis
Ampi/amoxycillin } improved by BLI	++	+	++	3+	++	+	–	Bronchitis/sinusitis/otitis media; combined with gentamicin in abdominal and urinary infections
Ticar/piperacillin	++	+	++	++	++	+	3+	Pseudomonal infections, often combined with aminoglycoside
Cephalosporins								
First: cephalothin, cefazolin, cephalexin	++	3+	++	0	+	++	–	Surgical chemoprophylaxis, skin/soft tissue infections, also combined with aminoglycosides
Second: cefuroxime, cefoxitin, cefamandol	3+	++	++	0	++	++	–	Mild–moderate mixed anaerobic–aerobic skin/soft tissue, respiratory and abdominal infections
Third: cefotaxime, ceftriaxone	++	+	++	0	3+	3+	+	Severe Gram-negative infections including meningitis
Third: ceftazidime	++	+	++	0	3+	3+	++	Pseudomonal and other Gram-negative rod infections
Other beta-lactams								
Aztreonam	–	–	–	–	3+	3+	3+	Severe systemic Gram-negative infections
Imipenem, meropenem	3+	3+	3+	++	3+	3+	3+	Severe systemic infections (not MRSA)
Vancomycin	–	MRSA	++	3+	–	–	–	Serious staphylococcal infections, especially MRSA; enterococcal infections
Aminoglycosides								
Streptomycin	–	–	–	–	–	3+	–	Tuberculosis, Gram-negative zoonoses (brucellosis, plague, tularaemia)
Gentamicin	–	–	–	–	++	3+	3+	Severe systemic Gram-negative infections
Tobramycin	–	–	–	–	++	++	3+	Pseudomonal infections
Amikacin	–	–	–	–	++	3+	3+	Otherwise-resistant Gram-negative rod infections
Chloramphenicol	3+	MRSA	++	++	3+	++	–	MRSA, rickettsial and rare infections
Lincosamides								
Clindamycin	3+	++	++	–	–	–	–	Anaerobic infection, toxoplasmosis
Macrolides								
Erythromycin	++	++	++	+	++	–	–	Chancroid, legionellosis, pertussis, Campylobacter infection
Roxithromycin	++	++	++	+	++	–	–	Erythromycin substitute
Azithromycin	3+	+	+	–	3+	–	–	Chlamydia, STDs, unusual infections
Clarithromycin	++	3+	3+	+	3+	–	–	Chlamydia, MAC, leprosy
Tetracyclines	+	+	+	+	+	+	–	Unusual bacteria: *Borrelia* spp., brucella, mycoplasma, rickettsia, vibrios
Nitroimidazoles								
Metronidazole, tinidazole	3+	–	–	–	–	–	–	Anaerobic infections, amoebiasis, giardiasis
Qunolones								
Norfloxacin	–	–	–	–	–	++	–	Urinary infections (2nd line)
Ciprofloxacin	–	+	+	–	3+	3+	++	Serious Gram-negative infections, (MRSA)
Rifamycins								
Rifampicin	++	MRSA	3+	3+	3+	++	+	TB, MRSA, leprosy, prophylaxis in meningococcal and haemophilus meningitis
Rifabutin								MAC
Trimethoprim and sulphonamides	–	+	+	–	+	++	–	Mild urinary and respiratory infections, nocardial infection, toxoplasmosis, PCP

Efficacy: **3+**, usually effective (> 80% strains usually sensitive); ++, moderate efficacy (50–80% strains usually sensitive); +, poor efficacy (25–50% of strains usually sensitive); –, minimal or no efficacy.
AnO, anaerobes (e.g. *Clostridia, Bacteroides* spp.); Sta, staphylococci; Str, streptococci; Enc, Enterococci; GNC, Gram-negative cocci (e.g. *Neisseria, Haemophilus* spp.); GNR, Gram-negative rods, especially enteric (e.g., *E. coli*); Ps, pseudomonads; MAC, *Mycobacterium avium* complex; MRSA, methicillin- (and multi –) resistant *S. aureus; PCP, Pneumocystis carinii* pneumonia; STD, sexually transmitted diseases; BLI, beta lactamase inhibitor

Spectrum. This changes somewhat with the generations, but all are broad-spectrum, active against many Gram-positive, Gram-negative and anaerobic bacteria. Enterococci are resistant to all. In general, first generation are best against Gram-positive bacteria, second generation best against anaerobes and third generation best against Gram-negative rods. Major uses are given in Table 1.

Toxicity. Low toxicity is similar to the penicillins

Cost. This varies, ascending with the generations.

Table 2 **The CBA of antibiotics**

Clinical syndrome	Bacterial causes	Antibiotic(s) of choice
Pharyngitis	*S. pyogenes, C. diphtheriae*, (viruses)	Penicillin
Pneumonia		
Classical	*S. pneumoniae*	Penicillin*
'Atypical'	Numerous	Erythromycin plus ceftriaxone
Cystitis	*E. coli* and other enteric Gram-negative rods, *S. saprophyticus*	Amoxycillin +/– clavulanate, or cephalexin Trimethoprim, norfloxacin
Urethritis	*N. gonorrhoeae* *C. trachomatis*	Amoxycillin or ceftriaxone (PPNG) Doxycycline

*ceftriaxone if resistance
PPNG, Penicillinase-producing *N. gonorrhoeae*

ANTI-MICROBIALS: SPECIFIC ANTI-BACTERIALS II

Other β-lactams

Aztreonam is a mono-bactam, i.e. a single β-lactam ring. Its action and pharmacokinetics are like an injectable cephalosporin, but its spectrum is solely Gram negative, including many pseudomonads, i.e. like the aminoglycoside gentamicin. Its use is restricted by its high cost.

Imipenem and meropenem are carbapenems, structurally like penicillin except that carbon replaces the sulphur atom in the five-member ring. Their actions and pharmacokinetics are like an injectable cephalosporin, but their spectrum is wonderfully wide (Table 1, p. 201), though they are inactive against MRSA, *E. faecium*, and some Gram-negative rods. Their use is restricted by policy and high cost to serious systemic infections, especially before precise microbial diagnosis.

Vancomycin and teicoplanin

Vancomycin and teicoplanin are glycopeptides and are bactericidal to Gram-positive bacteria.

Pharmacokinetics Vancomycin is not absorbed from the gut, so it is given intravenously, by slow infusion to decrease the side effects of headache and flushing (the 'red man' syndrome). Teicoplanin can be given intramuscularly. Distribution of both is wide into fluids and tissues, but suboptimal into CSF and brain. Their half-lives are long, hence 12- or 24-hourly dosing. Excretion is renal and, because of toxicity, serum levels are usually monitored, though safe levels are not established!

Spectrum. Use is limited to severe Gram-positive (including MRSA) infections, and oral treatment of unresponsive *C. difficile*-associated colitis.

Toxicity. The main toxic effect is hearing loss, but phlebitis or neutropenia can also occur. Nephrotoxicity is uncommon with current preparations.

Cost is relatively high.

PROTEIN SYNTHESIS INHIBITORS

All act on the 30S or 50S bacterial ribosome. Only aminoglycosides are bactericidal.

Aminoglycosides

The aminoglycosides contain streptamine or a streptidine-containing aminocyclitol, with side chains that are modified to produce the individual drugs. Gentamicin is actually a mixture of three related molecules.

Pharmacokinetics. Aminoglycosides are not absorbed from the gut and so are injected for systemic use. They are not metabolised significantly and have a relatively long half-life, about 3 hours. Excretion is renal. Tissue penetration is relatively poor in bone, lung and sputum and does not occur in CSF and brain. Once- or twice-daily dosing has largely replaced the former 8-hourly dosing.

Spectrum. Aminoglycosides are bactericidal with a broad Gram negative spectrum (Table 1); however, anaerobiosis prevents their uptake into cells so they are ineffective against anaerobes. They are mainly used for enteric Gram-negative rods; gentamicin, tobramycin and amikacin are also active against pseudomonads. Streptomycin was used for tuberculosis but is now used mainly in unusual Gram-negative zoonoses including brucellosis, plague and tularaemia.

Toxicity. Aminoglycosides have both renal and oto-toxicity, which is related mainly to total dose. This is reflected later in peak than in trough levels, which should be monitored carefully, especially in the old, the underweight and those with renal or auditory impairment.

Cost. This is minimal for gentamicin (including monitoring), higher with tobramycin and very high with amikacin.

Chloramphenicol

Chloramphenicol is a natural product that is now chemically synthesised. Action is on bacterial protein synthesis at the 50S ribosome, and is usually bacteriostatic only.

Pharmacokinetics. Chloramphenicol is a lipophilic drug; it is orally absorbed and has wide penetration including the interior of the eye, the CSF and brain. It is metabolised by the liver and excreted renally. It can also be given by injection.

Spectrum. Like its pharmacokinetics, the spectrum is also almost ideal, being extremely broad and including most bacteria, chlamydiae, rickettsiae and mycoplasmas. It is also cheap.

Toxicity. The use of this otherwise ideal antibiotic is limited by its toxicity. An unpredictable irreversible marrow aplasia giving aplastic anaemia occurs rarely (1 in 30 000) and is fatal without marrow transplantation. If liver function is impaired chloramphenicol levels can rise above normal, and dose-related reversible marrow hypoplasia can occur; newborns who fail to metabolise chloramphenicol adequately die from toxic complications (grey baby syndrome).

Lincosamides

Clindamycin is bacteriostatic, on bacterial protein synthesis at the 50S ribosome, like macrolides and chloramphenicol. It is orally absorbed, but is also injectable, and is widely distributed apart from CSF and brain. It is metabolised by the liver and excreted renally.

Spectrum. It is mainly used against anaerobes, staphylococci and as a reserve drug in toxoplasmosis. Its cost and toxicity limit its use.

Toxicity. Toxic effects include allergy, a metallic taste, and initiation of pseudomembranous enterocolitis associated with *C. difficile* overgrowth (p. 137).

Cost as well as toxicity limits its use.

Macrolides

The macrolides have an unusual 14- or 15-member macrocyclic lactone ring

Table 1 Simplified antimicrobial sensitivities

	Anaer e.g. clostr.	Streptococci	Staphylococci	GNR enteric	GNR pseudomonas
Penicillin G, V	3	3	3 or –	–	–
Cloxacillin / Vancomycin	2	2	3	–	–
Cephalosporin 1st generation	2	2	3 or –	1	–
2nd generation	2	2	3 or –	2	–
3rd generation	2	2	3 or –	3	1
Gentamicin Tobramycin Norfloxacin Aztreonam	–	–	–	3	3

Key: **3**, >80% strains sensitive; **2**, 50-80% strains sensitive; **1**, <50% strains sensitive; – Not active with sugars attached. There are now four macrolides in clinical use: erythromycin, roxithromycin, azithromycin and clarithromycin. Action is bacteriostatic, on bacterial protein synthesis at the 50S ribosome.

Pharmacokinetics. Erythromycin base is inactivated by gastric acid so is protected by enteric-coating or is given as a salt or ester. Intravenous use often causes thrombophlebitis. The half-life is about 2 hours, and excretion is hepatic; as a result, care is needed in liver failure but dosage is unaltered in renal impairment.

Spectrum. Macrolides are mainly used against Gram-positive and unusual bacteria: chlamydiae, legionellae and mycoplasmas (Table 1, p. 201).

Toxicity. These are very safe antibiotics with low toxicity.

Cost is low for erythromycin, higher for the three newer drugs.

Tetracyclines
The tetracyclines have a basic structure of four fused rings (Fig. 1, p. 200), with the usual side chain changes to produce different members of the family. Action is bacteriostatic, on bacterial protein synthesis at the 30S ribosome (like aminoglycosides, which however are bactericidal).

Pharmacokinetics. Tetracyclines are orally absorbed yet also injectable: they are widely distributed including CSF and brain. Metabolism occurs in the liver and excretion is renal.

Spectrum. The spectrum of tetracyclines is very broad, including most pathogenic bacterial genera (except *Pseudomonas)*, plus chlamydiae, mycoplasmas and rickettsiae, but is not deep, with many resistant bacterial strains. The use of tetracyclines as a growth promoter added to livestock feed increased the numbers of resistant strains. Use is, therefore, chiefly for unusual bacteria.

Toxicity. This is low, apart from deposition in immature bone and teeth (causing discoloration), and the usual allergy or gut intolerance.

Cost is low.

NUCLEIC ACID SYNTHESIS INHIBITORS

Nitroimidazoles
Metronidazole and tinidazole are nitroimidazoles, with a unique mode of bactericidal action, acting as electron acceptors and producing intermediate compounds toxic to bacterial DNA.

Pharmacokinetics. They are well absorbed, penetrate widely, are metabolised by the liver and excreted by the kidney.

Spectrum. Their spectrum includes almost all pathogenic anaerobes (except some cocci), microaerophilic bacteria and some parasites. Their principal use is in prophylaxis and treatment of anaerobic bacterial infections, and in amoebiasis, giardiasis and trichomoniasis.

Toxicity is minimal, and cost is now low.

Quinolones
Structurally, the quinolones (e.g. norfloxacin and ciprofloxacin) are fluoroquinoline carboxylic acid derivatives with two six-member rings, and distinctive side chains. Action is bactericidal by inhibiting bacterial DNA gyrase, hence preventing supercoiling of DNA.

Pharmacokinetics. Quinolones are well absorbed and distributed, but serum concentrations are low. Their half-lives are relatively long, about 4 hours, and excretion is renal.

Spectrum. This is chiefly Gram-negative bacteria including pseudomonads, with poor activity against Gram-positive and anaerobic bacteria, although improved in the newer drugs.

Norfloxacin is used mainly in urinary and gut infections, while ciprofloxacin is used chiefly in serious systemic Gram-negative infections. Numerous other quinolones are available in different countries.

Toxicity can affect the CNS with headache, mood changes and fits.

Cost is moderate to high.

Trimethoprim and sulphonamides
Sulphonamides are derived from the single ring compound sulphanilamide, which is structurally similar and competes with an intermediate in the synthesis of folic acid, para-aminobenzoic acid. They are commonly used in combination with trimethoprim, which inhibits the next step to tetrahydrofolic acid in folic acid synthesis. Cotrimoxazole is sulphamethoxazole plus trimethoprim.

Pharmacokinetics. All have good absorption, wide distribution, long half-lives (to 10 hours) and renal excretion.

Spectrum. Chiefly Gram-negative rods were sensitive, but resistance is now common. Uses are mild urinary or respiratory infections, and unusual infections including *P. carinii* pneumonia, nocardiosis, chancroid and typhoid fever.

Toxicity. This is mainly allergy with rash and fever. Megaloblastic anaemia is uncommon and reversed by folinic acid.

Cost is low.

Rifamycins
The rifamycins are semi-synthetic derivatives of a natural product, rifamycin B.

Pharmacokinetics. All are well absorbed orally, penetrate widely including the CNS, and are cleared by liver metabolism and excretion mainly in bile.

Spectrum. Because rifamycins enter cells they are used against intracellular organisms such as mycobacteria. Rifampicin is a first line drug for TB and is also used for leprosy, resistant (especially MRSA) staphylococcal infections (with fusidic acid) and as prophylaxis in contacts of meningococcal and haemophilus meningitis. Rifabutin is used in combination with other drugs (e.g. ethambutol, clarithromycin) in the treatment and prophylaxis of atypical mycobacterial infections (MAC) in AIDS.

Toxicity. Unlike some drugs used in TB, rifampicin has few toxic side effects.

Anti-microbials: specific anti-bacterials

- Spectrum determines use, which is modified by toxicity and cost.
- Penicillins, cephalosporins and other beta-lactams are safe, effective bactericidal antibiotics of great use in a wide range of infections, limited by developing resistance.
- Aminoglycosides particularly gentamicin are very effective against many Gram-negative infections but need careful dosing and monitoring because of toxicity. Resistance is a lesser problem.
- Metronidazole is a very useful drug against most anaerobes, while clindamycin is less reliable and more costly.
- Older broad-spectrum bacteriostatic antibiotics including chloramphenicol, cotrimoxazole and tetracyclines are now mainly used for unusual bacteria, chlamydiae, mycoplasmas or rickettsiae.
- Quinolones are costly and used mainly in special infections.
- Vancomycin is reserved for serious resistant Gram-positive infections.

ANTI-MICROBIALS: SPECIAL ANTI-MICROBIALS

ANTI-MYCOBACTERIAL DRUGS

All forms of mycobacterial infections need prolonged, combined treatment with two or more drugs because:

- mycolic acids in their cell walls make them impermeable to many drugs
- mycobacteria grow slowly, requiring long-term treatment
- some mycobacteria are intracellular pathogens so agents must enter human cells
- antibiotic-resistance is common and most infections are mixed containing some resistant strains
- incidence in the immunodeficient is growing and here natural defences are impaired or absent.

Tuberculosis

Streptomycin is an aminoglycoside; it is bactericidal, causing abnormal proteins to be synthesised. It is not absorbed orally, penetrates the CNS poorly, is excreted by the kidney and has auditory toxicity. It has been a first-line drug against TB since the 1950s but is now less used as therapy is usually entirely oral.

Rifampicin is a red rifamycin which blocks nucleic acid synthesis by binding to RNA polymerase. It is well absorbed orally, widely distributed including the CNS, metabolised by the liver and excreted mainly in bile. It is also excreted in sweat, tears and urine, which in compliant patients it colours red (warn patients lest they panic on 'bleeding'). It is a first-line bactericidal drug for TB and leprosy.

Isoniazid (INAH) is synthetic isonicotinic acid hydrazide which is bactericidal, probably by inhibiting mycolic acid synthesis. It is well absorbed orally and widely distributed in the body, including the CNS. It is used only in mycobacterial infections, chiefly TB, and toxicity is neurological (prevented by routine pyridoxine) or hepatic.

Ethambutol is a synthetic mycobacteriostatic drug, probably inhibiting RNA synthesis. It is well absorbed and well distributed. It is used only in mycobacterial infections, chiefly TB. Toxicity includes optic neuritis, so dosage and regular vision/fundus reviews are critical.

Pyrazinamide is a synthetic tuberculocidal drug which is absorbed orally and penetrates both the CNS and cells, including macrophages. Hepatotoxicity was previously common with high doses, but the value of the drug is now more evident in safer, lower doses.

Reserve drugs: cycloserine, ethionamide, prothionamide and viomycin are used if resistance or intolerance exist to first-line drugs.

Leprosy

Dapsone is a synthetic sulphone, very similar to sulphonamides. It is cheap, given orally, usually well tolerated and was widely used alone for decades until widespread resistance forced the present combined therapy with rifampicin (see above).

It is also is used in toxoplasmosis, usually with pyrimethamine.

Table 1 **Classes of anti-fungal drugs**

Target	Chemical class	Mode of action	Drug	Spectrum
Cell membrane				
Function	Polyenes	Membrane leakage by ergosterol binding	Amphotericin B Nystatin	Wide Local candidiasis
Synthesis	Azoles	Inhibit erogosterol synthesis	Clotrimazole, miconazole Ketoconazole Fluconazole Itraconazole	Medium, local Broad Broader Broadest (e.g. includes *Aspergillus* spp.)
Nucleic acid synthesis	Pyrimidines	Inhibit DNA synthesis by conversion to 5-fluorouracil	Flucytosine	Cryptococcosis, combined in systemic candidiasis
	Benzofurans	Inhibit DNA synthesis (may also inhibit cell-wall chitin synthesis)	Griseofulvin	Dermatophytes

Atypical mycobacteria

M. avium complex (MAC). Rifabutin, ethambutol, clarithromycin and ciprofloxacin are suppressive.

M. marinum. Cotrimoxazole (pp. 170, 203) is used.

ANTI-FUNGAL DRUGS

Anti-fungals can be classified like antibacterials, by target site and chemical groups (Table 1).

Nystatin is a polyene acting on fungal cell membrane function. It is not absorbed orally or parenterally so is only used locally on *Candida* spp. infections of skin, mouth, vagina or bladder (by instillation).

Amphotericin B is the standard parenteral drug for systemic fungal infections. It is poorly distributed but low concentrations in blood, CSF and urine do not correlate with efficacy. Excretion is biliary, and slow: renal toxicity is considerable. Administration and dosage are specialised and debatable. It is also used in amoebic meningoencephalitis and leishmaniasis.

Flucytosine is a pyrimidine which inhibits fungal DNA synthesis through the action of its product 5-fluorouracil. It is absorbed orally but is also given parenterally. It is widely distributed and mainly excreted unchanged by the kidney. Activity is mainly against *Cryptococcus* and *Candida* spp. Toxicity is chiefly dose related, including gut intolerance, abnormal liver function and bone marrow suppression, especially in AIDS and/or renal impairment.

Azoles include clotrimazole, miconazole and econazole (all used locally), and ketoconazole, fluconazole and itraconazole, (used systemically for systemic disease). All inhibit ergosterol synthesis in fungal cell membranes. All are well absorbed orally; fluconazole is also given intravenously. Ketoconazole is active against *Candida* spp., the four systemic mycotic fungi, but not *Aspergillus*, *Cryptococcus* or *Mucor* spp. Fluconazole is, in addition, active against *C. neoformans*, while itraconazole also has some activity against some *Aspergillus* and *Mucor* spp. Toxicity is low, but drug interactions with warfarin, isoniazid, rifampicin, cyclosporin or phenytoin can be dangerous; in vitro, ketoconazole and amphotericin B are antagonistic.

ANTI-PARASITIC DRUGS

Parasites are eucaryotic organisms so it is harder to find drugs that are selectively toxic to the pathogen. Table 2 lists the major anti-protozoal drugs.

Table 2 **Major anti-protozoal drugs**

Disease	Anti-protozoal drug
Amoebiasis	Metronidazole/tinidazole, chloroquine, diloxanide, emetine
Giardiasis	Metronidazole/tinidazole, furazolidone
Trichomoniasis	Metronidazole/tinidazole
Amoebic meningoencephalitis	Amphotericin B
Cryptosporidiosis	Paromomycin
Malaria	Primaquine, chloroquine, doxycycline, mefloquine, primaquin, pyrimethamine, quinine
Toxoplasmosis	Pyrimethamine, sulphamethoxazole, spiramycin, dapsone
Pneumocystosis	Pentamidine, sulphamethoxazole/trimethoprim, atovaquone
Leishmaniasis	Pentamidine, antimony compounds,[a] amphotericin B
Trypanosomiasis	
African	Pentamidine, antimony compounds,[a] nifurtimox
American	Nifurtimox, benznidazole

[a] Includes suramin, melarsoprol and tryparsamide

Table 3 **Major anti-helminthic drugs**

Disease	Anti-helminthic drugs[a]
Ascariasis (roundworm), threadworms, hookworms	Mebendazole,[a] albendazole, flubendazole,[a] pyrantel pamoate[a]
Trichuriasis	Mebendazole, albendazole, flubendazole
Strongyloidiasis	Thiabendazole
Cutaneous larva migrans, toxocariasis, trichinosis	Mebendazole, albendazole, flubendazole, thiabendazole,[b]
Filariasis	Diethylcarbamazine, ivermectin
Hydatids, taeniasis and other cestodes	Mebendazole, albendazole, flubendazole, praziquantel, niclosamide
Schistosomiasis or gut flukes	Praziquantel, oxamniquine
Liver or lung flukes	Praziquantel

[a] Mebendazole, albendazole, flubendazole, thiabendazole and pyrantel pamoate should not be used in pregnancy.
[b] Local application in cutaneous larva migrans.

Anti-malarials

Chloroquine is a 4-aminoquinoline which is well absorbed orally and is also given intravenously. It concentrates so much in liver, spleen and CNS that loading doses are not needed. It has been a major drug for malaria prophylaxis and treatment since the 1950s, but resistance in *P. falciparum* is now so widespread that it is only useful in Central America and parts of the Middle East. It does not eradicate the pre-erythrocytic liver stage (p. 72).

Mefloquine is a newer drug, a quinolinemethanol, which is well absorbed orally, concentrated in the liver and then excreted in the faeces, though with a very long half-life of 17 days. Resistance to it is rare but increasing in southeast Asia. It has troublesome neurological and cardiac toxicity, so careful use is essential.

Primaquine is an 8-aminoquinoline which is well absorbed orally, widely distributed (including the liver) and rapidly metabolised, being undetectable in 24 hours. It is used in a 14-day course for radical cure of *P. vivax* and *P. ovale* malaria. Haemolysis is common, especially in glucose 6-phosphate dehydrogenase deficiency.

Quinine, used for over 400 years, is a natural alkaloid from cinchona bark. It is given orally or intravenously (**not** intramuscularly) and is metabolised in the liver, with some renal excretion and a half-life of 18 hours. It is schizontocidal only, so must be used with other drugs such as doxycycline, pyrimethamine +/or sulphadoxine. Dosage is critical, and ECG and blood pressure should be monitored for cardiotoxicity.

Other anti-protozoal drugs

Antimonials. These include suramin (for prophylaxis and early treatment) and melarsoprol (for meningoencephalitis, 'sleeping sickness') in trypanosomiasis, and pentavalent compounds such as sodium antimony gluconate used in a leishmaniasis. All are given by slow intravenous injection, are toxic and require special care and knowledge.

Metronidazole, tinidazole: see p. 203.

Pentamidine is a diamidine which binds to DNA. It is given intramuscularly or intravenously to treat Gambian trypanosomiasis and *P. carinii* pneumonia (PCP), or by inhalation for PCP prophylaxis. As it does not enter the CNS, it is useless for the neurologic stage of trypanosomiasis. Common toxic effects include hypoglycaemia, hypotension, renal impairment and rashes.

Anthelmintics

Benzimidazoles. Thiabendazole is an unpleasant old drug which commonly causes vomiting but has a very broad spectrum and is effective in strongyloidiasis and both cutaneous and visceral larva migrans. Mebendazole, albendazole and flubendazole are newer drugs with few side effects. **Mebendazole** is little absorbed so used against intestinal nematodes. **Albendazole** and **flubendazole** are well absorbed and widely effective against most intestinal and tissue nematodes (not filaria) (Table 3).

Diethyl carbamazine (DEC) is a piperazine derivative, well absorbed orally, well distributed and metabolised in 48 hours before renal excretion. In use since the 1950s, it kills all human microfilariae but is less active against adult worms, especially *O. volvulus*. Repeat courses are often needed. Minor side effects are common, including allergic itchy rash and transient worsening of disease symptoms.

Ivermectin is a relatively new macrolide antibiotic, orally well absorbed and particularly active against *O. volvulus* microfilariae. It has minimal side effects and almost abolishes infectivity for 6–12 months after one dose. Its effect on adult worms is not well established.

Praziquantel is a prazino-isoquinoline stereoisomer mixture! It is an important new drug active against all three species of schistosomes, most other cestodes and also most flukes. It is well absorbed and causes fatal tetanic contractions in the helminth but only mild nausea or abdominal pain in some patients.

Anti-microbials: special anti-microbials

- The major anti-tuberculous drugs are isoniazid, rifampicin, ethambutol and pyrazinamide: treatment usually begins with three drugs (four if resistance is likely), then decreases to two for many months. Resistance is an increasing problem.
- Leprosy is now usually treated with dapsone and rifampicin.
- Superficial mycoses are treated with nystatin or an azole, while invasive, systemic or disseminated mycoses need amphotericin B, often with a second drug if possible.
- A wide range of drugs is needed for protozoal and helminthic diseases, and treatment remains unsatisfactory for many.

VACCINES AND IMMUNISATION

Boring immunisation schedules obscure exciting stories including Jenner's popularisation of vaccination to prevent smallpox, Pasteur's courageous production of rabies and anthrax vaccines, the dedicated sub-culture of the bacillus *M. bovis* for 10 years by Calmette and Guerin to produce BCG, the conquest of polio by the Salk and Sabin teams, and today's genetic engineering of safe, specific, effective vaccines.

Definitions

Immunisation is the artificial production of immunity to an infection: it can be **passive** immunisation by administering preformed antibody, or **active** immunisation by stimulating the host to produce protective antibody, cell-mediated immunity, or both. Simultaneous and combined immunisation is noted below.

Vaccination is a specific form of immunisation using vaccinia (the virus causing mild cowpox) to protect against variola (causing virulent smallpox). **Vaccine** is now a general term for any preparation containing one or more **immunogens**, i.e. substances stimulating active immunity. These are usually the protective antigens of the microbe itself or, rarely, a closely related one (e.g. vaccinia for variola).

Immunogenicity is the ability to produce immunity, usually qualified as poor, good or excellent.

Adjuvants are substances which enhance the immune response. They include:

- aluminium (widely used) and other salts
- bacterial products: killed *B. pertussis* in DPT is both an adjuvant to the toxoids and an immunogen itself
- cytokines (interleukins 1 and 2): experimental
- special delivery systems: antigen on small spheres such as liposomes of phospholipid (also experimental).

VACCINE TYPES

Traditional types of vaccine (Table 1) are:

Live attenuated bacteria or viruses

BCG, Measles–Mumps–Rubella (MMR) and polio are included in this category. Production is by the selection of mutants. Advantages usually include high, long-lasting immunity and few side-effects. Disadvantages include potential risks from inadequate attenuation, reversion to virulent wild type, contamination by other viruses and persistent infection especially if unknowingly given to an immunocompromised patient. However, genetic engineering of deletion mutants lacking specific virulence genes is now possible and will remove most risks.

Inactivated ('killed') bacteria or viruses

Production is by chemical inactivation, for example with formaldehyde. Advantages include safety because of non-infectivity, stability, and relative ease of production. Disadvantages include lower immunogenicity and hence repeated doses.

Microbial components

Viral proteins or bacterial polysaccharides are used as antigens. Production is by extraction of pneumococcal, meningococcal or *H. influenzae* type b (Hib) capsular polysaccharides (the last conjugated to other proteins to increase immunogenicity); by purification of plasma from hepatitis B surface-antigen chronic carriers; or by recombinant DNA technology in yeasts (hepatitis B). Advantages include production of serotype-specific (Hib) or multivalent vaccines. Disadvantages include the exacting safety measures to remove live infectious material.

Inactivated toxin (toxoid)

Production is by inactivation, usually by formaldehyde, of the bacterial toxin. Advantages include long-lasting immunogenicity and the ability to combine several immunogens, e.g. diphtheria and tetanus toxoids and killed *B. pertussis* as 'DPT' 'triple antigen'. Disadvantage is the restriction to toxin-mediated disease.

Heterologous vaccines

These use immunisation across species, i.e. using vaccinia, an animal virus sharing antigens with smallpox to immunise humans.

Experimental vaccines

Cloned or synthetic peptides. Production of the peptide in *E. coli*, yeast, insect or mammalian vectors by cloning genes produces a range of potentially immunogenic peptides (or glycosylated proteins). These are tested for potent T-cell and B-cell 'epitopes' to trigger T- and B-cell responses. Advantages are the wide range of microbes potentially suitable, the specificity and the safety. Disadvantages are the complex technology, wide-ranging testing, unsuitability for carbohydrate or glycolipid antigens and, often, the need to increase immunogenicity. The last can be achieved by attachment to larger carriers such as tetanus toxoid or polylysine to which eight antigen peptides have been attached: the 'octopus' molecule.

Microbial vectors for cloned genes. Production is by using an existing vaccine as the expression vector (e.g. BCG, attenuated salmonellae) with the inserted gene(s) as a polyvaccine, which in the patient multiplies to deliver immunising protein or peptide. Advantages are the potential for polyvaccines, even a one-shot vaccine. As BCG induces cell-mediated immunity, it could in theory carry antigens for other persistent intracellular organisms including *Brucella*, *Listeria*, *Rickettsia*, *Chlamydia*, *Histoplasma*, *Leishmania*, *Toxoplasma* spp., malaria and many

Table 1 **Major current vaccine types and usage**

Type	General use	Limited use	Developmental
Live attenuated bacteria	BCG	Typhoid (oral), tularaemia	Cholera, shigellosis
Live attenuated virus	Measles, mumps, rubella, polio (oral)	Yellow fever	Varicella, RSV, CMV, HSV, rotavirus
Inactivated bacteria	Pertussis	Anthrax, cholera, plague, Q fever, typhoid (inj)	Gonorrhoea
Inactivated virus	Polio (injectable)	Hepatitis A, rabies	
Microbial components	*H. influenzae* b[a], influenza	Hepatitis B, meningococcal,[a] pneumococcal[a]	
Toxoid	Diphtheria, tetanus		Cholera, botulism
Protozoal			Malaria

[a]Polysaccharide.

viruses! Disadvantages are the complex technology and the unsuitability for carbohydrate or glycolipid antigens.

Anti-idiotype vaccines

Production of anti-idiotype vaccines is by making a first antibody to the antigen, then making numerous second antibodies to the first and testing to find a second antibody (anti-idiotype) resembling the antigen. This protein second antibody can be used as a surrogate antigen. Advantages are safety, specificity and the ability to make an immunogen from carbohydrate or glycolipid antigens, which cannot be cloned or synthesised as above. Disadvantages are the complex technology and the wide-ranging search required to find suitable anti-idiotypes.

PASSIVE IMMUNISATION

Passive immunity is produced by giving preformed specific or non-specific immunoglobulin. The effect is therefore, immediate but temporary.

Specific immunoglobulin is from convalescent or immunised donors or, sometimes, from immunised horses (e.g. against gas gangrene or diphtheria). It is used postexposure in the non-immunised for botulism, gas gangrene, hepatitis B, rabies, snake or scorpion bite and tetanus; in varicella-zoster exposure in the immunocompromised; and in treatment of diphtheria, gas gangrene and tetanus.

Non-specific immunoglobulin is from pooled normal plasma from blood donors. It is used either in normal hosts before travel (against hepatitis A if vaccine is unavailable) or postexposure (measles) or monthly in antibody immunodeficiencies, such as agammaglobulinaemia, or in bone marrow transplants (for CMV).

Combined active and passive immunisation is used in postexposure prophylaxis in the unvaccinated for rabies or tetanus (Table 2) or at birth to babies of hepatitis B carrier mothers.

Simultaneous administration of several vaccines by different syringes at different sites is usually satisfactory, except for yellow fever and cholera vaccines, which reduce antibody response to each other.

WHEN AND WHO TO VACCINATE

Immunisation is a life-long commitment:

- immunisation schedules for childhood and adolescence are published for each country; schedules vary depending on disease incidence and vaccine costs
- adults should have ADT (tetanus) and oral polio vaccine (OPV) every 10 years from 15 to 65 years, pneumococcal vaccine at age 65 and influenza vaccine yearly thereafter, particularly with pulmonary or cardiac disease
- specific occupations need specific vaccines, including hepatitis B (health workers), anthrax, plague, Q fever, rabies, and tularaemia

- special-risk groups needing specialist knowledge are the armed forces, college students, the homeless, prisoners and pregnant mothers
- Travellers need routine, required and specific vaccines, such as hepatitis A, meningococcal, typhoid, cholera and yellow fever vaccines, depending on the area (p. 188).

Contraindications to vaccination

Contraindications and adverse reactions vary with each vaccine, but there are three general rules.

1. Live vaccines should not be given in pregnancy or to immunocompromised patients, except for measles vaccine to HIV-positive children.
2. Conversely, inactivated or component vaccines are safe. Splenectomised patients should have pneumococcal and meningococcal vaccines, before splenectomy if possible for best antibody response.
3. The adverse effects of all available vaccines are many times less than their benefit.

Future needs

Major infections with no satisfactory vaccine include hepatitis C, sexually transmitted diseases (especially gonorrhoea, HIV and syphilis), malaria, schistosomiasis, trypanosomiasis and other parasitic diseases (and viral infections including CMV, HSV and RSV). In all, antigenic variation, poor immunity even after natural infection and the threat of reversion or latency with living attenuated vaccines are obstacles slowly being overcome.

Table 2 **Tetanus prophylaxis in managing wounds**

1. The wound itself must be cleaned, debrided and managed appropriately
2. Toxoid alone is dependable up to 10 years after full immunisation
3. Immunoglobulin is added for tetanus-prone wounds

Tetanus immunisation history	Minor, clean wounds		All other wounds	
	Tet toxoid, CDT[a] or ADT[a]	Tetanus immunoglobulin	Tet toxoid, CDT[a] or ADT[a]	Tetanus immunoglobulin
3 doses or more, and <5 years from last dose	No	No	No	No[b]
3 doses or more and 5–10 years from the last	No	No	Yes	No[b]
3 doses or more and >10 years from the last	Yes	No	Yes	TPW only
Less than 3 doses or uncertain	Yes	No	Yes	Yes

[a]CDT or ADT is preferred to tetanus toxoid, to boost immunity to diphtheria also.
[b]In some countries, immunoglobulin is advised for tetanus-prone wounds (penetrating, necrotic or neglected) in this category. TPW, tetanus prone wounds

Vaccines and immunisation

- Immunisation is a safe, cost-effective way of preventing many severe health- or life-threatening diseases and should be used in childhood and adolescence, in adult life and in special groups including health workers and travellers.
- The adverse effects of all available vaccines are many times less than their benefit
- Mild upper respiratory tract infection is not a contraindication to immunisation: more serious disease with fever above 38°C should only defer immunisation by 10 days.
- Each medical consultation is an opportunity to consider immunisation.

CLINICIAN AND LABORATORY: MICROBIAL DETECTION AND IDENTIFICATION

Microbiology laboratories help clinicians directly in four ways.

1. Detecting and identifying the microbes causing infections
2. Measuring host antibody response when the microbes cannot be detected
3. Guiding therapy
4. Assisting infection control.

Laboratories also help indirectly by teaching and research.

Fig. 1 *C. neoformans* (Indian ink).

The laboratory, the clinician and the patient are all helped when the clinician provides written notes about the specimen, the sufferer and the suspected pathogen(s). This entails:

Fig. 2 *M. tuberculosis* (ZN stain).

- specimens correctly collected, legibly labelled and swiftly sent to the laboratory, so pathogens are present
- clinical conditions, differential diagnoses and administered anti-microbials noted on the request slip
- possible unusual pathogens noted, so special techniques can be used to find them.

Detecting and identifying the microbes causing infections usually requires four steps:

1. Specimen collection, transport and processing
2. Direct detection methods
3. Culture
4. Identification.

SPECIMEN COLLECTION, TRANSPORT AND PROCESSING

Specimens should be collected from the site of infection and the sites of any spread. Specimens may be swabs (good), fluids (better) or tissue (best).

Swabs. Eye, ear, nose, throat, wound, ulcer and genital swabs are relatively easy to obtain but provide small volumes which dry easily, so stains may be poor and cultures negative.

Fluids. Samples of pus, urine, sputum, bile, CSF and blood are better, as volumes are greater and organisms survive transport better; stains or cultures are more likely to be positive.

Tissue. Samples of tissue are best as organisms are usually most concentrated and most viable, though tissue is most difficult to collect.

Processing specimens. While specimens are sometimes processed directly at the bedside (blood films for malaria, cough plates in pertussis, fungal scrapings or clippings, genital discharges), most are transported to the laboratory for processing, as quickly as possible to preserve the pathogens. Transport media can be used if delay is unavoidable but these often interfere with microscopy and stains. *When in doubt about specimen collection, transport or processing, ask the laboratory.*

DIRECT DETECTION METHODS

Direct methods include microscopy, stains and detection of antigen, nucleic acid or metabolic products. These only take minutes or a few hours and can provide a microbe-specific diagnosis for many fungi and parasites, and for those bacteria with distinctive morphology, staining, antigens or metabolic products.

Direct microscopy

Direct microscopy is used for larger organisms with distinctive shapes. Techniques include:

- **wet mount** in saline for genital and gut parasites
- **KOH** for fungi in skin scrapings or sputum
- **Indian Ink** to highlight cryptococci in fluids (Fig. 1)
- **dark field (dark ground) microscopy** (DGM) for syphilis.

Stains

Stains are used to show and differentiate microbes:

- **Gram's stain** is the quick, easy, usual stain for common bacteria, showing their shape and type of cell wall (p. 3).
- **Acid-fast stains** (Ziehl–Neelsen, Kinyoun, Auramine–rhodamine) are used to find mycobacteria (Fig. 2), Nocardia and, oddly, *Cryptosporidium* spp.
- **Special stains** for organisms not staining with the above:
  — toluidine blue or PAS for fungi and *P. carinii*
  — silver stains for fungi, legionellae, *P. carinii*, rickettsias and treponemes
  — Giemsa and Wright's stains for malaria and babesiosis (overwhelming bacteraemia may also be seen in blood films)
  — acridine orange for staining DNA, even in damaged microbes
  — methylene blue for faecal leucocytes.

Antigen detection

Antigen detection is also known as immunodetection and uses a specific antibody to detect a microbial antigen.

- **Latex agglutination** uses antibody coupled to latex particles which visibly agglutinate with antigen (e.g. capsular polysaccharide) on a slide, detecting *N. meningitidis*, *H. influenzae*, *S. pneumoniae*, *S. pyogenes* or *Cryptococcus neoformans* (Fig. 3).
- **Co-agglutination** uses protein A on *S. aureus* to attach to the Fc part of an antibody and thus orientate the antigen-specific Fab fragment outward to agglutinate with antigen, as in latex agglutination
- **Direct immunofluorescence** (DIF) uses fluorescein-tagged antibody, detecting, for example, chlamydia.
- **Immunocytochemistry** uses specific antibody on tissue sections.

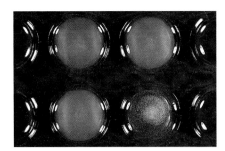

Fig. 3 Antigen detection (latex agglutination) of *C. neoformans*.

Fig. 4 Anaerobic jar components.

- **Radioimmunoassay** (RIA) is being replaced by *ELISA* (enzyme linked immunosorbent assay). Both use a solid phase such as a microtitre plate with antibody which binds antigen. This is then measured by the binding of a second ligand either labelled by radioactivity (RIA) or an enzyme detected by colour change (ELISA).
- **Monoclonal antibody** is being increasingly used.
- **Counter immunoelectrophoresis** (CIE) is now little used.

Nucleic acid detection

Nucleic acid detection is achieved in several ways:

- **Polymerase chain reaction** (PCR) multiplies even one segment of DNA a million times in a few hours. It is very specific and is so sensitive that contamination is a problem if technique is not fastidious; it is now commercially available
- **Nucleic acid probes** use labelled single-stranded DNA or RNA to probe for single-stranded target nucleic acid, by several techniques; although very specific, they are often no more sensitive than antigen detection tests
- **Plasmid fingerprinting** separates plasmids by agarose gel electrophoresis and is used epidemiologically
- **Restriction enzyme** analysis shows defined DNA nucleotide sequences; it is also used in epidemiological investigation.

Metabolic product detection

Metabolic product detection includes *gas liquid chromatography* (GLC). It detects the different fatty acids produced by different anaerobic bacteria in pus or cultures, helping rapid identification.

CULTURE

Culture of microbes in specimens takes at least some hours and more usually one or more days; it can be more specific than the non-cultural methods and gives live organisms for full identification and sensitivity testing. Each specimen type is cultured in (several) particular media chosen to grow the likely pathogens. Media may be liquid to give maximum growth (e.g. from swabs) or solid to separate different organisms. Media may be enriched, selective, indicator, specific or a combination.

Enriched media. Enriching substances to encourage growth include blood agar (BA) or chocolate agar (CA), which is blood agar heated to 60°C to release nutrients for fastidious bacteria.

Selective media. These vehicles use inhibitory substances to inhibit growth of some bacteria (and may be combined with indicator and/or enrichment also). They include:

- MacConkey agar (Mac) or deoxycholate citrate agar (DCA), which contain the inhibitor (bile), plus lactose and an indicator to grow enteric bacteria and differentiate between lactose and non-lactose fermenters (p. 40)
- Sabouraud's dextrose agar, with low pH to grow fungi and inhibit other organisms
- Lowenstein–Jensen agar, with glycerol to grow mycobacteria, and malachite green to inhibit other bacteria.

Indicator media. These use pH indicators in Mac and DCA, or can use antiserum (e.g. Hayward's medium for the Nagler test for clostridia).

Special media. Fastidious organisms require special media, including Robertson's cooked meat medium (RCM or CMM) for growing anaerobes, or special media for chlamydia, mycoplasma, rickettsia and viruses.

Incubation. Cultures may be incubated in special atmospheres: CO_2 for *Haemophilus* and *Neisseria* spp., anaerobic for clostridia, *Bacteroides* spp., etc., or microaerophilic for *Campylobacter* spp. (Fig. 4). Special temperatures are also sometimes required (e.g. 42°C for *Campylobacter* spp. or 30°C for *M. ulcerans*).

IDENTIFICATION

Provisional identification is often possible through

- Gram (or other) stain
- organism's shape (coccus or rod), pattern (chains, pairs) and characters (capsule, spores, intracellular position)
- growth requirements (aerobic, anaerobic, fastidious, unaffected by bile).

Definitive identification is usually then by either serological methods, including antigen detection as above, or biochemical methods. Biochemical methods include:

- detection of enzymes such as coagulase, catalase, oxidase
- substrate utilisation, especially sugars such as lactose, sucrose, glucose
- metabolism of sugars oxidatively (aerobically) or by fermentation (anaerobically).

Microbial detection and identification

- Clinicians must provide correctly collected and legibly labelled specimens for proper processing.
- Clinicians must notify clinical conditions, differential diagnoses and antimicrobials administered for reliable results.
- Direct detection depends on microscopy, stains, antigen detection (immunodetection), nucleic acid detection or metabolic product detection.
- Culture on enriched, selective, indicator and/or specific media gives further information plus live organisms for full identification and anti-microbial sensitivity tests.
- Identification if not definitive from the above information is made by biochemical and serological tests.

CLINICIAN AND LABORATORY: ANTIBODY RESPONSE AND GUIDING THERAPY

MEASURING HOST ANTIBODY RESPONSE

Measuring host antibody levels (called 'titres') using a specific antigen is the converse of microbial antigen detection; these major analyses are part of serology.

Antibody analysis is used only when the microbe or an antigen cannot easily or safely be found (e.g. *Legionella, Brucella, Mycoplasma, Rickettsia* infections, or systemic mycoses) for it has three major disadvantages: firstly, a specific antigen must be used, so the specific disease must be suspected; secondly, 2–4 weeks pass before IgM antibodies are detectable; and, thirdly, IgG may be present already from previous infection or immunisation, hence the usual need for paired acute and convalescent sera to show an antibody rise (see below).

Methods include agglutination (Fig. 1), flocculation, precipitation, haemagglutination and its inhibition (HAI), complement fixation tests (CFT), enzyme-linked immunosorbent assays (ELISA), fluorescent antibody and neutralisation tests.

Clinical uses are:

- diagnosis of acute infection, usually by antibody rise in two 'paired' sera, one acute and one convalescent; sometimes an elevated IgM in a single acute serum sample is used or a single elevated convalescent antibody if a response is short-lived, e.g. legionellosis

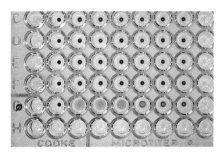

Fig. 1 **Antibody detection by particle agglutination.**

- determining the immune status, either before immunisation (e.g. rubella) or after exposure (needle-sticks, hepatitis B and C, HIV, etc.).

GUIDING THERAPY

SENSITIVITY TESTING

There are two steps in testing for antibiotic sensitivity. First, a precise laboratory method measures whether the organism is affected by ('susceptible' to) chosen concentration(s) of the test antibiotic. One concentration in disc methods gives only an approximate susceptibility, while dilution tests with a series of concentrations give an exact MIC (minimum inhibitory concentration: the lowest concentration inhibiting microbial growth) and can be continued to give the MBC (minimum bactericidal concentration: kills the microbe). Secondly, with rules based on experience, an imprecise prediction of the likely clinical response is given:

S: susceptible or sensitive, i.e. likely to respond
R: resistant, i.e. unlikely to respond
I: intermediate, may respond to very high dose or where antibiotic is concentrated, e.g. in urine.

Methods use known amounts of antibiotic and include those discussed below.

Diffusion tests (6–18 hours)
Disc diffusion uses antibiotic in paper discs placed on a lawn of organism on special agar (Fig. 2). The antibiotic diffuses outwards giving a gradient of decreasing concentration, so zone size is related to degree of susceptibility. The test is cheap, easy and accurate if standardised.

Stokes' adaptation of disc diffusion uses a sensitive control organism, either on half the plate or concentrically around the periphery (Fig. 3). It helps control disc quality.

The **E-test** uses a graded concentration on a strip, so the MIC is read where the junction of growth–no growth intersects the strip (Fig. 4). It is exact, relatively expensive and useful for unusual organisms or antibiotics.

Fig. 2 **Sensitivity test: disc method.**

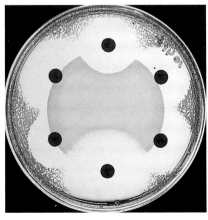

Fig. 3 **Sensitivity test: J Stokes method.**

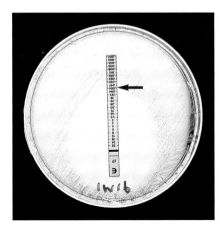

Fig. 4 **Sensitivity test: E-test strip:** *Nocardia* sp./**imipenem.**

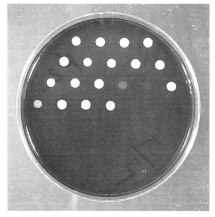

Fig. 5 **Agar dilution, control.**

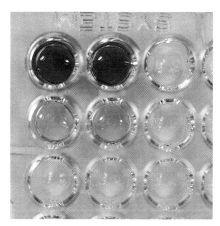

Fig. 6 **Beta-lactamase test (nitrocefin).**

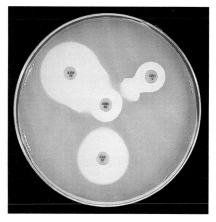

Fig. 7 **Extended spectrum β-lactamase test (ESBL).** Note deformed zone around third-generation cephalosporin induced by adjacent clavulamic acid disc.

Dilution tests (4–18 hours)

Agar dilution uses antibiotic diluted uniformly through special agar, testing about 30 different organisms per plate (Fig. 5). It is exact and economic when testing at least 25 organisms daily but usually takes 16–18 hours.

Broth dilution uses antibiotic diluted in broth in tubes or microtitre plates. It also is exact and economic in larger laboratories; it is the basis of several common automated systems, taking 4–6 hours.

Killing curves (6–18 hours)

Some automated methods use broth, and measure killing curves over time instead of the simple 'growth–no growth at one point in time' information provided by diffusion or dilution methods.

Synergy tests (usually overnight)

Tests to measure whether two antibiotics show synergy (at least four times greater effect than expected by addition), indifference (no effect on each other) or antagonism (decreased effect) were done in tubes but now usually use a chequer-board of broth dilutions in microtitre plates, with one antibiotic in doubling dilutions across and the other down.

Enzyme detection

Beta-lactamase is easily detected (e.g. in some strains of *H. influenzae* and many enteric Gram-negative rods) by a special cephalosporin, nitrocefin, which changes colour in a few hours when the lactam ring is broken by beta-lactamase (Fig. 6).

Extended-spectrum beta-lactamases (ESBL) destroy third-generation cephalosporins and are now found in numerous enteric pathogens in hospitals. They are easily shown in 12–18 hours in the laboratory when the clavulanate in an amoxycillin–clavulanate disc inactivates the ESBL and, thus, causes expansion of the zone around a nearby cefotaxime (or other third generation cephalosporin) disc (Fig. 7).

ANTIBIOTIC AND OTHER ASSAYS

Antibiotic levels can be measured overnight by bio-assay, measuring zones around four wells in agar, three containing known concentrations of the antibiotic and one with the patient's serum. This is cheap, slow and inexact.

Chemical methods are now common, exact, quick (20–60 minutes) and affordable. 'Peak' levels taken 10–30 minutes postdose, and 'trough' levels taken predose have been the usual samples for amino-glycosides and vancomycin. Efficacy is considered to be mainly related to the peak level, while toxicity is more related to the trough and the area under the (level–time) curve (AUC); this is being re-assessed with once-daily dosing, particularly of gentamicin.

Serum bactericidal titre is the killing power of the patient's serum against the organism causing their endocarditis or other serious infection. Cure is believed more likely if serum diluted 1:8 kills their organism, but this view is challenged so this overnight test is now uncommon.

GIVING ADVICE

Giving advice on appropriate tests to do and the meaning of particular test results is an important function of medical microbiologists and of senior scientists. They also can give advice on appropriate therapy, especially if infectious disease physicians are not available.

ASSISTING INFECTION CONTROL

This is detailed on page 191.

Antibody response and guiding therapy

- Host antibody response is used to diagnose infection only when the causative microbe is not easily or safely detected by microscopy or culture. Inherent disadvantages are the need for specific antigens, the slowness of antibody response and the possible presence of IgG antibody from previous infection or immunisation, so paired acute and convalescent sera and IgM tests are preferred.
- Antibody tests are mainly used to find the person's immune status before immunisation or after exposure.
- Antimicrobial sensitivity tests attempt to predict a patient's response from a laboratory test without reference to the host defence mechanisms so cannot be 100% reliable.
- Anti-microbial levels, usually in serum, rarely in CSF, can be used to estimate both likely efficacy and possible toxicity. There cannot be a single level which separates safety from toxicity.
- Giving advice on appropriate tests, the meaning of results, appropriate therapy and infection control are important functions of clinical microbiologists and infectious disease physicians.

INDEX

Note: organisms are discussed on pages 28–85, diseases on pages 86–195

A

Abdominal infections, 136–47
  abscesses, *see* Abscesses
  rare, 147
  tropical, 138
Abscesses
  breast, 186
  Brodie's, 178
  CNS, 90–1, 129
  dental, 102, 103
  intrabdominal, 142–3
    kidney, 150
    liver, 140, 143, 145–6
    pelvic, 159
    perinephric, 150
    psoas, 135, 143
    tubo-ovarian, 158, 159
  orbital, 97
  periodontal, 103
  peritonsillar, 105
  pulmonary, 116–17
  subperiosteal, 97
  vulval, 161
Absidia spp., 70
Acanthamoeba spp., 74
Acid-fast rods, 54–5
Acid-fast stain, 208
Acinetobacter spp., 37
Actinobacillus actinomycetemcomitans, 47
Actinomyces spp., 56
  CNS infection, 90–1
  mycetoma, 174
  orofacial infection, 108
  osteomyelitis, 181
Acute phase proteins, 21
Adherence, 25
Adhesins, 15
Adhesion (in phagocytosis), 20
Adjuvants, vaccine, 206
Adrenal tuberculosis, 153
Aerococcus viridans, 49
Aeromonas spp., 45
  hydrophila, 169, 170
Aflatoxins, 8
African eye worm (*Loa loa*), 80, 99
African trypanosomiasis, 77, 93, 98, 131
Agar dilution, 211
Aggressive pathogens, 14
Airborne transmission, 18
Albendazole, 205
Alimentary tract, *see* Gastrointestinal tract
Allergy, atopic, 24
Alveolar bone infections, 102
Amastigote
  leishmaniasis, 76, 121
  trypanosomiasis, 77
American cutaneous leishmaniasis, 76
Amino acid metabolism, 5
Aminoglycosides, 201, 202
Amoebae, 10, 74
Amoebic liver abscesses, 140, 145
  dysentry, 140
Amphotericin B, 204
Anaerobic bacteria, 48–9
  CNS infection, 90
  Gram-positive, 56
  Gram negative, 46
  metabolism, 4

Anaerobic cellulitis, crepitant, 168
Anaerobic streptococcal myositis, 169
Anaphylactic hypersensitivity, 24, 25
Ancylostoma braziliense, 81, 176
Ancylostoma duodenale, 79, 141
Aneurysms, mycotic, 126, 127
Angina
  Ludwig's, 106, 107
  Vincent's, 104, 105
Angioneurotic laryngeal oedema, 106
Angiostrongylus cantonensis, 89
Animal(s), *see also* specific animals
  as reservoirs, *see* Zoonoses
  as sources of infection, 18
  wounds infected from, 170
    bite, 171
Animal-loving dermatophytes, 9, 68
Anopheles spp., 72
Anthelmintics
  Ancylostoma braziliense, 81
  Ancylostoma duodenale, 79
  Ascaris lumbricoides, 79
  B. malayi, 80
  Diphyllobothrium latum, 83
  Dracunculus medinensis, 81, 177
  Echinococcus spp., 83, 146
  Enterobius vermicularis, 78
  Fasciola hepatica, 84
  Fasciolopsis buski, 84
  H. nana, 83
  L. loa, 80
  Onchocerca volvulus, 80
  Opisthorchis sinensis, 84
  P. westermani, 85, 93
  Schistosoma spp., 85, 131
  Strongyloides stercoralis, 79
  Taenia saginata, 82
  Taenia solium, 82
  Toxocara canis/T. cati, 81, 133
  Trichinella spiralis, 81, 185
  Trichuris trichiura, 78
  W. bancrofti, 80
Anthrax, 33, 168
Anti-bacterials, *see* Antibiotics
Antibiotics (antibacterials), 196–204, *see also* specific (types of) antibiotics
  assays, 211
  associated enterocolitis, 137
  characteristics, 199
  classification, 198
  development, 198
  elderly patients, 195
  GP patients, 194
  intestinal/diarrhoeal disease caused by, 137
  pharmaco-kinetics, 199
  resistance, 199
  sensitivity tests, 210–11
    Enterobacteriaceae, 40
    N. gonorrhoea, 155, 157, 159
  target site classes, 198–9
Antibody, 22–3, 192, *see also* Immunoglobulins, Serology
  defects, 192–3
  levels, measurement, 210
Antibody-dependent cytotoxicity, 24, 25
Anti-fungals, 204
  Aspergillus spp., 62
  B. dermatitidis, 66

Candida spp., 63
Coccidioides immitis, 67
Cryptococcus neoformans, 64
cutaneous mycoses, 69, 173
H. capsulatum, 65
P. brasiliensis, 67
subcutaneous mycoses, 69, 173
zygomycosis, 6, 70
Antigen detection, 208–9
Antigenic variation, 25
Antihelminthics, *see* Anthelmintics
Anti-idiotype vaccines, 206
Anti-malarials, 72, 93, 130, 205
Antimicrobials, 198–205, *see also* specific types
  general properties, 198–9
Antimonials, 205
Anti-mycobacterials, 204
  atypical mycobacteria, 55, 204
  M. leprae, 55, 92, 204
  M. tuberculosis, 54, 116, 119, 124, 153, 204
Anti-protozoals, 205
  B. coli, 75
  Cryptosporidium spp., 73
  D. fragilis, 75
  E. histolytica, 74
  G. lamblia, 75
  Leishmania spp., 76, 177
  N. fowleri, 74
  Plasmodium spp., 72, 93, 130, 205
  Toxoplasma gondii, 73, 185
  Trichomonas vaginalis, 75
  Trypanosoma brucei, 77, 83, 131
  Trypanosoma cruzi, 77, 125, 131
Antisepsis, 186
Aphthous stomatitis, 101
Apicomplexa, 10
Arachnida, 12
Arthritis, infective, 182–3
Arthropoda, 12, 71
Ascarid, dog/cat, 81
Ascaris lumbricoides, 79, 140–1
  travellers, 189
Ascomycotina, 6
Asepsis, 186
Aseptic meningitis, 86, 87
Asexual reproduction
  fungi, 8
  Plasmodium spp., 72
Aspergillus spp., 62, 63
  disseminated infection, 135
  flavus, 62
    toxin, 8
  fumigatus, 62
  necrotising infections, 169
  niger, diagnosis, 7
  respiratory infections, 120
Aspiration pneumonia, 114
Athlete's foot (tinea pedis), 172
Atopic allergy, 24
Attack, microbial, 14, 18–19
Autoclaving, 197
Autotrophs, 4
Azoles (incl. imidazoles)
  anthelmintic, 205
  antibacterial, 201
  antifungal, 204
  antiprotozoal, 205
Aztreonam, 201, 202

B

B-cells, 22, 23
  defects/dysfunction, 193
  functional assessment, 25
Bacilli
  gram-negative, 40–1
  gram-positive, 56
Bacillus spp., 33
  anthracis (and anthrax), 33, 168
    toxin, 15, 33
  cereus, 33, 138
    crepitant cellulitis, 169
Bacteraemia, 128–9
  in endocarditis, 126
  hospital-acquired, 190, 191
  metastatic spread by, 103
Bacteria, 2–5, 28–61, *see also genera/species*
  chemotherapy, *see* Antibiotics
  classification, 2
  CNS infection
    focal, 90–1
    meningeal, 86
  diarrhoeal disease, 138–9
  energy, 4
  growth, 4
  latent pathogenic, 17
  lymph node infection, 123
  metabolism, 4–5
  nomenclature, 2
  nutrition, 4, 25
  overgrowth (bowel), 137
  size and structure, 3
  skin infections, 164–9
    tropical and rare, 174–5
  urinary tract infection, 148
  in vaccines, live attenuated and inactivated, 206
  wound infections, 170–1
Bacteriuria, 148
Bacteroides spp., 48–9, 49
Bairnsdale ulcer, 174–5
Balanitis, 160
Balantidium coli, 75, 140
Bancroftian filariasis, 80, 177
Barcoo rot, 107
Bartholinitis, 156
Basidiomycotina, 6
BCG vaccination, 119
Beef tapeworm, 82
Bejel, 174, 175
Benzimidazoles, 205
Beta-lactam antibiotics, 198, 200–2
Beta-lactamases, 211
Bifidobacterium spp., 56
Biliary infections, 148–9
Biochemical tests
  bacteria
    Bacteroides spp., 48
    Campylobacter spp., 44
    E. coli, 41
    Fusobacteria spp., 48
    N. gonorrhoeae, 36
    N. meningitidis, 37
    Pseudomonas spp., 46
    staphylococci, 29
  fungi, *C. neoformans*, 64
Bite infections, 171
Black piedra of hair, 68, 172
Bladder infection (incl. cystitis)
  bacterial, 148, 148–9, 155
  Schistosoma spp., 153
Blastomycosis
  N. American (*Blastomyces dermatitidis*
    infection; blastomycocis), 9, 66, 67
    respiratory infection, 121
  S. American, *see* Paracoccidioidomycosis
Blepharitis, 96

Blood protozoa, 11, 76
Boils, 166
Bone/bone marrow infection, 178–81
Bordetella spp., 38–9, 39
  bronchioseptica, 39
  parapertussis, 39, 109
  pertussis, 38, 109
Borrelia spp., 52–3
  burgdorferi (Lyme disease), 50, 52, 53, 89,
    124, 132
    CNS infection, 89
    myocardial infection, 124
  of relapsing fevers (*B. recurrentis* etc.), 52,
    53, 71, 132
Botulism, 35, 92, 138
Bowel, *see* Gastrointestinal tract
Bradyzoites, 72
Brain, *see* Central nervous system
Branhamella catarrhalis, 37, 109
Breast infections, postnatal, 186
Brodie's abscess, 178
Bronchial infection, 110–11
  C. diphtheriae, 107
Bronchiectasis, 111
Bronchitis, acute, 110, *see also*
  Laryngotracheobronchitis
Broth dilution, 211
Brucella spp., 50, 184
  CNS infection, 89
  osteomyelitis, 181
Brugia malayi, 80, 177
Bubonic plague, 51, 184, 185
Bullous impetigo, 164
Burns, infection, 171
  streptococcal, 164
Buruli ulcer, 174–5

C

C-reactive protein, 21
Calymmatobacterium granulomatis, 47, 163
Campylobacter spp., 44–5, 138
Canaliculitis, 97
Cancer, immune defects, 193
Cancrum oris, 101, 108
Candida spp., 62–3
  cutaneous infection, 173
  disseminated infection, 135, 173
  vaginitis, 161
Capnocytophaga canimorsus, 171
Capsules, 3, 25
  staphylococcal, 28
Carbohydrate metabolism in infection, 27
4-Carbon compounds, conversion to 3-carbon
  compounds, 4
Carbuncle, 29, 166
Carcinogens and normal flora, 17
Cardiac infections, 124–7
Cardiobacterium hominis, 47
Caries, 102
Caseation, tuberculosis, 54
Cat parasites
  A. braziliense (cat hookworm), 81, 176
  Toxocara cati, 81, 99, 133
  Toxoplasma gondii, 72–3
Cat scratch fever, 171
Cattle tapeworm (*T. saginata*), 82
Cavernous sinus thrombosis, 97
Cavitation, tuberculosis, 54, 118
CD4+ T-cells, 23
CD8+ T-cells, 23
Cell-mediated hypersensitivity, 24, 25
Cell-mediated immunity, 23, 192
  defects, 193
Cell membrane, bacterial, 3
  function inhibitors, 199
Cell wall, bacterial, 3, 25

synthesis inhibitors, 198, 200–1
Cellular defence mechanisms
  non-specific, 20
  specific (immune system), *see* Cell-mediated
    immunity
Cellulitis
  C. perfringens, 34–5
  crepitant anaerobic, 168
  gangrenous, 169
  orbital, 97
  staphylococcal, 167
  streptococcal, 165, 166–7
  synergistic necrotising, 169
Central nervous system, *see also genera/species*,
    86–93
  diffuse infections, 88–9
  focal infections, 90–1
  meningitis, 86–7
  T. solium, 82
  tropical/rare infections, 92–3
Cephalosporins, 200–1
Cerebral infections, *see* Central nervous
    system
Cervical lymphadenitis, 108
  tuberculosis, 108, 118
Cervix, uterine
  cytologic metaplasia, 157
  infection (cervicitis), 156–7
Cestodes (tapeworms), 12, 82–3, *see also*
    genera/species
  intestinal/diarrhoeal disease, 82–3, 140, 141
  structure, 12
Chagas' disease, 77, 98, 124, 131
Chalazion, 96
Chancre, 52
Chancriform ulcers, 168
Chancroid, 38, 162
Chemical(s)
  as disinfectants, 197
  peritonitis caused by, 142
  as sterilants, 196
Chemotaxis (in phagocytosis), 20
Chemotherapy
  bacterial infections, *see* Antibiotics
  fungal infections, *see* Antifungals
  helminthic infections, *see* Anthelmintics
  protozoal infections, *see* Antiprotozoals
Chemotrophy, 4
Chlamydia spp., 58–9
  pneumoniae, 59, 105
  psittaci, 59
  trachomatis, 58–9, 99
    cervicitis, 156, 156–7
    lymphogranuloma venereum, 162–3
    perihepatitis, 147
    prostatitis, 154, 155
    salpingitis/pelvic inflammatory disease,
      158, 159
    urethritis, 154
Chloramphenicol, 201, 202
Chloroquine, 205
Cholangitis, 144
Cholecystitis, 144
Cholera, 44, 136, 139
Chorioretinitis, 100
Chromobacterium spp., 47
Chromoblastomycosis, 69, 176
Chronic obstructive pulmonary disease
    (COPD), 110–11, 114
Ciliates, 10, 75
Ciliophora, 10
Ciprofloxacin, 201, 203
Citrobacter spp., 41
Claviceps purpurea toxin, 8
Clindamycin, 201, 202
Clostridia spp., 34–5
  botulinum, 35, 92, 138
  difficile, 35, 137

myonecrosis, 169
perfringens, 34–5, 138, 169
  salpingitis/pelvic inflammatory disease, 158
  toxins, 15, 34
septicum, 35
tetani, 35, 93
  prophylaxis, 207
other species, 35
Co-agglutination, 208
Coagulation, 21
Cocci
  Gram-negative, 36–7, 49, 50
  Gram-positive, 30–1, 49
Coccidioides immitis, 9, 66–7, 67
  respiratory infection, 121
Cocco-bacillus, Gram-positive, 38, 51
Coliform bacilli, 40–1
Colitis, pseudomembranous, 35, 137
Colonization, 14
Common vehicle transmission, 18
Complement, 21
  defects, 192
Condylomata lata, 163
Confirmatory tests, *see genera/species*
Congenital defects in defence, 192–3, 208–9
Congenital infections/fetal infections, 186–7
Conjugation, bacteria, 5
Conjunctivitis, 97, 194
Contact, 28
Contamination, 14
Control of infection, *see also genera/species*
  GP patients, 194
  hospital-acquired infections, 191
  metazoa, 13
  protozoa, 11
  wounds, 170
Corneal infections, 96–7
Corynebacteria spp.
  diphtheriae, *see* Diphtheria
  minutissimum, 32, 68, 174
  other, 32, 106, 107
Cotrimoxazole, 203
Coxiella burnetii, 61
Cross-infection of neonates in hospitals, 187
Croup (laryngotracheobronchitis), 106, 109
Crustacea, 12, *see also* Trematodes
Cryptococcus neoformans, 9, 64, 88
  CNS infection, 88
  necrotising infection, 169
  respiratory infection, 121
Cryptosporidium spp., 73, 140
Culture, 209, *see also other genera/species*
  Bacteroides spp., 48
  Brucella spp., 50
  Chlamydia spp., 58
  Clostridia perfringens, 34
  Corynebacteria diphtheriae, 32
  E. coli, 41
  enteric bacteria, 40
  Fusobacteria spp., 48
  Haemophilus influenzae, 38
  Helicobacter pylori, 45
  Legionella spp., 39
  Listeria monocytogenes, 32
  meningitis, 87
  N. gonorrhoeae, 36, 155, 157, 159
  N. meningitidis, 37
  Pseudomonas spp., 46
  staphylococci, 28
  streptococci, 30
  V. cholerae, 44
Cutaneous diphtheria, 107
Cutaneous *E. histolytica*, 176
Cutaneous larva migrans, 81, 176
Cutaneous leishmaniasis, 76, 177
Cyst(s)
  CNS, 91
  hydatid, 82–3, *see also* Hydatid disease

protozoan
  E. histolytica, 74
  G. lamblia, 75
Cystic fibrosis, 111
Cysticercosis, 82
Cystitis
  bacterial, 148–9, 155
  schistosomal, 153
Cytokines
  in immunity, 23
  in infection, 27
Cytotoxic T-cells, 23
Cytotoxicity, antibody-dependent, 24, 25

D

Dacrocystadenitis, 97
Dacrocystitis, 97
Dapsone, 204
Defence/resistance, host, 14, 20–5, 192–3
  1st line of, 19, 192
  2nd line of, 20–1, 192
  3rd and 4th lines of, 22–5
  evasion, 25
  fungal infection, 9
    H. capsulatum, 65
  impaired and ruptured defences, 19, 192–3
    elderly, 195
    hospital-acquired infections, 190
  metazoan infection, 13
  non-specific, 20–1, 192
  protozoan infection, 110
  specific defence, 22–5, *see also* Immunity
  in travellers, preparing, 188
Dental infections, 102
Dermal leishmaniasis, post kala azar, 177
Dermatology, *see* Skin
Dermatophytes, 68–9, 172–3
Desert (veldt) sore, 107
Developing countries (incl. tropics), infections, *see also* Travellers
  abdominal, 138
    diarrhoeal disease, 138
  CNS, 92–3
  ocular, 98
  orofacial, 108
  sexually-transmitted diseases, 162–3, 188–9
  skin, 174–7, 188–9
  soft tissues, 174–7
  systemic, 130–1
  urinary tract, 152–3
Di George syndrome, 193
Diabetes
  foot infections, 169
  ischaemic-neuropathic osteomyelitis, 179
Diagnosis, 208–9, *see also genera/species*
  fungal infection, 7
  helminths, 13
  hypersensitivity, 25
  immunodeficiency, 25
Diarrhoeal disease, 136–41
  general features, 136–7
  travellers, 136, 189
Dientamoeba fragilis, 75, 140
Diethylcarbamazine, 205
Diffusion tests, antibiotic, 210
Digestive tract, *see* Gastrointestinal tract
Dikaryomycota, 6
Dilution tests, antibiotic, 211
Dimorphic fungi, 7
Diphtheria (*C. diphtheriae* infection), 32, 33, 106–7
  bronchitic, 107
  cutaneous, 107
  differential diagnosis, 107
    epiglottis, 106
  laryngeal, 107, 109

myocardial, 124
nasal, 106
nasopharyngeal, 107
pharyngeal, 106–7
throat, 104, 106
toxin, 15, 32
Diphyllobothrium latum, 83
Direct immunofluorescence microscopy, 209
  Chlamydia trachomatis, 155, 157, 159
Dirofilaria immitis, 81
Disc, intervertebral, infection, 183
Disc diffusion (antibiotics), 210
Disinfection, 196, 197
Diverticula/diverticulitis, 142, 143
Dog hookworm, 81, 176
Dog roundworm, 81, 99, 133
Dracunculus medinensis, 80, 176–7
Drinks, precautions with, 188
Drugs
  antimicrobials, *see antibacterials, etc.*
  immunosuppressive, 193
  intravenous abuse, endocarditis in, 126
Duke Endocarditis Service, 127
Dwarf tapeworm, 83
Dysentery, 136
  amoebic, 140
  bacillary, 41

E

E-test, 210
Ear infections, 94
  travellers, 188, 189
Echinococcus spp., 82–3, *see also* Hydatid disease and Hydatid cysts
Ecology
  fungal, 9
  metazoans (incl. helminths), 13
  protozoans, 11
Ecthyma gangrenosum, 169
Ectoparasites, 71
Eikenella corrodens, 47
Elderly patients, 195
Embden–Meyerhof–Parnas pathway, 4
Emboli, septic, 128, 129
Embolisation in endocarditis, 126
Empyema, 117, 170
Encephalitis, African trypanosomiasis, 93, *see also* Meningoencephalitis
Endocarditis, infective, 126–7
Endodontic infection, 102
Endometritis, 158
Endophthalmitis, 101
Endotoxin (lipopolysaccharide; LPS), 3
  damage caused by, 25
  shock caused by release, 26–7
Energy
  metabolism
    bacteria, 4
    humans in infection, 27
  sources, bacteria, 4
Entamoeba spp.
  histolytica, 74, 140, 145
    cutaneous disease, 176
    diarrhoeal disease, 74, 140
    liver abscesses, 140, 145
  other species, 74
Enteric bacteria, 40–1
Enteric fever, 42, 133
Enterobacter spp., 43
Enterobacteriaceae, 40–3
Enterobius vermicularis, 78, 141
Enterococcus spp., 30–1
Enterocolitis, 136
  antibiotic-associated, 137
  Y. enterocolitica, 50, 139, 147
Enterohaemorrhagic *E. coli*, 138

Enteroinvasive *E. coli*, 138
Enteropathogenic *E. coli*, 138
Enterotoxigenic *E. coli*, 138
Enterotoxin, 15, 136, 138, 139
 C. perfringens, 34
Entner–Doudoroff anaerobic pathway, 4
Entomophthoromycosis, 70
Environmental reservoirs, *see* Reservoirs
Enzyme (antibiotic-degrading) detection, 211
Eosinophilic meningitis, 89
Epidemiological information, 18
Epidermophyton spp., 68, 69
Epididymitis, 160
 chlamydial, 154
Epiglottitis, 106
Ergot alkaloids, 8
Erysipelas, 164–5
Erysipelothrix rhusiopathiae, 33, 170
Erythema marginata, 165
Erythema multiforme, *M. pneumoniae*, 112, 113
Erythema nodosum, 165
Erythrasma, 32, 68, 174
Erythromycin, 201, 203
Escherichia coli, 40–1, 138
Ethambutol, 204
Ethylene oxide, 196
Eubacterium spp., 56
Eumycotic mycetoma, 69, 174, 176
Exophiala wernecki, 68, 172
Exotoxins
 damage caused by, 25
 pyrogenicity, 26
 shock caused by, 26
Extradural abscesses, 91
Eye infections, 96–100, 194
 tropical, 98
 travellers and, 188, 189

F

Facial infection, tropical/rare, 108
Fallopian tube infection (salpingitis), 156, 158–9
Fascial space infections, deep (head/neck), 103, 105
Fasciitis, necrotising, 165, 169, 191
Fasciola hepatica, 84
Fasciolopsis buski, 84, 147
Fat metabolism in infection, 27
Feet, *see* Foot
Fetal infections, 186–7
 CNS, 93
 pneumonia, 112
Fever/pyrexia, 26–7
 travellers, 188
 of unknown origin, 134–5
Fibrinolysin, 21
Filamentous fungi, 7
Filarial worms, 80
 cutaneous and systemic infection, 177
 ocular infection, 99
Filtration, sterilisation by, 196
Fimbriae, 3
Fish tapeworm, 83
Fish-tank granuloma, 170, 174, 175
Fitz-Hugh Curtis syndrome, 147
Flagella, bacterial, 3
Flagellates, 10, 75
Flatworms, *see* Cestodes; Trematodes
Flavobacterium spp., 46–7
Flora/microflora, normal, 14, 16–17
 defined, 16
 fungi as, 9
 harmful roles, 17
  opportunistic infection, 8, 17
 skin, 16
 useful roles, 16

Flubendazole, 205
Flucytosine, 204
Flukes, *see* Trematodes
Fluoroquinolones, 201, 203
Folliculitis, 164
Food, precautions with, 188
Food poisoning, 136
 B. cereus, 33, 138–9, 138
Foot infections, *see also* Mycetoma
 bacterial mycetoma (Madura foot), 174
 diabetic, 169
 eumycotic mycetoma, 69, 174, 176
 fungal, 172
Formaldehyde, 196
Formalin, 196
Fournier's gangrene, 169
Francisella tularensis (tularaemia), 50, 51, 98–9, 185
 oculoglandular tularaemia, 98–9
Fungaemia, 128–9
Fungi, 6–9, 62–70, *see also genera/species*
 classification, 6
  clinical, 7
 diagnosing infection, 7
 drugs acting against, *see* Antifungals
 ecology, 9
 host defence, *see* Defence
 morphology, 7
 nomenclature, 6
 pathogenicity, 8, 17
 reproduction, 8
 structure, 6–7
 subcutaneous infection, 7, 69, 176
 superficial/cutaneous infection, 7, 62–3, 68–9, 162–3, 176
 systemic/deep infection, 7, 64–7
  CNS infection, 90, 91
  lymph node infection, 123
  osteomyelitis, 180–1
  respiratory infection, 121
Furuncles, 166
Fusobacteria spp., 48–9

G

Gallbladder inflammation, 144
Gangrene
 Fournier's, 169
 gas, 34, 169
 synergistic, 168
 vascular, infected, 169
Gangrenous infections (in general), 168–9
Gangrenous stomatitis, 101, 108
Gardnerella vaginalis, 161
Gas-forming infections, 168–9
 gas gangrene, 34, 169
Gastroenteritis, 136
 Salmonella spp., 42
Gastrointestinal tract (predominantly intestine)
 infections, 136–41, *see also specific regions/disorders*
  bacterial, 138–9
  cestodes, 82–3, 140, 141
  GP patients, 194
  nematodes, 78–9, 140, 141
  protozoal infection, 11, 74, 75, 140
  trematodes, 84–5, 140, 141, 147
 normal flora, 16
Gay bowel syndrome, 137
General paresis of the insane, 88–9
General practice, 194
Genetics, bacterial, 5
Genital infections, 156–63, *see also* Sexually-transmitted diseases
 in GP patients, 194
Gentamicin, 201–2

Geographic distribution, *see also* Developing countries, Travellers
 fungi, 9
 metazoa, 13
 protozoa, 11
Ghon focus, 118
Giardia lamblia, 75, 140
Gingivitis, 102
Glomerulonephritis, acute, 152
Glue ear, 94
Glutaraldehyde, 196
Glycolysis, 4
Gonococcus (*N. gonorrhoeae*) and gonorrhoea, 36
 arthritis, 183
 cervicitis, 156
 perihepatitis, 147
 pharyngeal, 104
 salpingitis/pelvic inflammatory disease, 158, 159
 urethritis, 154
Gram-negative organisms, 2
 bacilli, 40–1
 cocci, 36–7, 49, 50
 endotoxin, 3, 25, 26–7
 rods, 38–9, 42–3, 46–7, 48–9, 50–1
 structure, 3
Gram-positive organisms, 2
 anaerobes, 56
 bacilli, 56
 cocci, 30–1, 49
 cocco-bacilli, 38, 51
 oxidase-positive, 36–7, 44–5
 rods, 32–5
 structure, 3
Gram stain, 2, 208, *see also Bacterial genera/species*
 arthritis (infective), 183
 pneumonia, 113
 Streptococci spp., 30
Granuloma, fish-tank, 170, 174, 175
Granuloma inguinale, 162–3
Granulomatosis infantiseptica, 33
Granulomatous hepatitis, 147
Growth, bacteria, 4
Guinea worm (*D. medinensis*), 80, 176–7
Gum (gingival) infection, 102
Gumma, 89
Gut, *see* Gastrointestinal tract

H

Haematogenous osteomyelitis, acute, 178
Haemoflagellates, 76
Haemolytic streptococci, 30
Haemophilus spp., 38, 39
 aphrophilus, 38
 ducreyi, 38, 162
 haemolyticus, 38
 influenzae (incl. type b), 38
  epiglottic infection, 106
  meningeal infection, 87
  pneumonia, 113, 115
  vaccination, 113
 parainfluenzae, 38
Haemorrhagic rash, meningococcal, 37, 128, 129
Hafnia spp., 43
Hair
 piedra, 68, 172
 tinea, 173
Hair follicle infection, 164
Hansen's disease (leprosy), 54–5, 174
 chemotherapy, 55, 92, 204
 congenital, 186
 ophthalmopathy, 98
 peripheral neuropathy, 55, 92

Head louse, 71
Heart infections, 124–7
Heat
  disinfection by, 197
  sterilisation by, 196–7
Helicobacter pylori, 45
Helminths (worms), 12–13, 78–85, *see also*
      Cestodes; Nematodes; Trematodes *and
      genera/species*
  chemotherapy, *see* Anthelmintics
  diagnosis, 13
  ecology, 13
  host resistance, 12–13
  pathogenicity, 12–13, 17
  reproduction, 12
  structures, 12
Helper T-cells, 23
Hepatic disease, *see* Liver
Hepatitis, granulomatous, 147
Herpes simplex, 162
Heterophyes heterophyes, 141
Heterotrophs, 4
Histoplasma capsulatum (Histoplasmosis), 9,
      64–5, 66
  respiratory infection, 121
History-taking, pyrexia of unknown origin, 134,
      135
HIV infection
  immune system in, 193
  pyrexia of unknown origin and, 134
Homosexual men, gut infections, 137
Hookworms, 79, 141
  dog/cat, 81, 176
Hordeolum, 96
Hospital-acquired (nosocomial) infections,
      190–1
  neonatal, 187
  respiratory (incl. pneumonia), 112, 114–15,
      190, 191
  surgical wound, 17, 170, 190–1
Hospital-acquired (nosocomial) pyrexia of
      unknown origin, 134
Host–microbe relationships, 14–15, *see also*
      Humans
  H. capsulatum, 65
  host damage, 25
  host resistance/defence, 14, 20–5, 192–3, *see
      also* Defence
Human(s)/people, *see also* Host-microbe, and
      Immunocompromised persons
  bites from, 171
  fungi as normal flora of, 9
  metabolism, 16, 27
  as reservoirs, 18
  as sources of infection, 18
Human immunodeficiency virus, 134, 193
Humoral defence mechanisms, 21
Hydatid disease
  CNS, 91
  kidney, 152
  liver, 146
  lung, 116, 120
  soft tissues, 177
Hydatid worm, *see Echinococcus* spp.
Hymenolepis nana, 83
Hypersensitivity, 24
  fungal spores, 8

I

Icterohaemorrhagic fever (Weil's disease), 184
Imidazoles, 201, 204–5, *see also* Azoles
Imipenem, 201, 202
Immigrants, 189
Immune complex-mediated hypersensitivity,
      24, 25

Immunisation, 22, 206–7
  active, 206–7, *see also* Vaccination
  active and passive combined, 207
  definitions, 206
  passive, 206, 207
  travellers, 188
  when and who to immunise, 207
Immunity, 22–5, 192–3, *see also* Defence
  disorders, 24–5, 192–3
  microbial factors impeding, 14
  normal flora and, 16
Immunocompromised persons (and
      opportunistic infections), 14, 17, 36,
      192–3
  fungal infections, 8
    candidiasis, 173
  necrotising infections, 169
  pneumonia, 115
  stomatitis, 101
Immunocytochemistry, 208
Immunodeficiency, 24–5
  diagnosis, 25
Immunodetection, 208–9
Immunofluorescence microscopy, 209
  chlamydia, 155, 157, 159
Immunogenicity of vaccines, 206
Immunoglobulins, 22–3, *see also* Antibody
  in passive immunisation, 207
Immunosuppressive drugs, 193
  transplant patients, 193
Impedins, 15
Impetigo, 164
  bullous, 164
Incidence rates, 18
Infection, defined, 14
Infective dose, 15
Infectivity, period of, 15
Ingested/orally-acquired pathogens, *see also*
      gastrointestinal tract
  gastroenteritis, 136, 138–9
  homosexuals, 137
  neonates, 187
  protozoal and parasitic, 140–1
Ingestion (in phagocytosis), 20
Injury, bowel, 137, empyaema and trauma
      including bites, 170–7
Insects, 12
  protozoa carried by, 76–7
  skin infections, 71
Interleukin-1, 27
Intervertebral disc infection, 183
Intestine, *see* Gastrointestinal tract
Intracranial infections, *see* Central nervous
      system
Intracranial thrombophlebitis, 122
Intravenous drug users, endocarditis, 126
Invasiveness, 15
Iridocyclitis, 100, 194
Iritis, 193, 194, 196
Irradiation, 100, *see also* Radiation
Ischaemic-neuropathic osteomyelitis, acute, 179
Isoniazid, 204
Ivermectin, 205
Ixodes spp., 19, 53

J

Joint infection, 182–3
  tuberculous, 181, 182
Jones criteria, rheumatic fever diagnosis, 125

K

Kala azar, 76, 130, 177
Kallikrein, 21

Kawasaki disease, 132
Keratitis, 96–7
Kerion, 173
Kidney infections, 148–51
  tropical and rare, 152, 153
Killing of microbes
  by antibiotics, killing curves, 211
  by host cells, 20
Kingella spp., 37
Klebsiella spp., 42–3
  necrotising infection, 169
  ozaenae, 42, 43
  pneumoniae, 42, 43
  rhinoscleromatis, 42, 43, 108
Krebs cycle, 4

L

Laboratory, 208–9
Lactobacillus spp., 56
Larva migrans
  cutaneous, 81, 176
  ocular, 99
  visceral, 81, 133
Laryngotracheobronchitis, 106, 109
Larynx
  angioneurotic oedema, 106
  infections, 106, 109
    C. diphtheriae, 107, 109
Latent pathogens, 17
Latex agglutination, 208
Legionella spp. (incl *L. pneumophila*), 39, 113
Leishman–Donovan bodies (amastigote), 76,
      121
Leishmania spp., 76, 130
  cutaneous leishmaniasis, 76, 177
  donovani (and kala azar; visceral
      leishmaniasis), 76, 130
    dermal leishmaniasis following, 177
  mucocutaneous leishmaniasis, 76, 177
Lemierre's disease, 105
Leprosy, *see* Hansen's disease
Leptospira interrogans, 50, 53, 184
Lice, *see* Louse
Lincosamides, 201, 202
Lipopolysaccharide, 3, 25–7
Listeria monocytogenes, 32–3, 33, 132
  congenital, 186
Liver
  abscesses, 143, 145–6
    amoebic, 140, 145
    pyogenic, 145
  cysts, 146
  tropical/rare infections, 147
Liver fluke, sheep, 84
Loa loa, 80, 99
Lobomycosis, 69, 176
Louse/lice, 71
  diseases transmitted by
    epidemic typhus, 60–1, 71
    relapsing fevers, 53, 71, 132
Ludwig's angina, 106, 107
Lumbar puncture (LP)
  acute meningitis, 87
  diffuse chronic CNS infection, 88, 89
Lung disease, *see also* Respiratory tract
  chronic obstructive (COPD), 110–11, 114
  underlying, pneumonias complicating, 114
Lung fluke, 85, 93, 121
Lungworm, rat, 89
Lupus vulgaris, 174
Lyme disease, 89, 124, 132, *see also Borrelia* spp.
Lymph node infection/lymphadenitis, 123
  cervical, 108, 118
  mediastinal, tuberculous, 118
  mucocutaneous lymph node syndrome, 132

Lymphadenitis, 123
Lymphangitis, 122–3, 165
Lymphocutaneous sporotrichosis, 176
Lymphocytes, 22, 23
    functional assessment, 25
Lymphogranuloma venereum, 58–9, 162, 163
Lysozyme, 21

M

Macrolides, 201, 202–3
Macronodular candidiasis, 173
Macrophages, 23, see also Phagocytosis
Madura foot, 174
Malaria (Plasmodium spp. infection), 72, 130–1,
    188–9
    antimalarials, 72, 93, 130, 205
    cerebral, 92
    travellers, 188–9
        prophylaxis, 188
Malassezia furfur, 68, 172
Malignancy, immune defects, 193
Malnutrition, 27
    immune impairment, 193
Mastitis, 186
Mastoiditis, 95
Maternal infections, 186
    babies acquiring, 187
Mebendazole, 205
Media, culture, 209, see also genera/species
Mediastinal lymph nodes, tuberculous, 118
Medical kit, travellers, 188
Mefloquine, 205
Meleney's gangrene, 168
Melioidosis, 46, 120
Membrane, cell, see Cell membrane
Memory cells, 22
Meningitis, 86–7, see also specific organisms
    eosinophilic, 89
    meningococcal, 37, 87
    syphilitic, 88
    tuberculous, 89
Meningococcus (N. meningitidis), 36–7, 194
    meningitis, 37, 87
    septicaemia, 37, 128, 129
Meningoencephalitis
    brucellosis, 89
    syphilitic, 88
Meningovascular syphilis, 88
Meropenem, 201, 202
Merozoites, 72
Metabolism
    bacteria, 4–5
        product detection, 209
    human, 16
        in infection, 27
Metagonimus yokogawai, 141
Metazoans, 12–13
Metronidazole
    as antibacterial, 201, 203
    as antiprotozoal, 205
Microbiology laboratory, 208–9
Microfilariae, 80
Microflora, see Flora
Microscopy, 208
    bacteria, see also genera/species
        Clostridia perfringens, 34
        Legionella spp., 39
        Listeria monocytogenes, 32
        N. gonorrhoeae, 36, 155, 157, 159
        N. meningitidis, 37
        staphylococci, 28
    fungi
        Coccidioides immitis, 66–7
        Cryptococcus neoformans, 64
        H. capsulatum, 65

P. brasiliensis, 67
helminths
    intestinal nematodes, 78, 79
    trematodes, 84, 85
meningitis, 37, 87
protozoa
    E. histolytica, 74
    N. fowleri, 74
Microsporum spp., 68, 69, 173
    canis, 68, 172
Miliary tuberculosis, 118
Mineral metabolism in infection, 27
Mites, 71
Moraxella (Branhamella) catarrhalis, 37, 109
Morganella spp., 43
Mosquito, Anopheles spp., 72
Mothers, 186–7, see also Maternal infections
Mouth, see Oral cavity
Mucocutaneous candidiasis, chronic, 173
Mucocutaneous leishmaniasis, 76, 177
Mucocutaneous lymph node syndrome, 132
Mucormycosis, 69–70, 91, 169
Mucous membranes (as barrier), 19, 192
    acquired breaches, 192
Mucoviscidosis, 111
Multicellular parasites, 12–13, see also
    genera/species
Murein, 3, 26, see also Peptidoglycan
Muscle
    necrosis, 169
    streptococcal infection, 169
    tropical pyomyositis, 174–5
Mycetoma, 69
    bacterial/actinomycotic, 56, 174
    eumycotic, 69, 174, 176
Mycobacteria spp., 54–5, 108
    atypical (in general), 55, 119
        cervical lymphadenitis, 108
        chemotherapy, 55, 204
    avium complex, 55, 204
    bovis, 54, 118, see also Tuberculosis
    leprae, 54–5, see also Leprosy
    marinum
        chemotherapy, 170, 204
        wound infection, 170
    structure, 3
    tuberculosis, 54–5, see also Tuberculosis
    ulcerans, 55, 174–5
Mycology and mycoses, see Fungi
Mycoplasma spp., 57
    pharyngitis, 104
    pneumonia, 113
Mycotic aneurysms, 126, 127
Mycotoxins, fungal, 8
Myocarditis, 124
Myonecrosis, clostridial/non-clostridial, 169
Myositis, anaerobic streptococcal, 169
    tropical pyomyositis, 174–5

N

Naegleria fowleri, 74
Nail
    paronychia, 164
    infections next to, 164
    tinea, 173
Nasal flora, 16
Nasopharyngeal diphtheria, 107
Natural killer cells, 20
Necator americanus, 79, 141
Necrosis, muscle, 169
Necrotising cellulitis, synergistic, 169
Necrotising fasciitis, 165, 169, 191
Neisseria spp.
    gonorrhoeae, 36, see also Gonococcus
    meningitidis, 36–7, see also Meningococcus

Nematodes (roundworms), 12, 78–81, see also
    genera/species
    intestinal, 78–9, 140–1
    structure, 12
    tissue, 80–1
Neonatal infections, 187
    pneumonia, 112, 187
Nervous system infection
    central, 86–93, see also Central nervous
        system
    peripheral, 92
Neurological infection, see Nervous system
    infection
Neurosyphilis, 88–9
Neutropenic pyrexia of unknown origin,
    134
Newborns, see Neonatal infections
Nitroimidazoles, 201, 203
Nocardia spp., 56
    CNS infection, 90–1
    necrotising infection, 169
    respiratory infections, 120
Noma, 101, 108
Norfloxacin, 201, 203
Nose
    diphtheria, 106
    flora, 16
    rhinocerebral zygomycosis, 91
    rhinoscleroma, 43, 108
    rhinosporidiosis, 69, 108, 177
Nosocomial problems, see Hospital-acquired
Nucleic acids
    detection, 209
    synthesis inhibitors, 198–9, 203
Nutrition
    bacteria, 4, 25
    human, 17
        impairment, 27, 193
Nystatin, 204

O

Ocular infections, 96–100, see also Eye
    infections
Odontogenic infection, 105
Oedema
    angioneurotic laryngeal, 106
    orbital, 97
Onchocerca volvulus, 80, 99
Oocysts, T. gondii, 72
Opisthorchis sinensis, 84, 147
Opportunistic pathogens
    bacteria, 16–17
    fungi, 8
Opsonins, 20
Oral cavity/mouth
    flora, 16
    infection, 101–3
        tropical/rare, 108
    pathogens ingested, 136–41, 187, see also
        Ingested pathogens
Orbital infections, 97
Orchitis, 160
Oriental sore (cutaneous leishmaniasis), 76,
    177
Osteomyelitis, 178–81
    acute, 178–9
    chronic, 178, 179, 181
    subacute, 178, 180
Otitis externa, 94
Otitis media, 94
Oxidase-positive gram-positive organisms,
    36–7, 44–5
Oxygen-dependent killing (respiratory burst),
    20
    evasion, 25

P

Pancreatitis, 143
Paracoccidioidomycosis (*P. brasiliensis*), 9, 66, 67
  orofacial infection, 108
  respiratory infection, 121
Paragonimus westermani (lung fluke), 85, 121
  CNS infection, 93
Parameningeal infection, 90
Paranasal sinus inflammation, acute/chronic, 95
Parasites, 10–13, 72–85, *see also specific types and genera/species*
  CNS infection, focal, 90
  intestinal/diarrhoeal disease, 136, 140–1
  lymph node infection, 123
  multicellular, 12–13
  ocular, 98, 99
  skin infections, 176–7
    skin-surface living parasites, 71
  unicellular, 10–11, 72–7, *see also* Protozoa
Paratrophy, 4
Paronychia, 164
Parotitis, 108
Pasteurella multocida, 51, 170–1
Pasteurisation, 196, 197
Pathogenicity, 14–15, *see also genera/species*
  bacteria, 16–17
  fungi, 8
    latent, 17
  metazoan parasites (incl. helminths), 12–13
    latent, 17
  protozoa, 11
    latent, 17
Pediculus spp., 71, *see also* Louse
Pelvic abscess, 159
Pelvic inflammatory disease, 158–9
Pelvic thrombophlebitis, 122
Penicillins, 200
Pentamidine, 205
Pentose phosphate pathway, 4
People, *see* Host; Humans/people
Peptidoglycan (murein)
  Gram-negative bacteria, 3
  Gram-positive bacteria, 3
  pyrogenicity, 26
Peptostreptococcus spp., 30, 49
Peptococcus spp., 49
Pericarditis, 124
Pericoronitis, 103
Perihepatitis, 147
Perinephric abscess, 150
Periodontal infections, 102–3
Peripheral nervous system infection, 92
Peritonitis, 142, 158, 159
Peritonsillar infection, 105
Pertussis, 38, 109
Phaeohyphomycosis, 68, 69, 176
Phagocyte defects, 192
Phagocytosis, 20, 21, 192
  microbial inhibitors, 15
Phagosome, 20
  survival within, 25
Pharmacokinetics of antibiotics, 199, *see also specific antibiotics*, 196–204
Pharyngeal infection, 104
  epiglottic vs, 106
Phialophora spp., 69, 176
Phosphoketolase pathway, 5
Phototrophy, 4
Phthirus pubis, 71
Phycomycosis, *see* Zygomycosis
Piedra, hair, 68, 172
Piedra hortai, 68, 172
Pig (pork) tapeworm, 82
Pig (pork) threadworm (*T. spiralis*), 81, 183
Pig-bel, 138
Pili, 3

Pinta, 52, 174, 175
Pinworm (US term for *E. vermicularis*), 78–9, 141
Pityrosporum orbiculare (and pityriasis versicolor), 68, 172
Plague bacillus (*Y. pestis*), 50, 51, 184–5
Plaque, dental, 102
Plasmids, 5
  fingerprinting, 209
Plasmin (fibrinolysin), 21
Plasmodium spp., 72, *see also* Malaria
Platyhelminthes, *see* Cestodes; Trematodes
Plesiomonas spp., 45
Pleura, tuberculosis, 118
Pleural cavity, pus collection (empyema), 117, 170
Pneumococcus, 30–1, *see also Streptococcus* spp.
Pneumocystis carinii, 115, 205
Pneumonia, 112–15, *see also* Respiratory tract infection, lower
Pneumonic plague, 51, 184
Polymerase chain reaction, 209
Pork, *see* Pig
Porphyromonas spp., 48
  asaccharolytica, 48, 49
Portal thrombophlebitis, 122, 145
Postanginal septicaemia, 105
Postpartum (puerperal) sepsis, 186
Pott's puffy tumour, 95
Praziquantel, 205
Pregnancy, 186–7
Prevalence rates, 18
Prevotella spp., 48
Primary pathogens, 14
Procaryotes, characteristics, 2
Proctocolitis, 137
Promastigote, 76
Propionibacterium spp., 56
Prostatitis, 148, 150–1
  chlamydial, 154
Prostheses
  adherence to surfaces, 25
  cardiac valve, infection, 126, 127
  prosthetic joint infections, 183
Protein
  metabolism in human infection, 27
  synthesis in bacteria, inhibitors, 198, 202–3
Protein A, staphylococcal, 28
Proteus spp., 43
Protozoa, 10–11, 72–7, *see also genera/species*
  chemotherapy, *see* Antiprotozoals
  ecology, 11
  intestinal, 11, 74, 75, 140
  nomenclature/classification, 10
  pathogenicity/host defence, 11
  reproduction, 10
  structure, 10
Providentia spp., 43
Pseudodiphtheria, 106, 107
Pseudomembranous colitis, 35, 137
Pseudomonas spp., 46
  aeruginosa, 46, 191
    necrotising infection, 169
    otitis externa, 94
  mallei, 46, 120
  pseudomallei, 46
Psoas abscess, 143
  tuberculous, 135
Pubic louse, 71
Puerperal sepsis, 186
Pulmonary disease, *see* Lung disease; Respiratory tract infection
Purine synthesis, 5
Purpura fulminans, 165
Purulent pericarditis, 124
Pus-forming (pyogenic/suppurative) disorders
  pyogenic liver abscess, 145–6
  pyogenic meningitis, 86

pyogenic osteomyelitis, 179
  staphylococcal infections, 28–9, 166–7
  streptococcal infections, 30–1, 164–5
  suppurative parotitis (acute), 108
  suppurative thrombophlebitis, 122, 145
Pyelitis, 148
Pyelonephritis, 148, 149
  acute/chronic, 148
Pylephlebitis, 122
Pyoderma, 164
Pyoderma gangrenosum, 168
Pyogenic disorders, *see* Pus-forming disorders
Pyomyositis, tropical, 174, 175
Pyothorax (empyema), 117, 170
Pyrazinamide, 204
Pyrexia, 26–7, 134–5, 188, *see also* Fever
Pyrimidine synthesis, 5
Pyrogens, 26
Pyruvate metabolism, 5

Q

Q fever, 61
Quinine, 205
Quinolones, 201, 203
Quinsy, 105

R

Radiation
  immunosuppressive effects, 193
  sterilisation employing, 196
Radioimmunoassay, 209
Rat bite fever, 171
Rat lungworm, 89
Reduviid bug, 77
Reiter's syndrome, 154–5
Relapsing fevers (*B. recurrentis* etc.), 52, 53, 71, 132
Renal infections, 148–53, *see also* Kidney
Reproduction
  fungi, 8
  helminths, 12
  protozoa, 10
Reservoirs (including environment), 18
  animals as, 18, 50–3, 72–3, 184–5, *see also* Zoonoses
  arthropods as, 71
  for fungi, 9
  hospital-acquired infections, 190
  for metazoa, 13
  for protozoa, 11
Resistance
  to antibiotics, 199
  to microbes, *see* Defence
Respiratory burst, 20, 25
Respiratory tract flora
  upper respiratory tract, 16
  wound infection by, 170
Respiratory tract infection, *see also* Lung disease
  GP patients, 194
  lower (incl. pneumonias), 110–21
    in abnormal host, 114–15
    atypical pneumonias, 57, 112–13
    congenital and neonatal, 112, 187
    elderly, 195
    fluke, 85
    hospital-acquired, 112, 114–15, 190, 191
    in normal host, 112–13
    in pre-existing lung disease, 110–111, 114
  travellers, prevention, 188
  upper, 94–5, 104–7, 109
Restriction enzyme analysis, 209
Retroperitoneal abscess, 143
Rheumatic fever/heart disease, 125

Rhinocerebral zygomycosis, 91
Rhinoscleroma, 43, 108
Rhinosporidiosis, 69, 108, 177
Rhizopus spp., 70
Rickettsia spp., 60–1
  lymph node infection, 123
  prowazeckii, 60–1, 71
  rickettsii, 61
  typhi, 61
Rifampicin, 203, 204
Rifamycins, 201, 203
Ringworm, 68–9, 172–3
River blindness worm (*O. volvulus*), 80, 99
Rocky Mountain spotted fever, 61
Rods
  acid-fast, 54–5
  Gram-negative, 38–9, 42–3, 46–7, 48–9, 50–1
  Gram-positive, 32–5
Roundworms, 12, 78–81, 140–1, *see also* Nematodes
Routes of spread/transmission, *see* Spread
Runyon classification, atypical mycobacteria, 55

S

Salmonella spp., 42, 138–9
  paratyphi, 42, 133
  typhi, 42, 133
Salpingitis/salpingo–oophoritis, 156, 158–9
Sandfly and trypanosomiasis, 76
Sarcomastigophora, 10
Sarcoptes scabei, 71
Scabies, 71
Scalded skin syndrome, 167
Scarlet fever, 165
Schistosoma spp., 85, 131
  bladder, 153
  cervix, 156
Scrofuloderma, 174
Secretory otitis media, 94
Sepsis, *see* Septicaemia
Septic thrombophlebitis, 122
Septicaemia/sepsis, 128–9
  defined, 128
  meningococcal, 37, 128, 129
  postanginal, 105
  puerperal, 186
  Salmonella spp., 42
  shock associated with, 26–7
Septicaemic plague, 51, 184
Serologic immunity, 22–3, *see also* Antibody
Serology, 209
  bacteria, 209
    Brucella spp., 50
    enteric bacteria, 40
    H. pylori, 45
    Legionella spp., 39
    Mycoplasma spp., 57
    N. gonorrhoeae, 36, 155, 157, 159
    N. meningitidis, 37
    Rickettsia spp., 60
    streptococci, 30
    Treponema pallidum, 52
    Ureaplasma spp., 57
  fungi
    C. immitis, 67
    H. capsulatum, 65
    P. brasiliensis, 67
  protozoa
    E. histolytica, 74
    T. gondii, 72
Serratia spp., 43
Serum bactericidal levels, 211
Sexual reproduction
  fungi, 8
  protozoans

Plasmodium spp., 72
T. gondii, 72
Sexually-transmitted diseases, 19, 154–63, *see also* Genital infections; Urinary tract infections
  bacterial, 38, 52
  balanitis, 160
  cervicitis, 156–7
  epididymitis, 160
  intestinal infection, 137
  prostatitis, 154
  protozoal, 75
  salpingitis/pelvic inflammatory disease, 158–9
  tropical and rare, 162–3
    travellers, 188, 189
  urethral infection, 154–5
  vulvovaginal infection, 161
Sheep liver fluke, 84
Shigella spp., 41, 139
Shock, septic, 26–7
  endotoxic, 26–7
  exotoxic, 26
Sinusitis, acute/chronic, 95
Skin, *see also* Cutaneous; skin infection
  congenital disease, 192
  as defensive barrier, 19, 192
    breaches, 192
  flora, 16
    wounds infected by, 170
Skin infection, 164–77
  bacterial, 164–9
  elderly, 195
  fungal, 7, 62, 68–9, 172–3, 176
  GP patients, 194
  normal flora, 16–17, 173
  opportunistic, 16, 17, 173
  parasitic, 176–7
  with pre-existing skin conditions, 171
  travellers and, 188, 189
  tropical and rare, 174–7
Sleeping sickness, 93, 131, *see* Trypanosomiasis
Slime production, staphylococcal, 28
Soft tissue infection, 164–9
  elderly, 195
  tropical and rare, 174–5
Soil-contaminated wounds, 170
Sources of infection, 18
  fungi, 9, 62
  hospital-acquired infections, 190, 191
  protozoa, 11
  zoonotic, *see* Zoonoses
South American blastomycosis, 66–7, 108, 121, *see also* Paracoccidioidomycosis
South American trypanosomiasis, *see* Trypanosomiasis
Species, microbial, disease and, 15
Specimen collection/transport/processing, 208
Spinal cord infection, syphilis (tabes dorsalis), 89
Spirillum minor, 47, 171
Spirochaetes, 52–3
  skin infections, 174, 175
Splenic abscess, 143
Spores
  bacterial, 3
  fungal, hypersensitivity, 8
Sporothrix schenckii, 9
Sporotrichosis, 69, 176
Sporozoa, 10, 72–3
Spread/transmission of infection
  bacteria, zoonotic, 50
  fungi, 62
  hospital-acquired infections, 190, 191
  metazoa, 13
  protozoa, 11
Sprue, tropical, 137
Staff
  neonatal infection from, 186

protection, 191
  wound infection, source for, 190–1
Stains, 208
  Gram, 2, 208, *see also* Gram stain
Staphylococcus spp., 28–9, 166–7
  aureus, 28, 29, 166, 167
    diarrhoeal disease, 139
    endocarditis, 126–7
    otitis externa, 94
    septicaemia, 128, 129
    CNS infection, 90
  epidermidis, 28, 29
    pneumonia, 114, 115
  saprophyticus, 28, 29
    skin infection/manifestations, 166–7
Stenotrophomonas spp., 46
Sterilisation, 196–7
Stokes' adaptation, 210
Stomatitis, 101
Streptobacillus moniliformis, 47, 171
Streptococcus spp., 30–1, 164–5
  agalactiae, 31
    anaerobic myositis, 169
    host defence and, 14
    meningitis, 86–7
    otitis, 94
  pneumoniae (pneumococcus), 30, 31
    pneumonia, 113
  pyogenes, 30, 31, 164, 165
    cellulitis, 165, 166–7
    glomerulonephritis, 152
    necrotising fasciitis, 165, 169
    rheumatic fever and heart disease, 125
    skin infections/manifestations, 164–5, 169
    throat infection, 104
  viridans, 31
Streptomycin, 201, 202, 204
Strongyloides stercoralis, 17, 79, 141
Stye, 96
Subcutaneous mycoses, 7, 69, 176
Subdural abscesses, 91
Subperiosteal abscess, 97
Subphrenic abscess, 143
Sulphonamides, 201, 203
Suppressor T-cells, 23
Suppurative disorders, *see* Pus-forming disorders
Surgical wound infections, 17, 170, 190, 191
Surveillance and hospital-acquired infections, 191
Synergy tests, 211
Syphilis, *see* Treponema spp.
  CNS, 88–9, 91
  congenital, 186
  cutaneous, 175
  endemic, 174–5
  primary, secondary, tertiary, 162–3
Systemic infections
  immune impairment in, 193
  mycoses, 64–7, *see also* Fungi
  ocular involvement, 98–9
  tropical and rare, 130–3

T

T-cells, 22, 23
  dysfunction/defects, 193
  functional assessment, 25
Tabes dorsalis, 89
Tachyzoites, 72
Taenia saginata, 82
Taenia solium, 82
Tapeworms, *see* Cestodes
Teeth infection, 102
Teichoic acid, 28
Teicoplanin, 202
Temporal bone, mastoid process infection, 95

Testis, inflammation, 160
Tetanus, 35, 93, 207, see also Clostridia spp.
Tetracyclines, 200, 201, 203
Thiabendazole, 205
Threadworm (E. vermicularis), 78–9, 141
  pig (T. spiralis), 81, 185
Throat infections, 104–5
Thrombophlebitis, suppurative/septic (venous
    infection with thrombosis), 122
  portal, 122, 145
Thrombosis
  cavernous sinus, 97
  venous infection with, 122, 145
Ticks
  Ixodes spp., 19, 53
  tick-borne relapsing fevers, 53, 132
Tinea, 68–9, 172–3
Tinea nigra, 68, 172
Tinea versicolor, 68, 172
Tinidazole
  as antibacterial, 201, 203
  as antiprotozoal, 205
Tissue nematodes, 80–1
Tissue protozoa, 11, 76
TNF, 27
Tonsillitis, 104
Tooth infections, 102
TORCH group/syndrome, 93
Toxic shock syndrome, 167
Toxicity of antibiotics, 199, see also specific
    antibiotics
Toxins
  bacteria, 15
    B. anthracis, 15, 33
    Clostridia botulinum, 35, 138
    Clostridia difficile, 35
    Clostridia perfringens, 15, 34, 138
    Clostridia tetani, 35
    Corynebacteria diphtheriae, 15, 32
    damage caused, 25
    in diarrhoeal illness, 136, 138, 139
    E. coli, 138
    extracellular, 25–6
    inactivated, as vaccines, 206
    intracellular, 3, 25–7
    Shigellae, 41, 139
    staphylococcal, 29, 139
    Vibrio cholerae, 15, 44, 139
    Vibrio parahaemolyticus, 44
  fungal, 8
Toxocariasis (T. canis/dog roundworm; T. cati;
    cat roundworm), 81, 133
  ocular, 99
Toxoid, 204
Toxoplasmosis, 17, 72–3, 185
  congenital, 186
  ocular, 98
Tracheitis, 109, see also
    Laryngotracheobronchitis, 106, 109
Trachoma, 99
Transduction, bacteria, 5
Transformation, bacteria, 5
Transmission, routes, see Spread
Transplant patients, immunosuppressive drugs,
  193
Traumatic wound infections, 170
  bite wounds, 171
  bowel, 137
Travellers, 188–9
  diarrhoeal disease, 136, 188
Trematodes (flukes), 12, 84–5, see also
    genera/species
  intestinal, 84–5, 140, 141, 147
  structure, 12
Treponema spp., 52
  carateum, 52, 174, 175
  pallidum (and syphilis), 52

pertenue (and yaws), 52, 98–9, 133
  ocular infection, 98–9
  skin infection, 174, 175
Tricarboxylic acid cycle, 4
Trichinella spiralis, 81, 185
Trichomonas vaginalis, 75, 161
Trichophyton spp., 68, 69, 172, 173
Trichosporon beigelii, 68, 172
Trichuris trichiura, 78, 141
Trigonitis, 148
Trimethoprim, 201, 203
Trophozoite, 10
  B. coli, 75
  D. fragilis, 75
  E. histolytica, 74
  G. lamblia, 75
  T. vaginalis, 75
Tropical sprue, 137
Tropical ulcer, 175, 189
Tropics, see Developing countries
Trypanosoma spp., 76–7, see also
    Trypanosomiasis
Trypanosomiasis
  African trypanosomiasis; sleeping sickness
    (T. brucei), 77, 93, 131
    CNS infection, 93
    ocular infection, 98
  Chagas' disease; S. American
    trypanosomiasis (T. cruzi), 77, 131
    myocarditis, 124
    ocular infection, 98
Tsetse fly, 77
Tuberculides, 174
Tuberculosis (primarily M. tuberculosis
    infection; occasionally M. bovis), 54, 89,
    118–19
  antimycobacterials, 54, 116, 119, 124, 153, 204
  extrapulmonary, 119
    adrenal, 153
    bone, 181
    cervical lymphadenitis, 108, 118
    CNS, 89, 91
    cutaneous, 174
    joints, 181, 182
    psoas abscess, 135
    renal, 153
  miliary spread, 118
  pulmonary, 118–19
Tubo-ovarian abscess, 158, 159
Tularaemia, 98–9, 185, see also Francisella
    tularensis, 50–1
Tumour necrosis factor (TNF), 27
Typhoid fever, 42, 133
Typhus
  epidemic louseborne (R. prowazecki), 60–1,
    71
  epidemic murine, 61

U

Ulcers
  Bairnsdale/Buruli, 174–5
  chancriform, 168
  corneal, 96–7
  tropical, 175, 189
Ultraviolet light, sterilisation employing, 196
Umbilical stump infection, 187
Ureaplasma spp., 57, 154, 155
Ureteritis, 148
Urethral syndrome, 148, 155
Urethritis, 148, 154–5, 159
Urinary tract infections, 148–55, see also
    Sexually-transmitted diseases
  elderly, 195
  GP patients, 194
  hospital-acquired, 190, 191

lower, 148
tropical and rare, 152–3
upper, 148
Uterine cervix, see Cervix, 156–7
UV light, sterilisation employing, 196
Uveitis
  anterior, 100
  posterior, 100

V

Vaccination (active immunisation), 22, 206–7
  contraindications, 207
  diphtheria, 107
  future needs, 207
  pneumococcus, 31
  pneumonia-causing organisms, 113
  travellers, 188
  tuberculosis, 119
  types of vaccine, 207
  when and who to vaccinate, 207
Vagina, normal flora, 16
Vaginitis, 155, 161
Vaginosis, bacterial, 161
Valve infection (cardiac), 126–7
Vancomycin, 201, 202
Vascular gangrene, infected, 169
Vasculitis, S. aureus sepsis, 128, 129
Vector-borne transmission, 18–19
Veillonella parvula, 49
Veldt sore, 107
Venous infection with thrombosis, 122, 145
Verotoxin, 138
Vertebral body osteomyelitis, 180
Vibrio spp., 44, 139
  alginolyticus, 44
  cholerae (and cholera), 44, 136, 139
    toxins, 15, 44, 139
  parahaemolyticus, 44, 139
  vulnificus, 44
Vincent's angina, 104
  postanginal septicaemia complicating, 105
Virulence, factors determining, 15, see also
    genera/species
Viruses
  infections
    myocardial, 124
    pericardial, 124–5
    pharyngeal, 104
  in vaccines, live attenuated and inactivated,
    206
Visceral larva migrans, 81, 99, 133
Visceral leishmaniasis, 76, 130, 177
Vulval warts, 163
Vulvo-vaginitis, 155, 162

W

Wall, cell, see Cell wall
Warts, vulval, 163
Water, wounds infected from, 170
Waterhouse–Friderichsen syndrome, 37
Weil–Felix test, 60
Weil's disease, 184
Whipworm, 78, 141
White piedra of hair, 68, 172
Whooping cough, 38, 109
Worms, see Helminths
Wound infection, 170–1
  bites and burns, 171
  streptococcal, 164
  surgical, 17, 170, 190, 191
  tetanus prophylaxis, 207
  traumatic, 170, see also Traumatic wound
    infections
Wuchereria bancrofti, 80, 177

X

Xanthomonas spp., 46

Y

Yaws, 98–9, 133, 174–5, *see also Treponema pertenue*
Yeasts, 7

Yersinia spp., 50, 50–1, 147
  enterocolitica, 50, 139, 147
  pestis, 50, 51, 184–5
  pseudotuberculosis, 50–1, 147

Z

Zoonoses, 18, 50–3, 184–5
  bacterial, 50–3

  protozoan, 72–3
Zoophilic dermatophytes, 9, 68
Zygomycosis (phycomycosis), 6, 70
  necrotising infections, 169
  rhinocerebral, 91
Zygomycota, 6